CELL INTERACTIONS

Proceedings of the third Lepetit Colloquium, held in London, November 1971

Editor:
LUIGI G. SILVESTRI, *Milan*

1972

NORTH-HOLLAND PUBLISHING COMPANY – AMSTERDAM · LONDON

Library of Congress Catalog Card Number: 72 76 601
ISBN North-Holland: 0 7204 4112 9
ISBN American Elsevier: 0 444 10366 4

Publishers:

NORTH-HOLLAND PUBLISHING COMPANY – AMSTERDAM
NORTH-HOLLAND PUBLISHING COMPANY, LTD. – LONDON

Sole distributors for the Western Hemisphere:

AMERICAN ELSEVIER PUBLISHING COMPANY, INC.
52 Vanderbilt Avenue, New York, N.Y. 10017

PRINTED IN THE NETHERLANDS

FOREWORD

Ladies and Gentlemen,

Before handing over to the chairman of the first session, it is my privilege to welcome you all here on behalf of the "Gruppo Lepetit". Those of you who have already attended some of our previous meetings will notice that the format of the Third Lepetit Colloquium has been changed. Previously we had a rather large number of speakers who gave formalized talks, on this occasion we have arranged a restricted number of lectures followed by informal workshops. This change reflects our intention to accept the criticisms which were raised against sessions too crowded. The new format was more advisable as the present meeting is largely multidisciplinary. We hope that the informality of the workshops on the interaction of different cell systems will help interaction of scientists as well. Anyhow, the present arrangements must be regarded as experimental. Our only ambition is to do something which could be of real usefulness to the scientific community. Please remember that we are open to every sort of suggestion that could help us to be helpful to people engaged in biomedical research.
I have another task to perform: to thank those people who helped us in the planning of this Colloquium, particularly I want to express our gratitude to Prof. Rita Levi Montalcini who gave us invaluably good advice. We regret that owing to her heavy engagements she has not been able to come. I want to thank also Prof. Benedetto Pernis whose suggestions have been essential for the realization of this meeting and Drs. Mario Crippa and Glauco Tocchini-Valentini.
I want to express our thanks to the President of the Royal Society of Medicine who has accepted our request to hold the meeting in this beautiful hall. I want to thank Mr. S. Hunt, Managing Director of Lepetit of England for arranging the meeting.
Finally let me express my gratitude to dr. E. Verwey and to dr. S. Riva for their collaboration in preparing this volume which allowed a prompt publication.

CONTENTS

INTRANEURAL INTERACTIONS

CELL INTERACTIONS IN DEVELOPING MAMMALIAN CENTRAL NERVOUS SYSTEM

RICHARD L. SIDMAN
Department of Neuropathology, Harvard Medical School
Boston, Massachusetts, USA

Abstract: Several recently analyzed examples of cell interactions in the developing mammalian brain will be considered. The general objective is to define gene-controlled events that determine the behavior of cells during genesis of a complex organ.

The first case concerns interaction of photoreceptor and pigment epithelial cells at late stages of development of the rat retina. Inbred RCS rats homozygous for the autosomal recessive gene retinal degeneration, rd, accumulate excess rhodopsin in an extra lamellar material located between the normal-appearing photoreceptor outer segments and the pigment epithelial cells during the third postnatal week. Subsequently the extra material and the entire population of photoreceptor cells degenerate. Electron microscopic autoradiography after pulse labeling with tritiated amino acids indicates that during the third week, protein turnover decreases markedly in outer segments while synthesis increases in pigment epithelial cells in mutant compared to normal rats. The pigment cells deposit excess protein in the extra lamellae; later, when rod-type proteins are no longer synthesized by either cell but protein degradation slowly continues, the degenerative phase of the disorder supervenes. The primary target of the rd gene in this intimately interacting cell system remains to be defined.

The second case concerns granule cell-Purkinje cell interactions in the mouse cerebellar cortex. In staggerer (sg/sg) mice, granule cell axons fail to form recognizable synaptic contacts on Purkinje cell dendritic spines. Granule cells do form normal synaptic contacts with basket and stellate cells and receive morphologically normal inputs via synaptic glomeruli. Purkinje cells in turn receive synaptic contacts from climbing, basket, and stellate axons, though the number of such contacts is reduced. The sg locus either controls specifically the granule-Purkinje cell relationship, or exerts some less direct effect on the Purkinje cells such that all of its inputs become reduced, the granule cell input most of all. Analysis of sg and other cerebellar mutants leads to two further conclusions about interrelations between these two neuron types: Purkinje cells acquire and maintain their basic form, including their dendritic spines, independent of granule cell inputs; granule cells, by contrast, appear markedly dependent on their target neurons, and degenerate if they fail to make a threshold number of synaptic contacts.

The third example concerns cell interrelationships in the developing cerebral cortex. Young neurons move outward from sites of origin at or near the ventricular surface by passing along radially-oriented "guide" fibers. Their somas take up layered positions in the cortical plate in an "inside-out" sequence relative to the times of cell origin. On route to the cortical plate, young neurons contact early-formed axons of thalamic neurons--one among many examples of the arrival of afferents at the very time their target cells are forming. In reeler (rl/rl) mutant mice, cortical neurons are generated in normal locations at normal times, but fail to attain a normal laminar distribution. A detailed ongoing anatomic analysis indicates tentatively that the early-generated and late-generated cortical layers are more or less reversed in position and that the disorder arises early in development, possibly in relation to abnormal distribution of early axonal contacts.

At later stages of development the mechanisms by which brain cells recognize their neighbors can be analyzed in dissociation-reaggregation tissue cultures. Cerebral cells of normal mouse embryos dissociated with trypsin will reaggregate and sort out to reestablish a cortical pattern. Corresponding cells of reeler embryos reestablish the mutant pattern. Thus, positional information at relatively late stages of cortex formation must be mediated directly by the cortical cells. Cell separation methods for obtaining purer sub-populations are under development, and surface "recognition" factors are being sought by immunological means.

1. INTRODUCTION

One of the many goals of developmental biology is to elucidate principles governing formation of complex nervous systems. The developing brain in man, the species of primary interest, is still relatively inaccessible by methods of cell biology, but important aspects of neural development have become approachable in other mammals and are now under intensive analysis in several laboratories. Since mammalian neuroembryology is still largely in a descriptive phase, it may be useful to consider a few of the conceptual and tactical approaches that seem likely to prove interesting and profitable. This conference provides a particularly suitable setting because it appears that the key advances now and for some time ahead are likely to be made in terms of the analysis of cell interactions.

During development, as in adulthood, the mammalian central nervous system is so heterogeneous in internal composition as to defy analysis by conventional chemical methods. It yields best to the Golgi and electron microscopic methods, preferably used in parallel on the same material. This approach is applicable to mammals in general, as exemplified in the recent revealing studies of cell relationships in the early mouse brain by Hinds and Ruffett (1971) and in the fetal monkey cerebellum and cerebrum by Rakic (1971a,1971b). As if the heterogeneity problem were not enough, the neuroembryologist also must contend with the fact that the properties and relationships of most cells in the young brain are changing with time. The autoradiographic approach, especially when used in parallel with methods that reveal cell shapes, adds a dynamic quality by allowing not only the determination of cell birthdays (defined as the final day of nuclear DNA synthesis in a given class of cells) but also a mapping of the rate and extent of cell rearrangements (reviewed in Sidman, 1970). The extraordinary display of varied and ever-changing cell relationships revealed by these morphological methods implies the existence of a series of cell interactions that determine the eventual form and function of the nervous system.

Three examples of the genetic manipulation of these cell relationships will form the substance of this paper. The first example, retinal degeneration (*rd*) concerns the relationship between photoreceptor and pigment epithelial cells in the rat retina and demonstrates altered synthesis of a specific cell protein as a prelude to virtually complete degeneration of photoreceptor cells. In the second example, staggerer (*sg*), a mutation in the mouse, the molecular aspects of the disease are completely unknown, but the mutation is expressed as a disorder of the relationship between granule cell and Purkinje cell neurons of the cerebellum. The third example, reeler (*rl*), also in the mouse, involves a widespread derangement of cell relationships in cerebellar and cerebral cortices and in a few deeper structures such that laminar organization and cell orientation are markedly disturbed. Consideration of this last disorder is leading us into a more detailed analysis of early embryonic cell relationships that influence patterns of histogenesis and into an *in vitro* analysis of recognition properties among developing neurons.

Modification of development by genes is chosen in preference to experimental intervention by means of other agents on the premises that the fundamental chemical defect is likely to be simple, the event controlled by the gene is likely to be a developmental turning point of some intrinsic importance (otherwise we would probably not recognize an abnormal phenotype), and the phenotypic consequences are likely to involve disordered relationships that would in most instances have been beyond the imagination or ability of the experimenter to obtain by deliberate design. On choosing the genetic approach, however, the investigator must be prepared to invest much effort and time in uncovering one after another in a series of secondary phenotypic expressions, often quite intriguing in their own right, while seeking the elusive primary genetic effect.

2. RAT RETINAL DEGENERATION, rd

The most obvious expression of this autosomal recessive mutation is selective and progressive degeneration of retinal photoreceptor cells, beginning at about 20 days after birth and reaching virtual completion by about 70 days. Detailed documentation was presented some time ago that in the third postnatal week, prior to overt cell degeneration, the mutant retina accumulates excess rhodopsin in membranous material organized into lamellar sheets and whorls located between the normal-appearing photoreceptor outer segments and the pigment epithelial cells (Dowling and Sidman, 1962). The absorption spectrum of the extra rhodopsin was normal, and visual function as measured by the electroretinogram was not measurably altered until the time of onset of cell degeneration.

In an important series of studies Droz (1963), Young (1967), and Young and Droz (1968) established that in normal animals, protein of the rod outer segment is continuously synthesized in the inner segment, incorporated into membranous discs at the base of the outer segment, displaced distally along the outer segment without decrement ("escalator-fashion"), broken off at the distal end, and probably digested by the pigment epithelial cells (Young and Bok, 1969). The labeled protein traced in these experiments is mainly opsin (Matsubara, Hiyata, and Mizuno, 1968; Hall, Bok, and Bacharach, 1969). In the adult rat, outer segment turnover takes about 9 days (Droz, 1963). During development, as the outer segments lengthen, the synthesis rate presumably exceeds the degradation rate. The two rates must balance once full outer segment length is attained.

These studies set the stage for an analysis of protein synthesis and turnover in the mutant retina. Herron et al. (1969) discovered that pigment epithelial cells in the mutant, in contrast to the normal, do not engulf the membranes shed from the distal ends of the rod outer segments, and hypothesized that the mutation affects primarily the pigment epithelial cell even though the photoreceptor cell is the one that actually degenerates. They reasoned that the failure of pigment epithelial cells to take in the shed rod discs leads to formation of an extracellular barrier that impedes nutritional exchange between the pigment epithelial cells and choroidal blood vessels on one side and the photoreceptor cells on the other. Thus cut off from nutrients, the photoreceptor cells die. Consonant with this view, electron microscopic evidence was obtained that pigment epithelial cells in the mutant lack phagosomes, which in the normal represent the phagocytosed rod disc membranes (Bok and Hall, 1971).

Meanwhile we had come to a different view of the disorder on the basis of similar autoradiographic data (Fig. 1).

The escalator-like displacement of rod discs is slowed in the mutant (O'Neil, 1967; independently recognized by Herron et al. 1969). If the disc degradation rate by pigment epithelial cells in the mutant were zero during the third week after birth, as suggested by Herron et al. (1969, 1971) and by Bok and Hall (1971) the net rhodopsin accumulation would be proportional to the synthesis rate. To reach the measured rhodopsin values of 100% greater than normal in the third week (Dowling and Sidman, 1962), synthesis would have to be maintained at or above the normal rate, and yet the autoradiographic measurements indicate that it actually falls below normal prior to postnatal day 17 and at all stages thereafter (Fig. 1 and LaVail, Sidman, and O'Neil, 1972). Furthermore, the degradation rate probably is greater than zero, since both the normal outer segments and the extra lamellar material do eventually disappear in the mutant. Therefore, the synthesis rate must be well above normal. The issue is from where does the excess protein come if not from the photoreceptor cell itself?

The autoradiographic data in Fig. 1 strongly suggest that the mutant pigment epithelium is the source, i.e., that the mutant pigment epithelial cell is behaving like a photoreceptor cell with respect to protein synthesis. Electron microscopic autoradiograms (LaVail, Sidman, and O'Neil, 1972) show that protein newly synthesized from tritium-labeled amino acids passes from the pigment

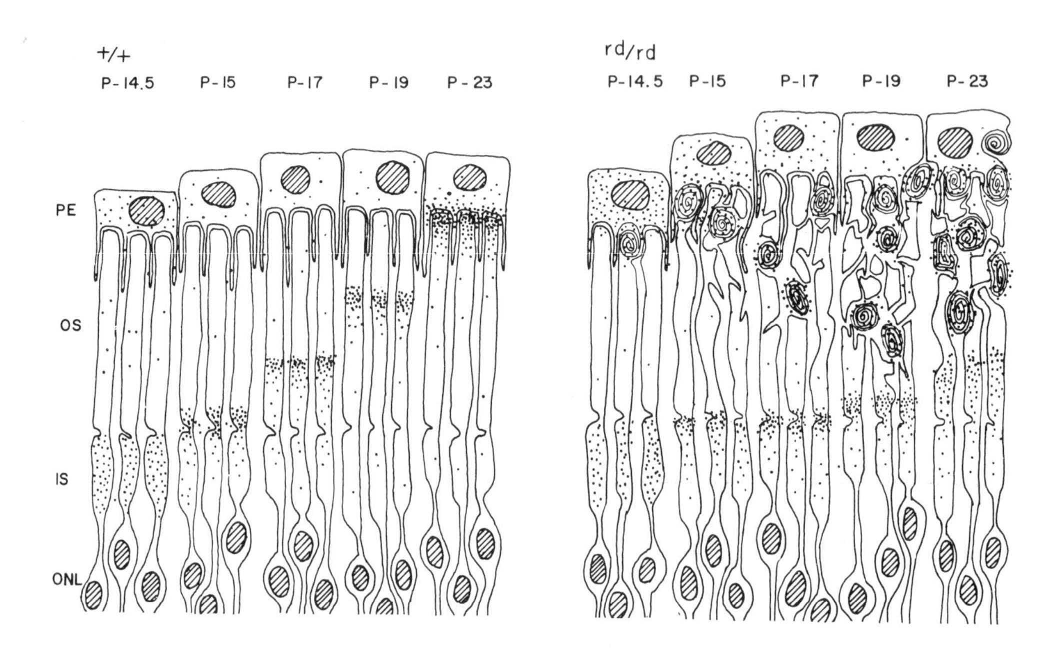

Fig. 1 Schematic drawing of autoradiograms of retinas from normal Fisher albino rats (+/+) and RCS mutant rats (rd/rd) which had been injected with H^3-amino acids on postnatal day 14 and killed at subsequent time intervals from 5 hours (P-14.5) to 9 days (P-23) as indicated. In normal retinas, newly-formed (labeled) outer segment (OS) discs are displaced outward in "escalator" fashion until 9-10 days later, when the discs are shed and at least partially ingested by pigment epithelial cells (PE). In mutant retinas, displacement of outer segment discs is slowed; at 9 days postinjection (P-23) most of the label is still present over the basal third of the outer segments. As early as 5 hour postinjection (P-14.5) and at all subsequent time intervals studied silver grains are present over the lamellar whorls adjacent to the pigment epithelial cells in the mutant retinas. Displacement of protein into the lamellar whorls from the pigment epithelium is maximal at approximately postnatal days 14 to 20. ONL, outer nuclear layer; IS, inner segments.

epithelial cell cytoplasm into the apical processes within one hour and into the lamellar sheets and whorls within 2 hours. The quantity of tritium so displaced is maximal at postnatal days 14 to 20, when the accumulation of rhodopsin-containing lamellae is greatest, and becomes less at later ages in parallel with the decline in rate of rhodopsin accumulation. Careful electron microscopic examination reveals that the extra lamellar whorls are composed of both pigment epithelial cell membranes and rod disc membranes (LaVail, 1972). It should be stressed that while the amount of excess protein synthesized by the mutant pigment epithelial cells correlates closely with the measured accumulation of rhodopsin, and no other source of the extra rhodopsin has been recognized, direct proof is lacking that the pigment epithelial product is actually rhodopsin.

The developmental and functional relationships are numerous between retinal pigment epithelial cells and photoreceptor cells (references cited in LaVail, Sidman, and O'Neil, 1972). The two structures arise from adjacent sectors of the primitive neuroepithelium of the embryonic optic vesicle. Under some circumstances a new neural retina can be regenerated from the pigment epithelium. In adulthood, derivatives of vitamin A cycle between the two cellular compartments when the eye is alternately dark-adapted and exposed to strong light. It comes as no great surprise, then, that the pigment epithelial cell should increase its protein-synthetic activity or take on a new synthetic function similar to that usually considered to be the exclusive prerogative of the photoreceptor cell.

Although this is one of the few instances in which the effect of a neurological mutation can be described in chemical terms, and despite the striking pathological response on the part of photoreceptor cells, it is still far from clear what is the primary effect of the rd gene, and whether it controls directly the pigment epithelial cell, the photoreceptor cell, or both.

3. THE STAGGERER MUTANT MOUSE, sg

What are the prospects for obtaining a genetic "handle" on synaptic organization in the central nervous system? Is there a gene for each class of synapses or is there more likely a limited number of genes which set development on particular paths such that subsequent synaptogenic events unfold on the basis of local or distant environmental influences? The problem is to choose among the 100 or more neurological mutants (Sidman, Green and Appel, 1965; O'Leary et al., 1968) for the ones that offer insight into this issue.

Staggerer (Sidman, Lane and Dickie, 1962) may be such a mutant. The mutation is autosomal recessive and lies in linkage group II close to the marker genes dilute (d) and short ear (se). The outstanding clinical features, recognizable by the end of the second postnatal week, are hypotonia and ataxia. The marker genes allow recognition of the staggerer mouse in a segregating litter at stages before the clinical signs become expressed. In mutant animals of such litters the cerebellum is grossly reduced in size within a few days after birth and thereafter. The granule cell neuron population in the cerebellar cortex is particularly decimated. Genesis and early maturation of granule cells appear to be qualitatively normal, but by postnatal day 14 large numbers of these cells are degenerating in the granular layer in the ventromedial parts of the cerebellar cortex. The cell loss then spreads in a dorsolateral direction until by day 33 almost no granule cells remain (D. Landis, 1971; Landis and Sidman, 1972b).

As in the case of the rd disease, however, the cell type which undergoes degeneration may not necessarily be the direct site of mutant gene action. A search for other phenotypic expressions has uncovered evidence of not only increased cell death and perhaps reduced rate of cell proliferation in the external granular layer of the neonatal animal, but also of a disorder of Purkinje cells. The latter is recognized by the small cell size, markedly electronlucent cytoplasm, small and closely packed dendritic branches, and lack of dendritic spines of the type normally contacted synaptically by granule cell axons (Sidman, 1968; Landis and Sidman, 1972b).

Analysis of synaptic organization in the postnatal developing cerebellar cortex by electron microscopy provides a clue to the role of the sg locus and again focuses our attention on a specific cell interaction, as summarized in the schematic diagram (Fig. 2).

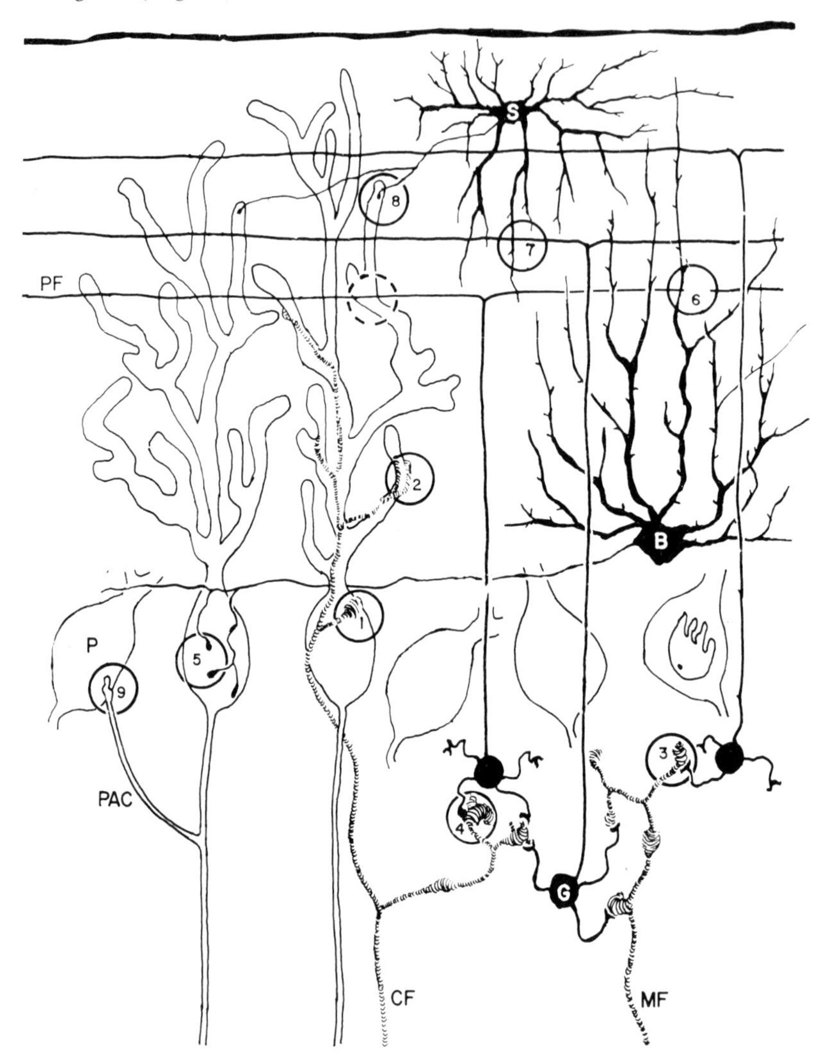

Fig. 2 Schematic drawing of the cerebellar circuitry in the staggerer mouse with each major class of synapses encircled and numbered: 1,climbing fiber-Purkinje cell soma; 2,climbing fiber-Purkinje cell dendrite; 3,mossy fiber-granule cell dendrite; 4,climbing fiber-granule cell dendrite; 5,basket cell axon-Purkinje cell soma; 6,parallel fiber (granule cell axon)-basket cell dendrite; 7,parallel fiber-stellate cell dendrite; 8,stellate cell axon-Purkinje cell dendrite; 9,Purkinje cell axon collateral-Purkinje cell soma. Synapses between parallel fibers and Purkinje cell dendrites (broken circle below synapse #8) were not encountered. Abbreviations: B,basket cell; CF,climbing fiber; G,granule cell; MF,mossy fiber; P, Purkinje cell; PAC, Purkinje axon collateral; PF,parallel fiber; S, stellate cell.

All but one of the major normal classes of synapses are represented in electron micrographs of the staggerer cerebellar cortex. Those present include the climbing fiber, basket cell axon, stellate cell axon, and Purkinje recurrent collateral inputs to the Purkinje cell, the mossy and climbing fiber inputs to the granule cell, and the granule cell inputs to basket and stellate neurons. Of these, the synapses on Purkinje cells are probably reduced in number, though clearly present.

The striking finding, however, is that not a single definite example was recognized of a synapse between a granule cell axon and a Purkinje dendrite (normally the most numerous and readily visualized synapses in the cerebellar cortex). This observation is all the more impressive since granule cell axons are organized into typical parallel fiber bundles, and do lie in abundance contiguous with Purkinje dendrites.

Three interpretations can be considered. 1) The failure of these synapses to form at any age could reflect a general synaptogenic insufficiency of the granule cell. This seems unlikely because the granule cell axon clearly engages in synapse formation with virtually all its normal targets except the Purkinje cell, and because its own dendrites in turn are contacted by the usual glomerular inputs, including the climbing fiber collaterals recently described in the rat by Chan-Palay and Palay (1971). 2) The disorder may be exactly what the synaptic analysis indicates, namely a direct and specific failure of the granule cell-Purkinje cell relationship. Such an intriguing possibility implies a one gene: one synapse principle. No independent evidence for or against this idea is yet available. 3) Perhaps the most plausible alternative is that the Purkinje cell may be the primary target of the sg gene via molecular mechanisms that have nothing to do directly with synaptogenesis. As a relatively late and nonspecific consequence, all synaptic classes on the Purkinje cell may be reduced, some more than others. The abnormal form and apparent slow maturation of the Purkinje cell from relatively early stages is consistent with this third possibility. A particularly important fact that points to the Purkinje cell itself is the failure of Purkinje cells in the mutant to develop the dendritic spines that in normal animals become the sites of innervation by granule cell axons. Evidence that Purkinje cells normally can form and maintain these dendritic spines independent of granule cell axons is considered in the discussion below on the reeler mutant. The more general argument, that the Purkinje cell abnormality might be a nonspecific response to the failure of synaptogenesis on the part of the granule cells, seems unlikely because Purkinje cells in the weaver mutant do not share the staggerer-type Purkinje cell defect even though granule cells in weaver die before establishing synapses (Sidman, 1968).

Two further hypotheses about normal causal interrelations among cells must be entertained if the staggerer phenotype is to be accounted for on the basis of a primary effect on the Purkinje cell. The first hypothesis is that Purkinje cells normally may influence cellular kinetics in the nearby external granular layer; in staggerer a diminution of this influence would lead to the observed early reduction in cerebellar size, to the decreased surface area (i.e., decreased absolute number of external granule cells), and to the increased number of pyknotic nuclei in the external granular layer. The second hypothesis is that survival of maturing granule cells might be dependent on the establishment of specific contacts with Purkinje cells; this would be the relatively unusual circumstance in which the presynaptic cell undergoes degeneration (see, for example, Cowan, 1970). Both hypotheses are consistent with data from other cerebellar mutants, particularly nervous (Sidman and Green, 1970; S. Landis, 1971) and reeler (see below), but the critical tests as yet have not been devised. In any case, whatever the primary gene action and the exact causal chain, it seems clear that granule cell-Purkinje cell interaction is central both to the staggerer problem and to genesis of the normal organization of cerebellar cortex.

4. THE REELER MUTANT MOUSE, rl

Reeler also is an autosomal recessive mutation and is genetically independent of other known cerebellar mutations. We maintain it on the C57BL/6J and C3H/HeJ inbred backgrounds and in F_1 hybrids between these two strains. On all genetic backgrounds, reeler shows a unique widespread abnormality of cerebral cortex as well as cerebellum, and poses the challenge of what might be the common denominator for all its phenotypic expressions in the central nervous system.

Based on evidence from Nissl, Golgi and autoradiographic material, it was proposed that reeler's disorder is fundamentally a problem in cell-to-cell recognition (Sidman, 1968). The main argument was that cells appeared to be randomly oriented and malpositioned in all those brain regions (and only in such regions) in which late-forming neurons must pass by earlier-formed cells in order to gain their proper positions. The hypothesis seemed to receive dramatic support from dissociation-reaggregation tissue culture experiments. A suspension of single cells dissociated from the cerebral cortex of normal mice during a narrow time span in late embryonic stages will aggregate to form small spheres. Within each sphere the cells sort out to establish a radially-symmetrical pattern resembling isocortex, with an outermost molecular layer, a cortical plate containing at least two broad laminae of neuronal somas, many with apical processes oriented radially outward, and an innermost fiber zone containing a few randomly oriented cell bodies enmeshed in a rich tangle of axons (DeLong, 1970). Cells in similar cultures from reeler littermates also aggregate and sort out, but come to display a different pattern of organization that is more like the *in vivo* pattern of embryonic reeler cortex, in that the aggregated cells are grouped into a "cortical plate" but unlike cells in the normal cultures, lack radial orientation and are not organized into obvious laminae (DeLong and Sidman, 1970).

These intriguing results have led us to undertake a deeper analysis of the reeler problem both on morphological and experimental planes. New Golgi and electron microscopic studies on the cerebellar cortex, histological studies of cell and fiber organization in the basal forebrain, and further tissue culture experiments will be reviewed below. The deeper we look the more elusive the primary abnormality seems to be, though it remains reasonable to expect that a simple story will yet emerge, since the disorder is based on an alteration at a single genetic locus. Cell recognition still appears to be the key issue, but we can now ask--recognition of what, on the part of which cells, and when?

Cerebellar Cortex. All parts of the reeler cerebellar cortex are abnormal (Sidman and Rakic, 1972a), but the severity of the disorder varies from area to area (Fig. 3). In places there is a shallow but distinct molecular layer composed mainly of parallel arrays of granule cell axons oriented at a right angle to Purkinje dendrites. The somas of basket and stellate cells are mixed, whereas in the normal they are clearly segregated. In other areas the majority of the granule cell somas lie external to the Purkinje somas instead of in the normal position internal to them, and the Purkinje dendrites lie among or even deep to the granule cells. In many areas Purkinje somas are scattered throughout the depths of the cerebellum except for the molecular layer and for the territories occupied by neuron somas and dendrites of the deep cerebellar nuclei. The more displaced the Purkinje cells lie from their normal position the more abnormal are their dendrites in volume, branching pattern and orientation while still retaining the distinctive morphology of Purkinje cells. In a few areas the cerebellar white matter extends almost to the external surface, and little is seen of the normal cortical cell arrangements.

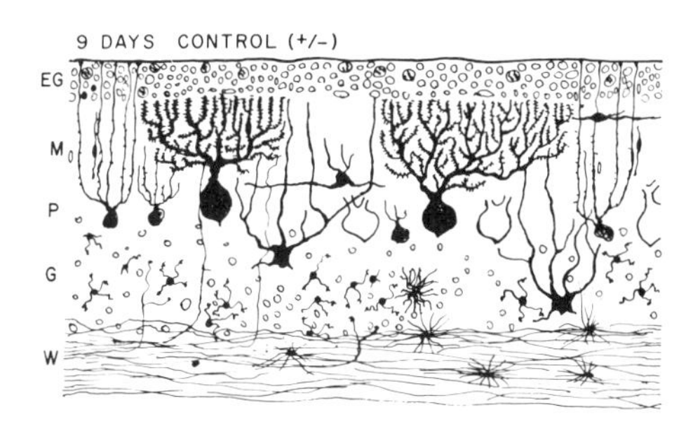

Fig. 3 Composite drawings from Golgi-impregnated preparations of normal and reeler littermates at 9 postnatal days. Abbreviations: G, granular layer; EG, external granular layer; M, molecular layer; P, Purkinje cell layer; W, fibers of the prospective white matter.

An outstanding feature of cerebellar and cerebral organization in general is that the regional topographic patterns are faithfully reproduced from reeler brain to reeler brain; that is, the mutant brain is constructed according to a tightly controlled blueprint, albeit a blueprint differing from the normal in some important developmental detail (Caviness and Sidman, 1972a; Sidman and Rakic, 1972a).

On electron microscopic study (Sidman and Rakic, 1972b) we were surprised to find that all the major classes of synapses typical of the normal cerebellum are also present in the mutant cerebellum. Most of them were identified not only in the relatively well-organized superficial zones of the cerebellar cortex, but also in the very abnormal deeper zone. Markedly aberrant Purkinje cells, for example, receive synaptic contacts on their somas and dendritic shafts from the terminals of axons which have the morphology of normal climbing fibers. As in the normal (Larramendi, 1969), these terminals are most prominent on the Purkinje somas at early postnatal stages and only later are predominantly distributed on dendrites. Scattered granule cells, another source of afferent axons, are also distributed among the displaced Purkinje cells (Fig. 3). These Purkinje cells develop dendritic spines, of which about 10% as measured in single sections, are contacted by processes that resemble granule cell axons except that they are single rather than assembled into bundles of parallel fibers. In addition, the aberrant Purkinje cells receive morphologically typical synaptic contacts of the types normally traced to basket and stellate cells. The identification of the several classes of synapses on the aberrantly located Purkinje cells, however, is only tentative, since axons might accept a synaptic configuration dictated by the post-synaptic neuron, as discussed for the normal cerebellum by Mugnaini (1970).

The electron microscopic data also suggest other developmental interactions of interest. While some of the dendritic spines on aberrant Purkinje cells do display synaptic contacts, as just described, most of them lack synapses and instead are enveloped completely in astroglial sheaths. Spine maintenance, then, is clearly independent of synaptic contact, a conclusion reached also from experimental studies involving selective granule cell destruction by virus (Herndon, 1971). Probably the initial formation of spines likewise is independent of synaptic contact by granule cell axons, both in the course of normal development (Larramendi, 1969) and in the reeler (Sidman and Rakic, 1972a,b).

In the molecular layer of the reeler cerebellum abnormalities of Bergmann glial cells allow further insight into the relationship between this class of glial cells and migrating young granule cells during development. Rakic (1971a) has examined postmitotic neurons migrating from the external granular layer past the Purkinje cell dendrites and soma to reach the granular layer in fetal and neonatal monkeys. These neurons consistently migrate along the radially-oriented processes of Bergmann glial cells and thereby find their way directly through the complex territory of the expanding molecular layer. These young granule cells do not leave the glial fiber to migrate along any of the other processes they meet in the molecular layer. Essentially the same relationship is seen in the normal mouse cerebellum (Landis and Sidman, 1972a). Is the relationship merely fortuitous? An answer comes from examination of the reeler cerebellum, where some of the Bergmann glial cells are inverted, with their somas at the external surface just beneath the pia and their fibers pointing obliquely inward (Fig. 3). Despite the reversed polarity and obliquity, granule cells migrate along these fibers and do not take the more direct but unguided radial route across the molecular layer (Sidman and Rakic, 1972a). This guidance mechanism has wider use; migrating cells follow radial fiber "struts" from the vicinity of the ventricular wall to the cortical plate several thousand microns away at late stages of cerebral development in the fetal monkey (Rakic, 1971b).

The general conclusion from the Golgi and electron microscopic studies of the reeler cerebellum is that although cells in the mutant show some major defect in alignment and orientation, they do associate in nonrandom ways, probably including the associations that lead to the formation of specific synapses.

Cerebral Cortex. A more precise definition of the reeler problem is emerging from a careful examination of the cerebral cortex in Nissl- and myelin-stained serial sections prepared by the method described in detail by Sidman, Angevine, and Taber Pierce (1971). The formulation referred to above, that those cells become misaligned which must migrate past previously-generated cells, failed to take into account the normal structure of the olfactory bulb in reeler (Caviness and Sidman, 1972a) or the relatively well-preserved laminar architecture of the hippocampus (Caviness and Sidman, 1972b). While it is not expedient to review here the details of these studies, it is worthwhile to state some of the general conclusions. Firstly, the cortical neurons in reeler are not randomly mixed. They lie in the normal cortical regions defined in terms of brain surface topography. Further, within a given region they group together into a few laminar compartments each of which contains a complement of cells apparently homologous with those in the same region (though not necessarily in the corresponding lamina) of the normal cortex. Secondly, and most strikingly, in the cortical areas which receive projections from the olfactory bulb, and in retrohippocampal and anterior hippocampal areas, the relative depths of these cell laminae from the external surface are reversed with respect to each other in reeler compared to the control. Thirdly, some major fiber bundles are markedly abnormal in position; for example, the anterior limb of the anterior commissure is displaced from its usual juxtaventricular location to a superficial site just beneath the lateral olfactory tract.

These impressive abnormalities in position of cell groups and fiber bundles suggest that the reeler disorder arises very early in embryonic life, perhaps at times when afferent axons are contacting recently-generated neurons on their way toward or into cortical structures. References to the idea that afferent fibers may exert a developmental influence on their target cells are cited in Cowan (1970) and Caviness and Sidman (1972b). Abnormalities in spatial or temporal aspects of this relationship of afferent fiber and migrating cell could underlie the reeler problem. Once the basic contacts are made, even if made in unusual locations, most other developmental events should proceed according to normal rules. The net result should be, as it is, a more or less successful grouping together of cortical cells of a given class, but into abnormally positioned laminae.

The positioning disorder is likely to be mild where the tissue normally features either a single-layered arrangement of cortical cell bodies (e.g. hippocampus) or a relatively simple and unidirectional afferent input (e.g. olfactory bulb), while it will be slightly more complicated where there are two or more inputs somewhat separated from each other (e.g., area CA1 of the hippocampus with its segregated perforant and alvear inputs) and very abnormal in areas where the cortex normally acquires additional laminar refinements and receives varied inputs terminating in spatial distinct territories (e.g., pyriform cortex, and especially many areas of isocortex). A direct examination of the early embryonic reeler brain for such predicted spatial or temporal anomalies of afferent inputs in relation to their target young neurons seems mandatory.

Further Tissue Culture Experiments. The series of dissociation-reaggregation experiments (DeLong, 1970; DeLong and Sidman, 1970) reviewed above, indicate that isolated, suspended cells prepared from developing cerebrum at late embryonic stages will aggregate *in vitro* and establish a radially symmetrical tissue pattern resembling immature cerebral isocortex. By whatever mechanism the cells had acquired their positions *in vivo* prior to the experiment, they are now able to express and respond to the information necessary for pattern formation independent of direct intervention by afferent axons or other environmental influences that would have been disrupted at the time of tissue dissociation. Cells of the cerebellum form different patterns, while cells from the hippocampal region aggregate and sort out to form even more complex patterns. The developmental stage at which the tissue is taken for dissociation appears to be both critical and somewhat different for different brain regions. Reeler and normal tissues treated identically form different patterns *in vitro*.

In light of the histological analysis of the reeler brain summarized in the previous section, it is pertinent to reconsider what the dissociation-reaggregation experiments might teach us about cell interactions in the brain. If the disorder in reeler does indeed originate at early developmental stages, long before the optimal stage for establishment of *in vitro* patterns, the thesis becomes tenable that the culture experiment brings into view an important cell recognition mechanism but not necessarily the one controlled directly by the reeler locus. Or it may be that the same molecular mechanism is used for recognition at the early embryonic stages when the reeler disorder arises and again at later stages to allow cells to attain their local "addresses". It is the latter information which presumably is still expressed and obeyed as cells sort out in the aggregates. The culture experiments do not necessarily tell anything about how the information was generated initially, but do indicate that a recognition mechanism is operative and indeed, represent an assay for it.

Certain apparent properties of this recognition mechanism may be artifacts of the culture system. For example, the limited temporal period in which patterned aggregation is obtained could reflect systematic variations in the size of aggregates as a function of age at dissociation, or some other methodological parameter, rather than a basic developmental property of brain cells. For this reason a detailed analysis of the technical variables in the culture system is underway. Furthermore, each aggregate contains a mixture of many cell types, a situation that complicates several key aspects of the biological analysis. It therefore becomes desirable to devise techniques for separating classes of cells that will retain their ability to interact *in vitro*.

One working hypothesis is that the recognition properties expressed so vividly in the aggregates may be mediated by cell surface constituents. As a byproduct of the dissociation technique, we now are able to use isolated viable brain cells for immunofluorescence, cytoxicity, and other immunological assays that should allow the identification and localization of developmental antigens at the stages when brain cells are engaged in critical interactions. The reaggregation culture technique in principle provides an assay for the biological significance of such antigens. In this way among others, neural science is joining immunology in the mainstream of cellular and developmental biology.

REFERENCES

Bok, D., and M.O. Hall, (1971), The role of the pigment epithelium in the etiology of inherited retinal dystrophy in the rat, J. Cell Biol. 49,664.
Caviness, V.S. and R.L. Sidman, (1972a), Olfactory structures of the forebrain in the reeler mutant mouse, J. Comp. Neur. (in press).
Caviness, V.S. and R.L. Sidman, (1972b), Retrohippocampal, hippocampal and related structures in the forebrain in the reeler mutant mouse, (submitted for publication).
Chan-Palay, V. and S.L. Palay, (1971), Tendril and glomerular collaterals of climbing fibers in the granular layer of the rat's cerebellar cortex. Z. Anat. Entwickl.-Gesch. 133,247.
Cowan, M., (1970), Anterograde and retrograde transneuronal degeneration in the central and peripheral nervous system, In: Contemporary Research Methods in Neuroanatomy. W.J.H. Nauta and S.O.E. Ebbesson, eds. (Springer-Verlag, New York) p. 217.
DeLong, G.R., (1970), Histogenesis of fetal mouse isocortex and hippocampus in reaggregating cell cultures, Dev. Biol. 22,563.
DeLong, G.R. and R.L. Sidman (1970), Alignment defect of reaggregating cell in cultures of developing brains of reeler mutant mice, Dev. Biol 22,584.
Dowling, J.E. and R.L. Sidman,(1962), Inherited retinal dystrophy in the rat, J. Cell Biol. 14,73.
Droz, B., (1963), Dynamic condition of proteins in the visual cells of rats and mice as shown by radioautography with labeled amino acids, Anat. Rec. 145,157.
Hall, M.O., D. Bok, and A.D.E . Bacharach, (1969), Biosynthesis and assembly of the rod outer segment membrane system. Formation and fate of visual pigment in the frog retina, J. Mol. Biol. 45,397.
Herndon, R.M., (1971), The interaction of axonal and dendritic elements in the developing and mature synapse, In: Cellular Aspects of Neural Growth and Differentiation, D. Pease, ed. (Univ. Calif. Press, Berkeley) p. 167.
Herron, W.L., B.W. Riegel, O.E. Myers, and M.L. Rubin, (1969), Retinal dystrophy in the rat--a pigment epithelial disease, Invest. Ophthal. 8,595.
Hinds, J.W. and T.L. Ruffett, (1971), Cell proliferation in the neural tube: An electron microscopic and Golgi analysis in the mouse cerebral vesicle, Z. Zellforsch. 115,226.
Landis, D., (1971), Cerebellar cortical development in the staggerer mutant mouse, J. Cell Biol. 51,159a.
Landis, D. and R.L. Sidman, (1972a) Electron microscopic study of cerebellar cortical development. I. Purkinje and granule cells in normal mice, (submitted for publication).
Landis, D. and R.L. Sidman (1972b), Electron microscopic study of cerebellar cortical development. II. Staggerer mutant mice, (submitted for publication).
Landis, S., (1971), Selective mitochondrial abnormality in Purkinje cells of nervous mutant mice, J. Cell Biol. 51,159a .
Larramendi, L.M.H., (1969), Analysis of synaptogenesis in the cerebellum of the mouse, In: Neurobiology of Cerebellar Evolution and Development. R. Llinas, ed. (Am. Med. Assoc., Chicago) p. 803.
LaVail, M.M., R.L. Sidman, and D.A. O'Neil, (1972), Photoreceptor-pigment epithelial cell relationships in rats with inherited retinal degeneration. Autoradiographic and electron microscopic evidence for a dual source of extra lamellar material. J. Cell Biol. (in press).

Matsubara, T., M. Miyata, and K. Mizuno, (1968), Radioisotopic studies on renewal of opsin, Vision Res. 8,1139.
Mugnaini, E., (1970), Neurons as synaptic targets, In: Excitatory Synaptic Mechanisms,P. Andersen and J.K.S. Jansen, eds. (Universitetsforlaget, Oslo), p. 149.
O'Leary, J.L., J.M. Smith, A.B. Harris, and R.R. Fox, (1968), Animal prototypes in hereditary ataxia, In: The Central Nervous System. International Academy of Pathology, Monograph No.9, O.T. Bailey and D.E. Smith, eds. (Williams and Wilkins Co., Baltimore) p. 124.
O'Neil, D.A., (1967), Protein synthesis and turnover studied autoradiographically in photoreceptor cells of rats with inherited retinal dystrophy, Master of Science Thesis, Boston College, Dept. of Biology, Boston, Mass.
Rakic, P., (1971a), Neuron-glia relationship during granule cell migration in developing cerebellar cortex. A Golgi and electron microscopic study in Macacus rhesus, J. Comp. Neur. 141,283.
Rakic, P., (1971b), Guidance of neurons migrating to the fetal monkey neocortex, Brain Res. 33,471.
Sidman, R.L., (1968), Development of interneuronal connections in brains of mutant mice, In: Physiological and Biochemical Aspects of Nervous Integration. Society of General Physiologists, Monograph, F.D. Carlson, ed. (Prentice-Hall, New Jersey) p.163.
Sidman, R.L., (1970), Autoradiographic methods and principles for study of the nervous system with thymidine-H^3, In: Contemporary Research Methods in Neuroanatomy. W.J.H. Nauta and S.O.E. Ebbesson, eds. (Springer-Verlag, New York) p. 252.
Sidman, R.L., J.B. Angevine, Jr. and E. Taber Pierce, (1971), Atlas of the Mouse Brain and Spinal Cord, (Harvard University Press, Cambridge) 261 plates.
Sidman, R.L., M.C. Green, and S.H. Appel, (1965), Catalog of the Neurological Mutants of the Mouse, (Harvard University Press, Cambridge) 82 pages.
Sidman, R.L. and M.C. Green, (1970), "Nervous", a new mutant mouse with cerebellar disease, Symposium of the Centre National de la Recherche Scientifique, Orleans-la-Source, France, In: "Les Mutants Pathologiques chez l'Animal, Leur Interet pour la Recherche Biomedicale," (CNRS, Paris) p. 69.
Sidman, R.L., P. Lane, and M. Dickie, (1962), Staggerer, a new mutation in the mouse affecting the cerebellum, Science 137,610.
Sidman, R.L. and P. Rakic, (1972a), Cell distribution and orientation in the cerebellar cortex of postnatal reeler mutant mice,(submitted for publication).
Sidman, R.L. and P. Rakic, (1972b), Synaptic organization in the cerebellar cortex of reeler mutant mice, (submitted for publication).
Young, R.W., (1967), The renewal of photoreceptor cell outer segments, J. Cell Biol. 33,61.
Young, R.W., and D. Bok, (1969), Participation of the retinal pigment epithelium in the rod outer segment renewal process, J. Cell Biol. 42,392.

Manuscript received after December 1, 1971

CONSTANCY OF NEURONS, PRECISION OF CONNECTIVITY PATTERNS AND SPECIFICITY AS A PRODUCT OF NEURON DIFFERENTIATION IN INVERTEBRATES

By

G.A. Horridge

Department of Neurobiology
Research School of Biological Sciences
The Australian National University
P.O. Box 475
Canberra. A.C.T. 2601 Australia

INTRODUCTION

There is a level of explanation of behaviour in terms of neurons at which the primary interest lies in which nerve cell one is talking about. This is the level of the actual circuit diagram. Most of "Interneurons" was written before the concept was clear but connectivity pattern was defined in the glossary as "the actual complete anatomical pattern of connexions between identified neurons" (Horridge, 1968). Subsequently the paper on accuracy of patterns of connexions began with the words "At one level of analysis the action of any small region of the nervous system is the result of a connectivity pattern" (Horridge and Meinertzhagen, 1970). The implication is that the connectivity pattern is the actual wiring diagram at a definite place and that each neuron can be individually recognized. The emphasis is upon the actual anatomical description rather than the idealized diagram in which individual neurons or particular locations are not distinguished at cell level. Conceptually we are far from the inferred idealized black box system derived from physiological pathways of excitation.

Function of the nervous system as the consequence of a particular connectivity pattern is a concept related to a more ancient difficulty, "The ancient form/function controversy is resolved only when the anatomy or structure is known in sufficient detail that the function becomes self-evident" (Horridge, 1968, page 421). But when we turn to the description of anatomical constancy in this detail we encounter first the problem of identifying the neurons so that we can recognize the same ones in a different animal, and secondly, the precision of the connectivity pattern when different individuals are compared. Quite soon experiments will be concerned with the effect of the genetic background on the pattern, and then the range of variation and its causes become relevant. The philosophical implications could be important. For example if individual differences in connectivity pattern can be shown to be inherited, and to be the cause of functional differences, they provide a substantial support for the inheritance of consistent behavioural differences such as ethologists describe.

In contemplating the required level of anatomical description, we find a tradition of regional neuroanatomy of tracts, centres, etc., and much electron microscopy of anonymous neurons but virtually no examples of the precision of the lay-out at single neuron level. It was pure sarcasm to write "a measure of the care which a research worker has taken is to be found in his account of the variability in a range of his specimens" (Bullock and Horridge, page 808) because at single cell level I could refer only to Hertweck (1931) who showed that 2-3% of the chordotonal sense organs of *Drosophila* larvae contain four instead of three sensory neurons.

The technical discovery of how to make serial 1μ sections was a response to the necessity to trace the projection of the photoreceptor cells on the second order neurons in insect eyes. More or less by chance, the technique which Meinertzhagen evolved to trace a few axons in the locust turned out more suitable for tracing several hundred axons simultaneously in the fly, which is more complex than the locust. There is no arborization but only a terminal bulb on these axons, so that we were almost at synaptic level in a study by light microscope. We found every axon to be perfectly projected (Horridge and Meinertzhagen, 1970).

Having stumbled upon a means of demonstrating the projection at single neuron level, and found the precision to be fantastic, the obvious next step is to investigate how widespread are these and related phenomena and how the patterns originate in growth. This is one aspect of the general problem of differentiation and mutual recognition of cells.

SPECIFICITY IN GROWTH SITUATIONS, CONSTANCY OF NEURON FORM, PRECISION OF CONNECTING PATTERN

Three types of approach are possible to the problem of the specificity of neurons as the ultimate and most complex example of specificity of cell interactions. These are the (1) observation of (a) precision of connections repeated within an animal and between animals of the same genotype (b) comparison of animals of differing complexity, (2) the observation of (a) the process of neuronal development and (b) the effects of genetic lesions and (3) the re-establishment of specific connections by (a) normal regeneration and (b) in situations where ganglia or target tissue have been transplanted. These are all well tried methods of approach in the analysis of development of organs and of cells and there are limitations in the conclusions that may be drawn from them.

(1a) Precision and constancy of neurons within the nervous system

In several groups of invertebrates, notably nematodes, annelids, arthropods and some molluscs, there are numerous types of neurons, all of which are constant enough to be recognizable. They are always found in their characteristic places and intermediates do not occur. From the time when Retzius and then Zawarzin applied vital staining by methylene blue, and Lenhossék and Ramon y Cajal applied the Golgi method to invertebrate ganglia, the constancy of the anatomy of cell bodies, axons and major dendrites and the large number of recognizably different neurons has been well known but hardly remarked upon until neuronal specificity became a popular topic in the last year or two.

Workers on annelids take the view that not only are the larger neurons of a given segment constant between individuals but up to 100 ganglia of different segments have similar complements of neurons. Almost as an aside, Retzius (1891) illustrates three variants of the giant neurons in the segmental ganglion of the leech Hirudo (Fig. 1). The interesting point about these, not mentioned by Retzius, is that the variation is symmetrical on the

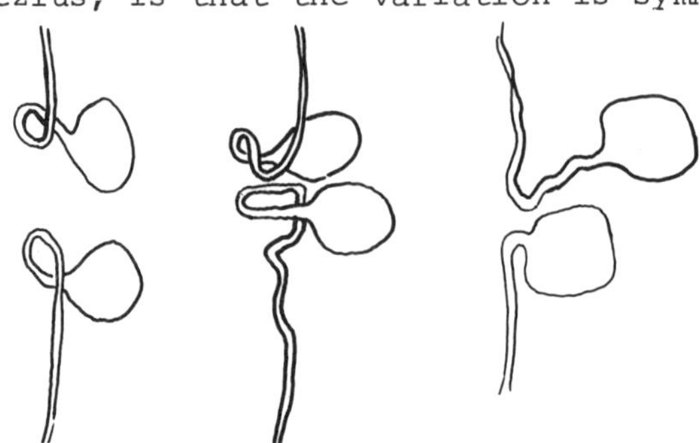

Figure 1. Three variant shapes of the origin of the axon in a pair of leech ventral ganglion cells (Retzius, 1891).

two sides, and he does not mention whether they are peculiar to single ganglia or general for all ganglia of one animal. There seems little point in following such detail unless the genetical background of the animals is known or constant.

In insect ventral ganglia about 50 neurons on each side, mostly motor neurons of that segment, are separately identifiable by their location and cell shape. By stimulating the cell with an internal microelectrode and recording the muscle response, each motoneuron can be identified with its peripheral muscle (Bentley, 1970). The constancy found in studies of this kind leads us to infer that the physiological methods are essential and at least as reliable as the morphological differences for identification of cells. But studies of neuron function alone are inadequate to identify neurons because they do not show how many of one type have to be distinguished.

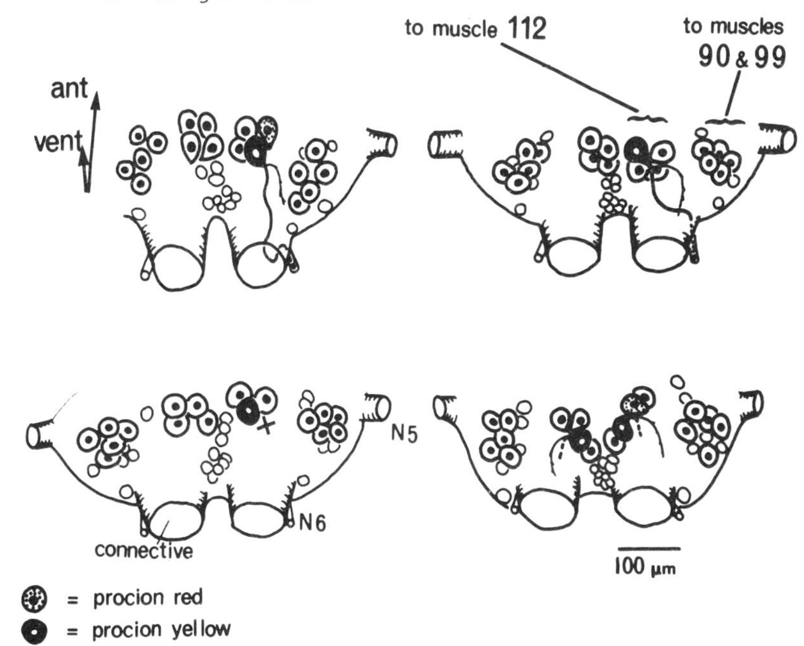

Figure 2. Mesothoracic ganglion of the locust Chortoicetes terminifera from the posterior ventral aspect. The cells with nuclei indicated are flight motoneurons. The medial group of four cells on each side innervate the metathoracic dorsal longitudinal muscle (112). Cells of this group have been injected with Procion dyes. The four preparations illustrated here show individual variation in the position of the cell bodies and in neuronal geometry. Note that one is missing at X. (Tyrer, 1972).

As yet there has been no test of cell constancy beyond the obvious largest cells, and no test of the precision of the connectivity pattern except in the fly eye and by vague inference from constancy of neuron function.

On the anatomical level "The constancy shows that the whole system, down to every cell, is predetermined directly or indirectly from the earliest stage, without plasticity or possibility of subsequent modification". But cell constancy does not imply synaptic constancy. However, the physiological evidence of recording the functions of these neurons shows that "The pattern of interneuron receptive fields must be a result of the previous existence of overlapping and inter-related subclasses, classes and superclasses of fibres some of which would and some of which would not connect during growth" (Quotations from Horridge, 1968, page 125). The constancy of the function mirrors the constancy of the anatomical form and suggests constancy of connectivity pattern. At present we simply do not know what fine detail we will be able to demonstrate as precise in the central neuropile.

On the sensory side a remarkable constancy of number and type of sensory neurons has been found where it has been sought. Small genetic differences in cuticular sense organs are well known within one species, and in *Drosophila*, for example, are inherited predictably. Bristle distribution is used as a species character in some insects and the bristles commonly are innervated by a sensory neuron (Stern, 1938).

Regularity in the layout of a sense organ (as between the same organ in different individuals) was one of the features which Young (1970) finds remarkable about the tibio-tarsal chordotonal organ of the mesothoracic leg of the cockroach. "Detailed study shows that individual neurons can be recognised on the basis of their size, shape and position. This implies that at the level of individual cells their morphological features are precisely determined in relation to their functional requirements". This conclusion about function could be reached because the anatomy and physiology of the sensory neurons was studied at single cell level with this precision in mind. There is no doubt that the sense organs of insects, and to a less extent of Crustacea and Annelida, will turn out to be differentiated in this precise way as more are examined.

On the histochemical side, recent results have also emphasized the constancy of invertebrate neurons. Specific methods of staining for transmitter-like amines, amine-oxidases, or neurosecretory products in invertebrates always show particular groups of neurons that are constant in number and position. In fact it would now be noteworthy to find an example, where individuals differ, anywhere among the molluscs or animals with ladder-like nerve cords.

A contribution from my own laboratory relates to the accuracy of a wiring diagram where there are many neurons in parallel pathways behind the compound eye. The eight primary photoreceptor neurons behind a single facet of a fly's eye look out in seven different directions because they lie at seven different points in the image plane of the lens. Cells 7 and 8 share the same axis and their axons have a separate projection which is known but not considered further here. Cells 1-6 send axons which terminate on second-order neurons of a ganglionic layer (the lamina) which lies immediately below the retina. The interest lies in the interweaving of these axons which ensures that the six differently numbered photoreceptors which look in one direction from behind six different facets all converge upon a group of second-order neurons (Fig. 3). Each group of second-order neurons receives one axon from a cell behind each of six different facets except near the equator of the eye. The pattern of photoreceptors 1-6 is not symmetrical and the pattern on the upper half of the eye is the mirror image of that on the lower half, with a sharp boundary between the two halves across the equator of the eye. The asymmetry

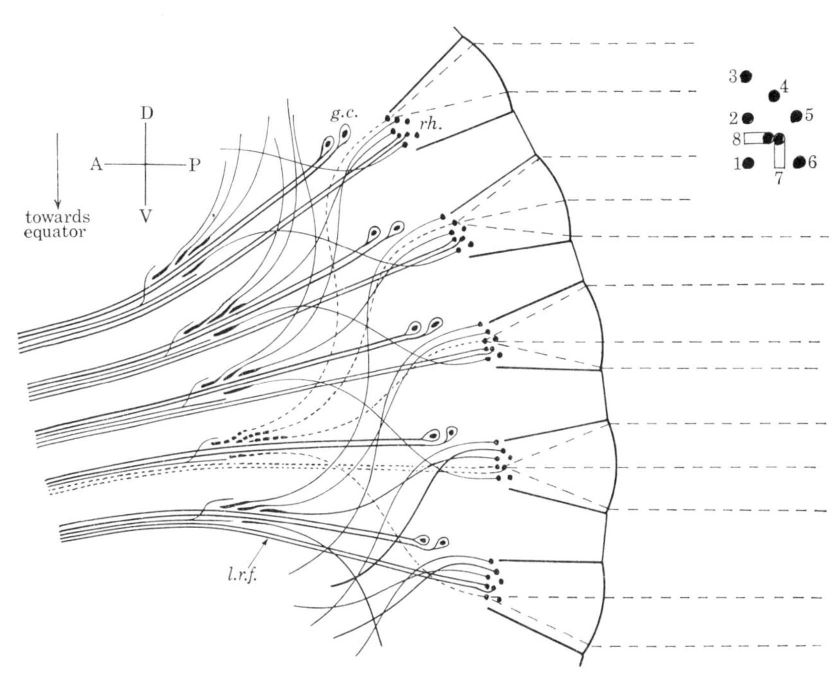

Figure 3. Interweaving of short retinula axons before they terminate upon the axons of the ganglion cells (g.c.) in the lamina cartridges. Light absorption occurs at the retinula cell rhabdomeres (rh.), which in cross section form a distinctive asymmetrical pattern (inset, top right). Axons of retinula cells which receive light from one particular direction (shown by the dashed rays) converge upon one cartridge and are drawn with dashed lines. On the equatorial side of each cartridge long retinula fibres (l.r.f.) of the two centrally situated retinula cells proceed directly to the medulla. The section is of the dorsal part of the left eye cut in vertical longitudinal section. In this plane no chiasma is seen in the bundles which run from the lamina to the medulla.

---

has the effect that four rows of second-order cell groups along the equator receive eight axons and two rows outside these receive seven axons. Even so the convergence of axons from receptors which look in the same direction is still perfect.

The first order axons form a single synaptic bulb without terminal arborizations, and this is known to be presynaptic from electron microscopy (Boschek, 1971). We were able to trace about 600 axons to their correct second-order cell without finding a single error although many of the axons had to cross the equator into a region of mirror-image symmetry. A small number of fibres (less than 5%) end in the wrong position of serial order but on the correct second-order neuron (Horridge and Meinertzhagen, 1970).

This result has a number of philosophical consequences. There is no requirement for such precision in a fly eye where every receptor has five others looking in the same direction and converging with it. The precision presumably exists because the growing sensory axons obey a few simple rules, laid down in their genetic code, as they grow to their connections in the embryonic optic lobe. The fact that axons 1-6 are reduplicated many hundreds of times, and that there is a complex interweaving, means that some mechanism must ensure that, for example, on axon number 3 is not mistaken for another axon number 3 from an inappropriate facet. Whatever rules of cell recognition define this growth pattern must be relatively simple but of a kind not previously known. One gene for each different connection is unreasonable.

Secondly the precision invites speculation as to whether all the sensory axons of the fly are equally precise in their choice of second-order neuron, whether all large neurons in the fly are precise in their connections, or whether all insects and possibly other groups of invertebrates are specified so exactly that every synapse is in the same location between the same neurons in every individual of the species (except for mutants acting on the nervous system).

Subsequently an individual fly was found with a dislocation in the equator so that two facets on one side of the equator show by their symmetry that they belong to the other side. By tracing all the fibres in this region Meinertzhagen (1972) found 17 errors the existence of which allows a deduction about the nature of the rules governing establishment of connections. All the erroneous fibres go in approximately the right direction relative to their asymmetry behind the facet and not relative to the axes of the eye in which they lie. They then join a second-order neuron and adhere to it on the side from which they have come. Therefore there need be no individual recognition signs which distinguish receptor axons from each other or second-order locations from each other. The rule could be that the photoreceptor axon grows oblivious of its milieu, with definite instructions governed by its position in its group of cells, and it stops when it meets any member of the right class of target neuron.

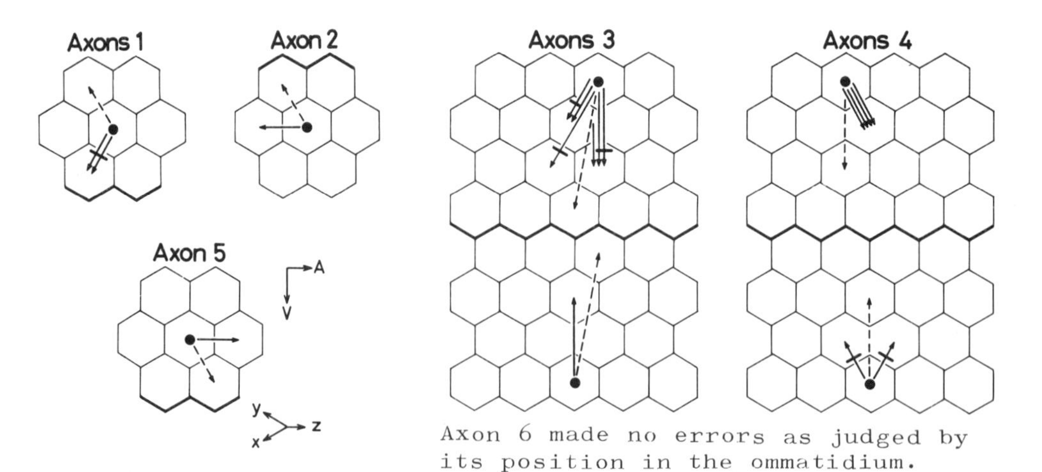

Figure 4. Retinula axons making incorrect projections to the lamina in an eye of the fly Calliphora. Axon paths 1 to 5 are shown as solid arrows from a facet of origin shown by a solid circle. Dashed arrows are the paths predicted on the basis of the retinula cell's position in the ommatidial pattern. The stronger line between facets shows the position of the equator and the cross line on some arrows shows fibres which in fact cross the midline (although not so drawn). (Meinertzhagen, 1972).

In vertebrates we can say little about cell constancy, specificity and precision because single neurons cannot be individually identified and found again in a different animal. The exceptional few examples of vertebrate giant fibres, notably the Mauthner cells, have been shown to be constant in anatomy and function down to synaptic level in any one species, and to show consistent species to species differences, so it is possible that the same is true for many vertebrate neurons, because Mauthner fibres are only giant interneurons. A most important task for the immediate future is the development of techniques and finding of a convenient vertebrate central nervous system to test whether corresponding neurons can be identified in different individuals. Present indications about this are equivocal. On the one hand we have constancy of behaviour patterns, stereotyped movements, constancy of anatomy at the regional level, and to some extent in the forms of dendrites. In some lower vertebrates, especially newts, the facility to regenerate central and motor connexions in sufficient detail for normal function to return also implies that the cells can distinguish each other, at least in the adult and this in turn points towards a stamped-in pattern of differences between cells or cell groups. Possibly transplants would show the pattern to be identical in every individual of the species.

On the other hand we have the growing evidence of abundant deaths of nerve cells during the development of the central nervous system in several kinds of vertebrates that have been examined (Hughes, 1968). About three quarters of the neurons of the developing ventral horn of amphibia die off at about the time that first limb movements appear. Reduction or increase in the peripheral field to be innervated appears to reduce or increase the number of neurons which survive. There is strong evidence that the surviving neurons are the ones which make connexions with muscles, and possibly the stamping-in of the adult pattern is derived embryologically from the differentiation of the muscles, not the neurons.

The recent change in attitude will now push the question of vertebrate connection patterns down to the level of individual cell constancy and synapse precision. At present we are quite unable to say whether the deaths of particular vertebrate central cells are programmed or whether there is a constant surplus and an accidental survival of some, i.e., whether the cells which we see as the survivors are the same individual cells in different animals of the one species. Furthermore we know almost nothing about the accuracy at cell and synaptic level even when adults are compared.

(1b) Comparison of animals of differing complexity

The most primitive members of an animal group do not have the fewest neurons. Examples such as the sea-anemones or the ballanoglossids have enormous numbers of small neurons which form nerve nets and appear to be of few neuron types. These animals have relatively simple behaviour patterns. As one turns to more advanced examples either within a phylum, or in different invertebrate phyla, one finds more neuron types. There are as yet no techniques by which we can describe complexity in the central nervous system, but in the more evolved examples there is an obvious greater complexity in that part of the dendritic and cell pattern which is constant from individual to individual. Parallel with this is a greater wealth of detailed behaviour, and we suppose (on very little evidence) a greater variety of functions of interneurons. This, together with the greater number of anatomical types of neuron and the greater predictability of exact dendritic form leads to the conclusion that evolution of more complex behaviour is the result of having more types of neurons, progressively increased specification of neuronal connexion patterns, and progressively more detailed specification of patterns of dendrites, synaptic properties and so on.

The same argument for evolution of greater specificity along with complexity can be advanced even within a single animal, in the cephalisation or regional specialization of a segmented nervous system.

In animals with many similar segments one finds in methylene blue or Golgi preparations exactly the same set of recognizable neurons in each segmental ganglion. The cell body and axon occupy the same position in the ganglion, the neuron is recognizable in each ganglion and only the details of its dendritic pattern differ in each ganglion, possibly by "accidents" of growth. Despite small differences we still expect the same functional pathways for cells of similar appearance.

Where particular segments have evolved specialized functions, more complex muscle groups, or more complicated behaviour, the neurons which are otherwise constant between segments are now different from their cousins in the other ganglia. At present we have no more than this anatomical indication of additional genetic effects on particular segments. With segmental differences in mind, we know nothing of neuronal development, and nothing about the detailed physiology, at cell level, of the functional differences between corresponding neurons in different ganglia.

If there are primitively no differences in growth between the set of developing neurons of one ganglion and those of another in a ladder-like nerve cord this sets a severe limit on the possible types of organization. Patterns like those of Fig. 5 A to D are readily grown without differences between segments. In patterns E and F, however, the neurons growing longitudinally down the system must be able to distinguish between the ganglia either by ganglion number or by time at which they arrive. If it were not so the higher centres would not be able to send command neurons terminating in a particular ganglion, or be able to receive sensory excitation in a way that allows discrimination between segments. We are completely ignorant of these problems of growth although most of the patterns in Fig. 5 (and others that are readily grown) have been found in the crayfish nerve cord by Wiersma or Kennedy and their associates.

I believe that the rules of connection between nerve cells are simple. Basically we have classes of cells which recognize each other. Within a segment the larger cells could be uniquely genetically determined and others could differentiate epigenetically upon them, with few extra rules. But where neurons are numerous, as in the compound eye, a few rules must determine the pattern for whole classes of cells.

1. Without segmental differences in target neurons.

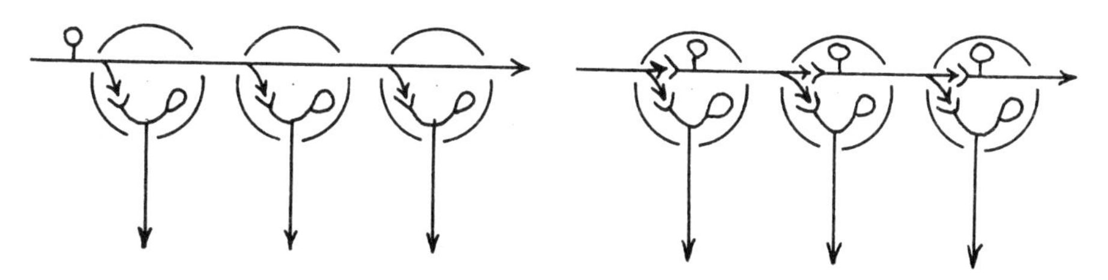

A. Premotor driving all ganglia. B. Chain of interneurons.

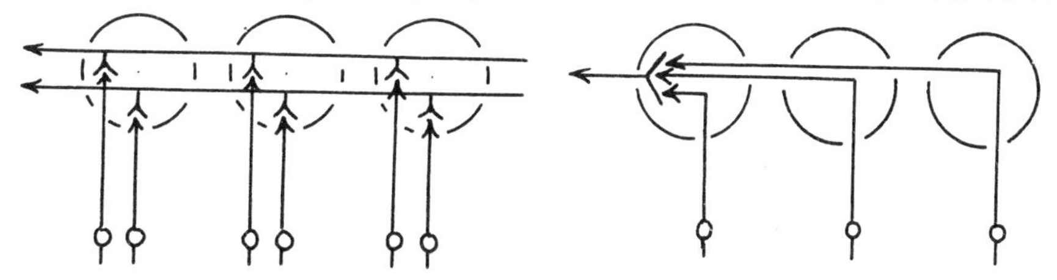

C. Sensory interneuron driven in either way from all ganglia.

2. Systems requiring segmental differences for growth.

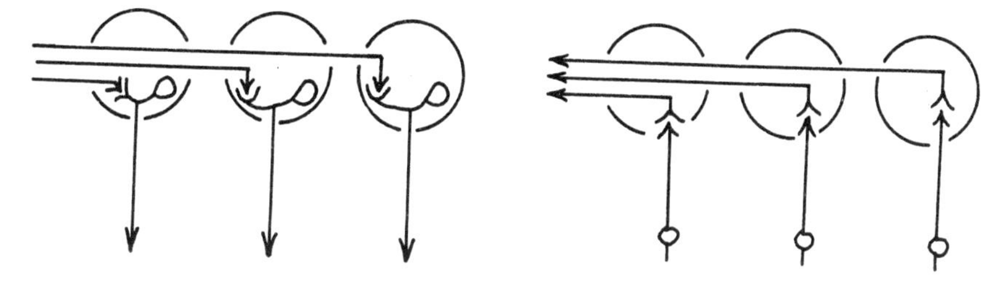

D. Separately driven segmental motoneurons. E. Separately driven sensory interneurons.

Figure 5. Neuron patterns in relation to rules of growth.

(2a) The process of neuronal development is hardly studied at the level of single identified cells in any invertebrate. Cell lineages of neurons are not known; all are small and sadly homogeneous when they first develop. Even the reason why one or two individual cells become giant fibres is not studied by the methods of molecular biology which are industriously applied to other differentiating systems. The reason is that the feature which makes the nervous system so interesting - the diversity of individual cells and specific connection patterns which reveal that every cell is unique - makes it an unattractive proposition for the biochemist.

(2b) Development can be interfered with by use of mutants, which act, to use Brenner's expression, in the differentiation dimension. The difficulty about mutants is that the ones of interest do not necessarily survive and one never knows at how many points the mutation is acting. In the nervous system a mutation is not likely to affect one attribute of one cell at a time. Others here deal with this field.

(3a) Regeneration of nerves into denervated tissue

Regenerating invertebrate axon stumps, advancing into their original tissue usually re-form their proper connections especially in young animals. This general statement has been inferred from numerous operations, for example on annelid worms (Faulkner, 1932), crustacea (Mapelli, 1931) and insects (Bodenstein, 1957; Drescher, 1960) over a period of many years, by observation of normal responses after functional recovery. This applies to sensory and motor axons and also to transections of the central nervous system. In some worms parts of the animal including whole ganglia are regenerated when lost.

In lower vertebrates, regeneration back to normal function can occur in all neuron types in fish and newts, and in some parts of higher vertebrates. But in higher vertebrates and particularly mammals there are failures to re-establish appropriate connections and erroneous connections can be formed. One of the ways of learning about mechanisms is to investigate these errors. Unfortunately in vertebrates there are usually too many parallel pathways, and the target organs lose their specificity when denervated, so that little has emerged at cell level.

Failure to re-establish appropriate connections in higher vertebrates does not necessarily imply that the neuron's specificity has been lost: a likely alternative is that scar tissue grows quickly and prevents the appropriate growth of axon stumps.

(3b) Regeneration after transplantation

In insects we have a favourable situation for working at single neuron level as the following examples illustrate.

The nerve cord of the cricket contains a number of giant axons which are excited by sensory cells in the anal cerci. The giant fibre impulses run forward and alert the anterior end of the animal when a slight air movement behind signifies danger. The cerci regenerate in the juvenile stages and the peripheral sensory neurons regrow *de novo*. When an anal cercus is transplanted to an abnormal situation further forward on the abdomen the sensory axons which grow from it to the central nervous system can still make functional connections with the giant fibres even though the synapses must be at an abnormal region of the target neuron. Specificity at cell level is not known in this case (Edwards and Sahota, 1968).

The motoneurons of insects are each individually recognizable by dendritic form, position size and shape of cell body and by the pathway of the axon to a definite part of a particular muscle. They are even fairly similar in different ganglia (Young, 1969). The three ganglia of the thorax, which send motoneurons to the leg muscles can be explored by micro-electrode until each motoneuron has been labelled with procion yellow, stimulated, and its muscle identified. A number of motoneurons have now been so identified with their muscles in the cockroach (Young, 1972) and in several species of locusts which can be cultured in the laboratory (Bentley, 1970; Tyrer, unpublished).

Young cockroaches regenerate lost legs and they also recover function when nerves regenerate after section. With the constancy

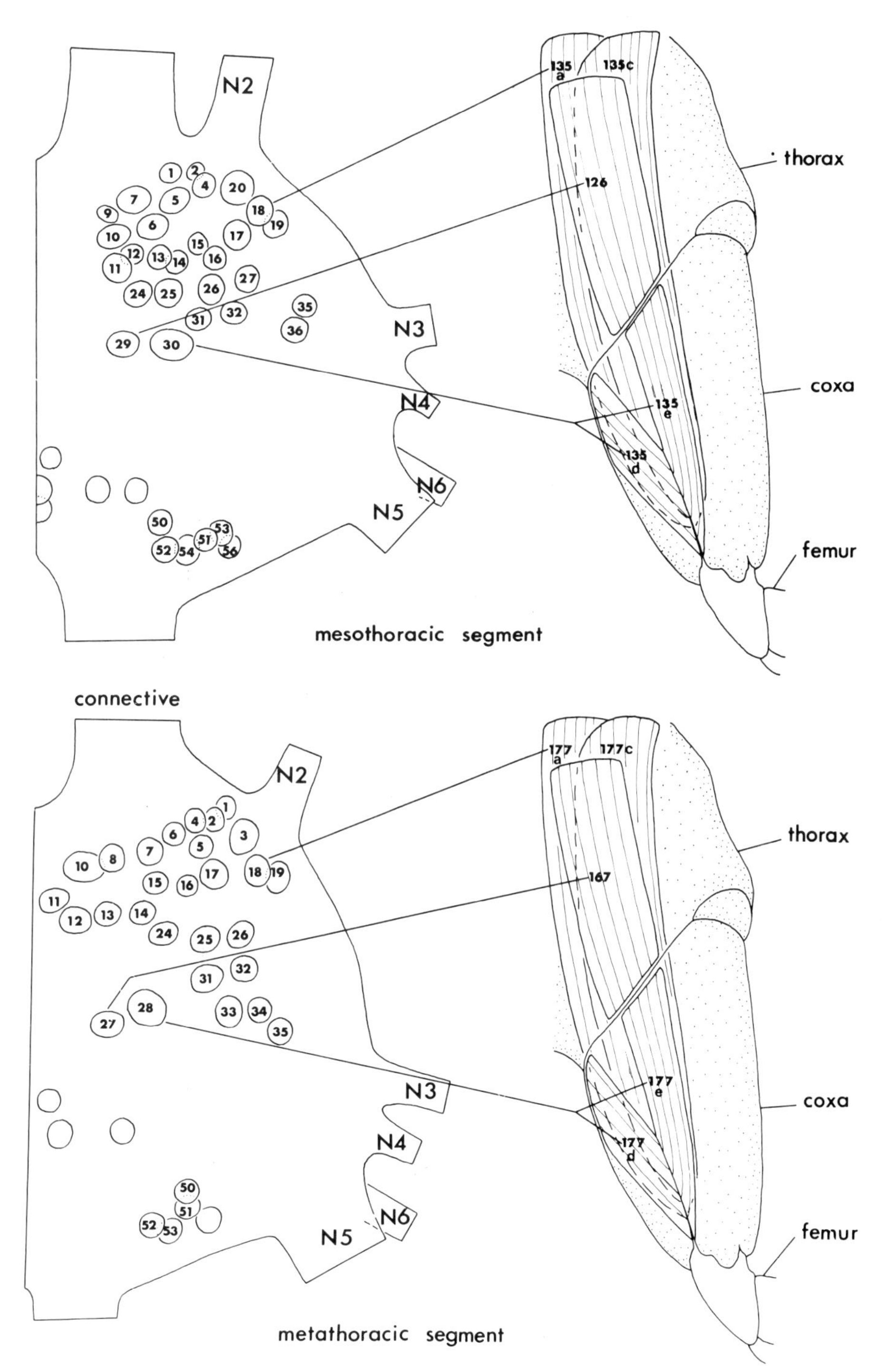

Figure 6. Specifity of motor neuron connexions: comparison of the meso- and metathoracic segments of the cockroach. For each is shown, left: a half ganglion with peripheral nerve trunks (N2 - N6) and identified motor neurons (numbered), and right: the proximal parts of the leg, with numbered muscles. Corresponding cell bodies in the two segments innervate corresponding leg muscles, as shown by connecting lines. (From Young, 1972)

of the normal system in mind, Dr Young in my laboratory has been examining the connection pattern in regenerates and transplants.

In the normal regenerated leg, which resembles in anatomy the leg it replaces, a normal innervation of a normal new set of muscles is the general rule. The same is apparently true when the nerve is cut and allowed to regenerate with the leg left in place. These conclusions rest at present mainly on the persistence of the central pattern of motoneurons and on the recovery of function, for the connections of only a few individually identified cells have been traced to their regenerated terminals.

A more interesting experiment, now in progress, is to transplant a metathoracic leg to the mesothoracic segment and then test the pattern of connections of motoneurons. Sure enough the mesothoracic central cells now innervate muscles in the metathoracic leg corresponding to the muscles they formerly innervated in their own leg. Several conclusions follow from this. First, the metathoracic leg muscles accept and distinguish between mesothoracic axons, so that in other experiments care must be taken to prevent unwanted invasion of transplanted legs. The only way to be certain of the connection between the identified cell body and the muscle fibre is to penetrate the one with a microelectrode and observe the contraction of the other. Secondly, this result shows that the mechanism of recognition of the various leg muscles by the corresponding axons is repeated segmentally.

The next stage is to modify and then control the process of recognition between motoneuron and target cell.

The preparation permits the same variety of experiments that have been done on vertebrates with the added advantage that the axons which grow in to the muscles are individually identified. This avoids the question of central cell death and possible re-innervation by a different group of motoneurons from those originally present. Conclusions of greater interest may be possible as the number of identified re-established connections at single cell level accummulate and errors begin to turn up.

In summary of this work, from the innervation pattern of transplanted parts placed in abnormal locations, it is possible to make a list of what cells another cell will and will not establish contact with. Each cell in the nervous system necessarily has a repertoire of possible target cells which will accept it and which it will innervate. This repertoire defines the limits of the mechanism of compatibility, which can therefore be studied by analysis of chemical and other factors which influence the repertoire. If the repertoire is a single neuron or group of muscle cells, which in turn will accept only their own innervating axon, one must know this before analysing the mechanism. If on the other hand the target muscle cells will accept any motor axon which in turn will innervate any muscle of that species, this must be known before the basic mechanism can be analysed.

I speak as if the mechanism can be worked out but all realize that this is only an aspect of the problem of differentiation of cell types and of the way in which organs are built up by the recognition of cell by cell. In the study of neurons we have cells with a fantastic diversity available, and these cells show that they recognize each other by forming synaptic connections or by rejecting them. We must first define the times and number of the choices which are available to a neuron, then observe by its own cellular behaviour that relevant differences exist for it, as distinct from biochemical or structural differences which we observe.

A general impression is that annelids and arthropods will turn out to be completely specified within each segment, but that interesting results which throw light on the origin of cephalization will follow from studying the innervation of transplants between different segments.

ACCEPTANCE OF ERRONEOUS INNERVATION

Errors sometimes arise in regeneration of axon stumps where the target cells have lost their original correct innervation for some time, and have then been re-invaded. From experiments of this type it is inferred that the target cells undergo a change which is called a loss of specificity, in that they accept other than their normal partner cells. This has been observed where target cells are (a) mammalian muscle innervated by motoneurons of other muscles (Sperry, 1945), (b) sympathetic ganglion cells innervated by spinal cells to the wrong ganglion (Guth and Bernstein, 1960), (c) cockroach giant fibres innervated by sensory neurons of the anal cercus of the wrong side of the animal (Edwards and Palka, 1971), (d) medulla units at the anterior end of spinal dorsal columns of the rat (Wall, 1971). An insufficient variety of experiments have been done for generalizations to be made about which individuals or even which classes of abnormal axons are accepted by denervated tissue. This loss of specificity is not necessarily universal but its existence means that the change in specificity can be investigated.

PARTNER RETURNS TO FIND IMPOSTER INSTALLED

An instructive situation arises when the proper axons arrive to find that their normal target cells have accepted abnormal axons after a period of denervation. All experiments have so far been performed on preparations which show only types of cells and are therefore not concerned with individually recognizable neurons or target cells.

In the cervical sympathetic chain of the cat preganglionic fibres normally innervate different postganglionic pathways so that stimulation of root T_1 causes dilation of the pupil and stimulation of root T_4 causes dilation of ear blood vessels.

When all the preganglionic fibres are cut there is an eventual regeneration back to normal. But if only fibres of T_1, T_2 and T_3 are killed and T_4 is left intact, it is possible for some months to dilate the pupil of the eye and the blood vessels of the ear by stimulation of T_4. Therefore T_4 fibres spread to abnormal connections. These abnormal connections persist until T_3 regenerates, when normal relations are again established (Guth and Bernstein, 1961).

This shows that the target cells prefer a certain input if they can get it, but failing the normal situation they lose specificity and accept abnormal partners. However they are able to throw off these abnormal connections when their proper partners return.

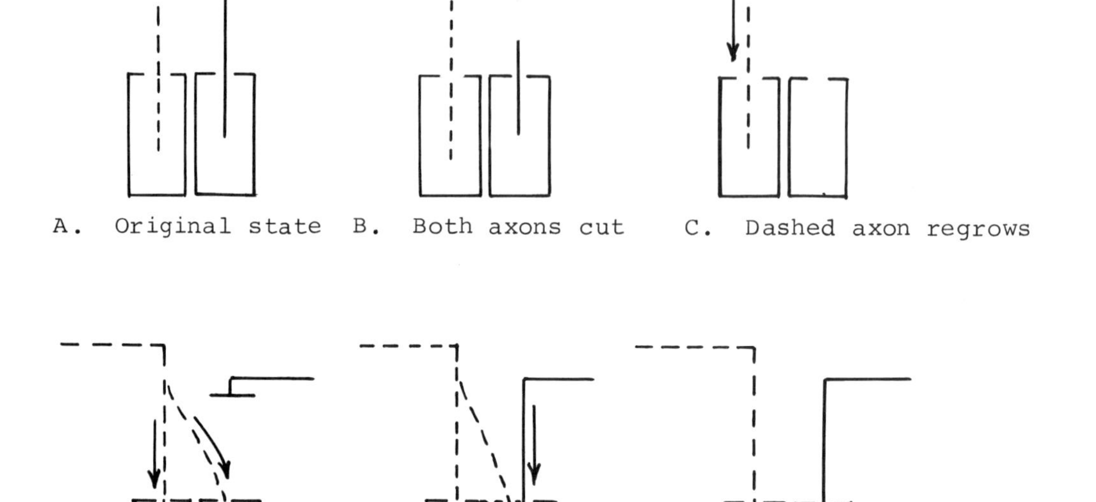

Figure 7. Regeneration of two axons to two target cells. A test with a choice brings out specifity not apparent in the no choice situation.

Recently Mark and collaborators (1970) in Melbourne have taken this one step further. The nerve to the superior oblique muscle is cut and prevented from growing again while this muscle is innervated by motor axons of the inferior oblique muscle. The neuromuscular junctions so formed persist but become non-functional when the normal innervation returns. The test of function was to stimulate the superior oblique nerve and look for responses of the superior oblique muscle. This shows that synapses can be switched off in function although persisting in structure, so that a problem in neuron specificity provides a possible synaptic mechanism for long-term plastic behaviour changes and learning.

CONCLUSION

Neurons and their target cells have a mechanism for recognizing their correct partners. The wide diversity of neuron types and of growth or regeneration situations means that analysis of neuronal specificity is a promising way of attacking the general question of how different cells recognize each other. With certain types of neurons we work with individually identified cells and we can present the growing axon with various choices of target cell under controlled conditions. The possibilities have been appreciated only in the last two or three years. The consequences cannot be predicted but they will perhaps include increased chances of recovery of function after damage to the human nervous system.

REFERENCES

Bentley, D.R. (1970). A topological map of the locust flight motor neurons. J.insect Physiol. 16, 905-918.

Bodenstein, D (1957). Studies on nerve regeneration in *Periplaneta americana*. J.exp.Zool. 136, 89-116.

Boschek (1971). On the fine structure of the peripheral retina and lamina ganglionaris of the fly, *Musca domestica*. Z.Zellforsch. mikrosk.Anat. 118, 369-409.

Bullock, T.H. and Horridge, G.A. (1965). Structure and function in the nervous systems of invertebrates. pp.1719. Freeman, San Francisco.

Drescher, W. (1960). Regenerationsversuche am Gehirn von *Periplaneta americana*. Zeit.Morph.Okol.Tiere 48, 576-649.

Edwards, J.S. and Palka, J. (1971). Neural regeneration. Delayed formation of central contacts by insect sensory cells. Science 172, 591-594.

Edwards, J.S. and Sahota, T.S. (1968). Regeneration of a sensory system: the formation of central connections by normal and transplanted cerci of the house cricket *Acheta domesticus*. J.exp.Zool. 166, 387-396.

Faulkner, G.H. (1932). The histology of posterior regeneration in the polychaete *Chaetopterus variopedatus*. J.Morph. 53, 23-58.

Guth, L. and Bernstein, J.J. (1961). Selectivity in the re-establishment of synapses in the superior cervical sympathetic ganglion of the cat. Exp.Neurol. 4, 59-69.

Hertweck, H. (1931). Anatomie und Variabilität des Nervensystems und der Sinnesorgane von *Drosophila melanogaster* (Meigen). Zeit.wiss.Zool. 139, 559-663.

Horridge, G.A. (1968). Interneurons, their origin, action, specificity, growth and plasticity. pp. 436. Freeman, San Francisco.

Horridge, G.A. and Meinertzhagen, I.A. (1970). The accuracy of the patterns of connexions of the first- and second- order neurons of the visual system of *Calliphora*. Proc.Roy.Soc.Lond. B. 175, 69-82.

Hughes, A.F.W. (1968). Aspects of neural ontogeny. pp. 249. Logos Press. London.

Mapelli, P. (1931). Processi rigenerativi del tessuto nervoso dell'*Astacus saxatilis*. Arch.zool. (Ital) Napoli 15, 127-156.

Mark, R.F., Marotte, L.R. and Johnstone, J.R. (1970). Reinnervated eye muscles do not respond to impulses in foreign nerves. Science 170, 193-194.

Meinertzhagen, I.A. (1972). Erroneous projection of retinula axons beneath a dislocation in the retinal equator of *Calliphora* (in MS). Canberra.

Retzius, G. (1891). Zur Kentniss des centralen Nervensystems der Würmer. Das Nervensystem der Annulaten. Biol.Untersuchungen. N.F.2, 1-28.

Sperry, R.W. (1945). The problem of central nervous re-organisation after nerve regeneration and muscle transposition. Quart.Rev. Biol. 20, 311-369.

Stern, C. (1938). The innervation of setae in *Drosophila*. Genetics *23*, 172-173.
Tyrer, M. (1972). Unpublished work on connections of individual motoneurons in the Australian plague locust *Chortoicetes*. Canberra.
Wall, P.D. and Egger, M.D. (1971) Formation of new connexions in adult rat brains after partial deafferentation. Nature Lond. *232*, 542-545.
Young, D. (1969). The motor neurons of the mesothoracic ganglion of *Periplaneta americana*. J.Insect Physiol. *15*, 1175-1179.
Young, D. (1970). The structure and function of a connective chordotonal organ in the cockroach leg. Phil.Trans.Roy.Soc. London B. *256*, 401-428.
Young, D. (1972). Unpublished work on connections of individual motoneurons to muscles of transplanted cockroach legs. Canberra.

ALTERATIONS IN SYNAPTIC EFFECTIVENESS ACCOMPANYING BEHAVIORAL MODIFICATIONS IN *APLYSIA*

ERIC R. KANDEL
VINCENT CASTELLUCCI
THOMAS J. CAREW
New York University Medical School and The Public Health Research Institute of the City of New York, New York, N.Y. 10016

In the marine mollusc *Aplysia* one can combine behavioral and cellular neurophysiological approaches to study the neural mechanisms of a reflex (gill withdrawal) that undergoes three simple behavioral modifications: habituation, dishabituation and sensitization. Most of the motor neurons, and a population of mechanoreceptor neurons and interneurons that mediate the reflex can be identified and studied intracellularly. (Kupfermann and Kandel, 1969; Peretz 1969; Castellucci, Kupfermann, Pinsker and Kandel, 1970; Kupfermann, Pinsker, Castellucci and Kandel 1971). As a result, a significant portion of the neural circuit of this reflexive behavior can be analyzed and the contribution of individual neurons to the total reflex and its modifications can be studied. We have so far used this preparation for two purposes: 1) to analyze the cellular mechanisms underlying habituation and dishabituation (Kupfermann et al. 1970, Castellucci et al. 1970) and 2) to examine the relationship of dishabituation to sensitization (Carew, Castellucci and Kandel 1971). Since this preparation shows both short-term and long-term habituation, it may also prove useful for studying the neural processes underlying the transition from short-term to long-term memory.

Habituation and dishabituation are often considered to be elementary forms of learning. Habituation is the decrease of a behavioral response that occurs when an initially novel stimulus is repeatedly presented. Spontaneous recovery of the habituated response occurs if the stimulus is withheld for relatively long periods of time. Dishabituation is the immediate restoration of a previously habituated response following the presentation of a novel or strong stimulus to another part of the animal.

A weak or moderate tactile stimulus applied either to the siphon or to the purple gland of *Aplysia* produces gill-withdrawal which is part of a more complex defensive withdrawal reflex. (Pinsker, Castellucci, Kupfermann and Kandel, 1970; Carew et al. 1971). When 5-10 tactile stimuli are repeatedly applied at intervals of two minutes or less, a progressive decrease (habituation) in the amplitude and duration of the gill-withdrawal reflex occurs. Recovery of reflex responsiveness following a single habituation session of ten stimuli requires a period of rest lasting from 10 minutes up to several hours. Immediate restoration of responsiveness (dishabituation) can be brought about, however, by presenting a strong stimulus to either the head or the tail of the animal. When habituation sessions are repeated over consecutive days, the defensive reflex undergoes long-term habituation lasting weeks. (Carew, Pinsker and Kandel 1971). Daily habituation training (10 stimuli/4 days) produces habituation of the reflex response which builds up progressively across days so that greater response decrement occurs on subsequent days. Thus, after 5 days, the mean duration of the reflex response (summed across ten trials) is only 20% of duration of the response exhibited on the first day. This habituation persists unchanged for at least a week and is only partially recovered after 3 weeks. As is the case for higher forms of learning, long-term habituation is sensitive to the temporal pattern of stimulation. To produce long-term alterations in reflex responsiveness, it is necessary to space the training trials (e.g. 10 stimuli/day for 4 days). Forty consecutive trials massed in a single day are no more effective in producing long-term habituation than are 10 trials in a single day.

We have so far only studied the neural mechanisms underlying the short-term behavioral modifications that accompany a single training session (10 stimuli) of habituation. In a semi-intact preparation intracellular recordings were obtained from identified motor neurons controlling gill movements. Habituation of the behavioral response to repeated tactile stimulation was found to be associated with a decrease in the amplitude of the complex excitatory postsynaptic potential and a reduction in the number of spikes produced in the motor neurons by the afferent input (Kupfermann, Pinsker, Castellucci and Kandel, 1970). This complex excitatory post-synaptic potential (EPSP) is produced by the activity of mechanoreceptor sensory neurons that make direct (monosynaptic) excitatory connections on to the motor neurons and indirect (polysynaptic) connections via excitatory interneurons. Dishabituation was associated with an increase in the previously decremented complex EPSP and an increase in the spike discharge of the motor cells. Peripheral effects such as sensory adaptation or motor fatigue, as well as the involvement of the peripheral nerve net work, were ruled out (Kupfermann et al 1970; 1971). In the isolated abdominal ganglion the alterations of EPSP amplitude produced by repeated electrical stimulation of an afferent nerve also paralleled the alteration of the EPSP produced by repeated tactile stimulation during the experiments in semi-intact preparations. Such EPSP decrement was found not to be associated with changes in the intracellularly measured input resistance of the motor neurons.

Since the complex EPSP involves both monosynaptic connections from sensory neurons to motor neurons and polysynaptic connections mediated via interneurons, the change could occur at several sites. We have so far only carried out a detailed analysis of the monosynaptic pathway. We found that an elementary, presumably monosynaptic, EPSP produced in a motor neuron by intracellular stimulation of a single sensory neuron, exhibited a prolonged decrement following 5-10 repeated stimuli that paralleled behavioral habituation; strong stimulation of another pathway produced a sudden increase in EPSP size (heterosynaptic facilitation) that paralleled dishabituation (Castellucci et al 1970).

These data indicate that at least part of the habituation and the dishabituation of the gill-withdrawal reflex is caused by changes in the synaptic effectiveness of previously existing excitatory synapses between the sensory and motor neurons. Our data do not, however, permit us to distinguish between pre- and post-synaptic contributions to these changes in synaptic effectiveness.

This preparation can also be used to examine the inter-relationship of different behavioral modifications. We have used it to examine the relationship of dishabituation (the increase in responsiveness of an habituated pathway following a novel stimulus), to sensitization (an increased responsiveness of non-habituated pathways following a strong stimulus). We found that habituation of the gill-withdrawal reflex elicited via stimulation of one part of the receptive field (siphon), in the intact animal, did not affect reflex responsiveness from stimulation of another part (mantle shelf) of the field. The effects of a strong stimulus were, however, more widespread, facilitating both habituated responses (dishabituation) and non-habituated responses (sensitization). Parallel results were obtained in the isolated ganglion. The complex EPSP produced in motor neuron L7 by stimulating one afferent nerve (siphon nerve) was relatively unaffected by decrement of the EPSP produced by repeated stimulation of another nerve (branchial nerve). However, strong stimulation of the connective from the head ganglia facilitated both the decremented and the non-decremented EPSPs produced by the two nerves. These experiments indicate that dishabituation of this reflex is not simply the removal of habituation, but an independent facilitatory process. Dishabituation can therefore be considered a special case of sensitization, a process which is spatially more widespread, in which a strong stimulus enhances reflex responsiveness of a number of reflex pathways independent of whether these have been previously habituated.

This preparation might also be useful in examining biochemical mechanisms of habituation and dishabituation. For example, the neural mechanisms of short-term habituation and of dishabituation do not appear to depend upon the synthesis of new proteins. These neural events were unaltered after protein synthesis had been inhibited by 97% for several hours (Schwartz, Castellucci and Kandel, 1971).

It would now be of obvious interest to apply these cellular approaches to the investigation of long-term habituation.

REFERENCES

Carew, T., Castellucci, V. and Kandel, E., (1971), An analysis of dishabituation and sensitization of the gill-withdrawal reflex in Aplysia, Intern. J. Neurosci. 2, 79.

Carew, T., Pinsker, H., and Kandel, E., (1971), Long-term habituation of a defensive withdrawal reflex in Aplysia, Science (in press).

Castellucci, V.F., Kupfermann, I., Pinsker, H., and Kandel, E.R., (1970), Neuronal mechanisms of habituation and dishabituation of the gill withdrawal reflex in Aplysia, Science, 167, 1445.

Kupfermann, I., and Kandel, E.R., (1969), Neural controls of a behavioral response mediated by the abdominal ganglion of Aplysia, Science, 164, 847.

Kupfermann, I., Pinsker, H., Castellucci, V.F., and Kandel, E.R., (1970), Neuronal correlates of habituation and dishabituation of the gill-withdrawal reflex in Aplysia, Science, 167, 1743.

Kupfermann, I., Pinsker, H., Castellucci, V., and Kandel, E., (1971), Central and peripheral control of gill movements in Aplysia, Science (in press).

Peretz, B., (1969), Central neuron initiation of periodic gill movements, Science, 166, 1167.

Pinsker, H., Castellucci, V.F., Kupfermann, I., and Kandel, E.R., (1970), Habituation and dishabituation of the gill-withdrawal reflex in Aplysia, Science, 167, 1740.

Schwartz, J.H., Castellucci, V. F., and Kandel, E.R., (1971), Functioning of identified neurons and synapses in abdominal ganglion of Aplysia in absence of protein synthesis, J. Neurophysiol. (in press).

REGENERATION AND CHANGES IN SYNAPTIC TRANSMISSION BETWEEN INDIVIDUAL NERVE CELLS IN THE CENTRAL NERVOUS SYSTEM OF THE LEECH

J. G. NICHOLLS and J. K. S. JANSEN*
Department of Neurobiology, Harvard Medical School
Boston, Massachusetts 02115 USA

Numerous experiments on invertebrates and vertebrates indicate that the connections of nerve cells are specific in the sense that synapses are established with certain cells but not with others. At present we have little information about the processes that enable nerve cells to find their targets, either during development or in regeneration after a lesion; nor is it clear how accurately each cell forms its connections. The relatively simple nervous system of the leech offers an especially favorable preparation for studying such questions at the level of individual identified cells of known function. Earlier work on chronically operated leeches has shown that the central nervous system can regenerate and that in addition the animals are capable of undergoing other changes in their nervous system to compensate for the effects of the lesion (Baylor and Nicholls, 1971). Thus, leeches moved in an obviously uncoordinated manner after the operation, but swam normally within a few weeks, whether or not regeneration had occurred.

In the present experiments reported it has been shown that:

1. Specific connections between individual nerve cells are reestablished by regeneration.

2. Some of the regenerated connections differ from those seen in normal animals.

3. Similar changes in the synaptic connections take place in parts of the nervous system not directly affected by the surgical lesion.

REFERENCE

Baylor, D.A. and J.G. Nicholls, (1971), Patterns of regeneration between individual nerve cells in the central nervous system of the leech, Nature (London) 232, 268.

*Present Address: Institute of Physiology, University of Oslo, Oslo, Norway.

INTERACTION BETWEEN NERVE CELLS DURING DEVELOPMENT AND IN THE ADULT. A DISCUSSION ON ENZYMATIC REGULATORY MECHANISMS OF NEUROTRANSMISSION.

EZIO GIACOBINI

Department of Pharmacology, University of Lund, Lund and Research Laboratories, AB Draco, Lund, Sweden

The establishment of interneuronal (synaptic) connections during the development of the nervous system may be regulated via mechanisms of neuronal interaction involving specific chemical affinities. One may infer that growing fibres carry their own identity and are able to recognize selectively other cells. Their "identification tags" may be of chemical nature as proposed by Sperry (1964) and could be identical with the specific proteins involved in the mechanism of chemical transmission (Giacobini, 1971 a, b) or could consist of specific surface structures (antigen-like ?) genetically controlled and capable of subtle recognition.

In order to test the first hypothesis, the initial appearance and the development of the transmitter's metabolizing enzymes have been investigated in "synapse-forming" and "asynaptic" groups of neurons belonging to sympathetic and spinal ganglia respectively, of the chick embryo (Giacobini, 1971 a).

Because of its complexity, the central nervous system does not lend itself to this kind of detailed study, we have therefore made our observations on autonomic ganglia. There is no reason to believe that the basal mechanisms of regulation in the ganglion cells should be different from those acting in the CNS.

The changes in the ratio between sympathetic and spinal ganglia for acetylcholinesterase (AChE), cholineacetylase (ChAc) and monoamineoxidase (MAO) activity indicate a relationship between development of the synaptic apparatus in the sympathetic ganglia and the peak activity of these enzymes, and suggest that specific enzymes related to the transmission mechanism may also be related to the "chemotactic guidance" governing the growth of new fibres (Giacobini, 1971 a).

On the other hand, in the adult autonomic neuron, the turnover of the transmitter seems to be related to its functional activity; the level of transmitter, however, remains constant unless synthesis is inhibited. The rate of synthesis of the transmitter in the neural tissue may therefore vary according to the level of nervous activity.

The interneuronal system of "chemical information" which regulates synthesis and inactivation of the transmitter is a result of an interaction with a) the synapse(-s) formed by the neuron with the periphery (peripheral synaptic output) and b) the synapse(-s) impinging on its cell body or processes (central synaptic output) as shown in the diagram.

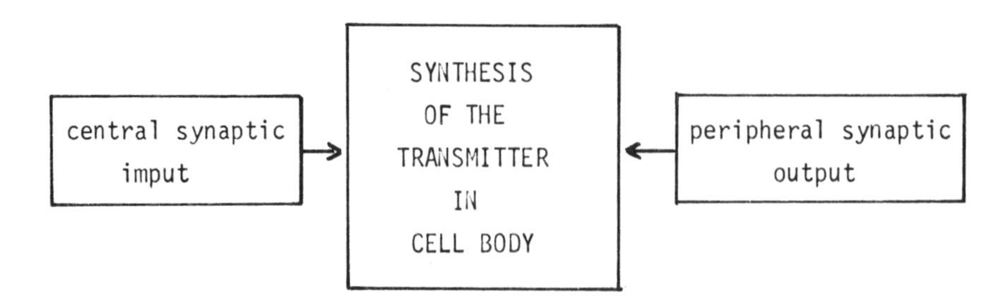

The role of synaptic connections and impulse activity in the maintenance of transmitter metabolism has been evaluated by studying the variation in activity of the enzymes synthetizing or inactivating the transmitter(-s) in adult sympathetic ganglia (Giacobini, 1971 b) under different conditions.

From these studies (Axelrod et al. 1970; Axelrod, 1971; Giacobini, 1971 b) it appears that the molecular mechanisms of regulation and interaction in the monoaminergic neuron are twofold and may involve transsynaptic induction of enzymes in the cell body or terminals according to the following:

A) rapid regulation of amine synthesis ⟶ via tyrosine hydroxylase (TH) activity (end product inhibition, variation in co-factor, endogenous inhibitors, precursor or substrate concentration).

B) slow regulation of amine synthesis ⟶ via new synthesis of tyrosine hydroxylase or dopamine-β-hydroxylase (DBH) or via increase or decrease in intraneuronal monoamineoxidase (MAO).

Other possible mechanisms are probably related to interactions between two or several different transmitters (e.g. noradrenaline and acetylcholine), or are hormonal in nature (Axelrod et al. 1970) (Fig. 2).

As reported in Fig. 1, the biosynthetic pathway for catecholamines is represented by four different enzymes, at least three of which (TH, DBH, PNMT) seem to be of "critical" importance (Axelrod, 1971). The biosynthesis of acetylcholine, however, is governed by one single enzyme, i.e. cholineacetylase.

In sympathetic neurons, the transmitter noradrenaline is synthetized and stored both in the cell body and in the terminals (Fig. 2).

As a direct consequence of such an arrangement, any condition which influences one of the "critical" enzymes (Fig. 1) in either the cell bodies located in the sympathetic ganglia, the nerve terminals, or the adrenal medulla (PNMT) is bound to produce some change in the rate of synthesis of the transmitter (catecholamines).

Moreover, since in the autonomic system several neurons are sequentially connected to each other, it is reasonable to think that a change produced in a neuron A or in the junction between neuron A and neuron B may also influence the next one, neuron C (Fig. 2). It is not unthinkable that this "interaction" might be exerted in both directions, i.e. A ⟶ C and C ⟶ A.

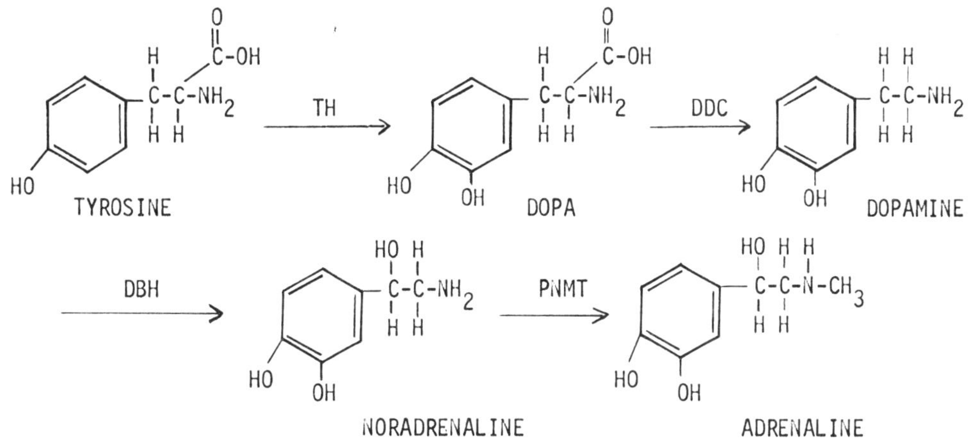

TYROSINE —TH→ DOPA —DDC→ DOPAMINE

—DBH→ NORADRENALINE —PNMT→ ADRENALINE

Fig. 1. Biosynthesis of catecholamines. TH = tyrosine hydroxylase, DDC = dopa-decarboxylase, DBH = dopamine-β-hydroxylase, PNMT = phenylethanolamine-N-methyl-transferase.

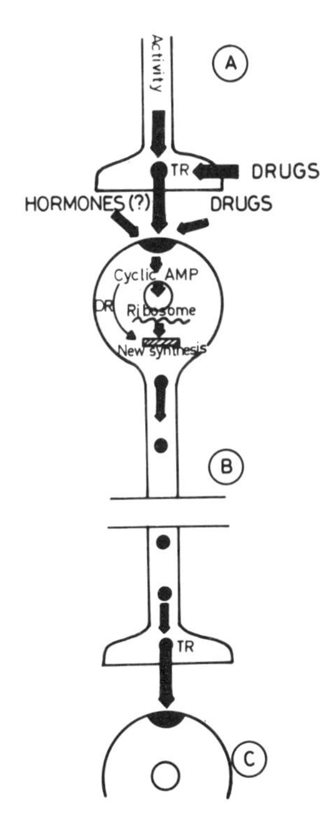

Fig. 2. Schematic diagram of synaptic connections between three neurons in the autonomic system pathway (A, B and C). Impulse activity triggers the release of the transmitter (TR) from the storage site. The action of the transmitter upon the receptor sets off a cyclic-AMP dependent mechanism for a new enzyme (trans-mitter-related) protein synthesis. The new formed enzyme migrates down the axon where it increases the synthesis of the transmitter.

Finally, the influence of the target organ on these mechanisms has to be taken into account.

On the basis of the results found in the literature (Tables 1 and 2), we shall now schematically consider four different groups of experiments which may illustrate the following conditions:

A/ increased synthesis of biosynthetic enzymes in the sympathetic neuron

B/ decreased activity of biosynthetic enzymes in the sympathetic neuron and its functional consequences.

C/ interference with postganglionic sympathetic function and its functional consequences.

D/ accumulation of the transmitter in the sympathetic neuron.

The literature related to these conditions is reported in Tables 1 and 2 and in the list of references.

A/ An increased synthesis of the rate limiting TH may be induced by a selective alteration (or depletion) of the postganglionic noradrenergic nerve endings by means of reserpine (Table 1). According to Thoenen et al. (1970) this induction is caused by a reflex increase in the activity of the sympathetic nervous system as follows:

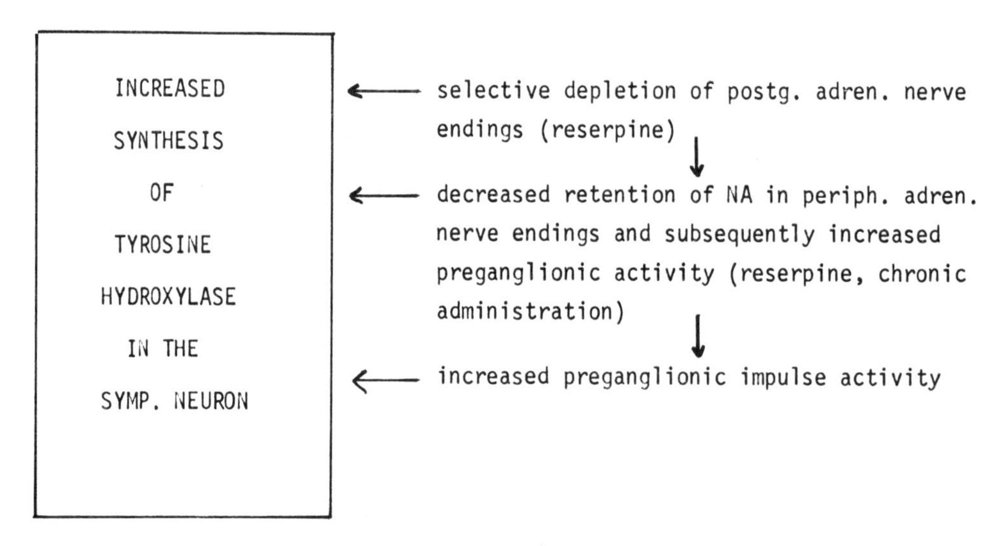

The neurally mediated increase in the TH activity in the adrenal gland and sympathetic ganglia can be prevented by inhibiting protein synthesis by the administration of actinomycin D or cycloheximide (Mueller et al. 1969 a) (Fig. 3). Furthermore, the increase in the TH activity in the rat superior cervical ganglion, produced by reserpine, is abolished by decentralization or administration of the nicotinic blockers chlorisondamine and pempidine (Mueller et al. 1970) (Fig. 3). Muscarinic blockers are ineffective.

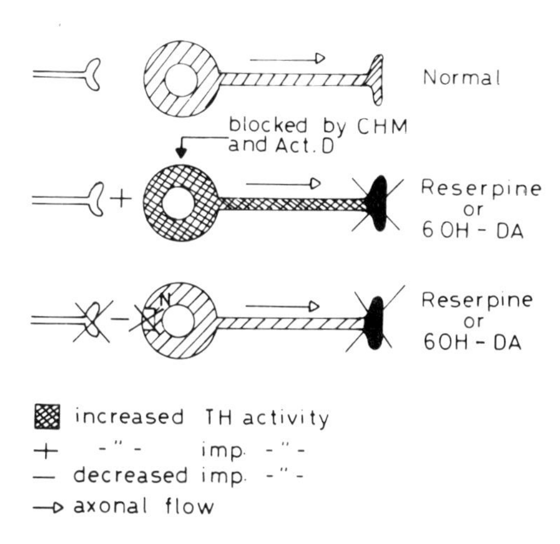

Fig. 3. Changes in tyrosine hydroxylase activity related to the effect of reserpine or 6-OH DA. N = nicotinic receptor; CHM = cycloheximide. See text.

It should be emphasized that each of the above agents (6-OH DA or reserpine) may induce a sustained and long lasting increase in sympathetic nerve activity. This increase might mediate an induction of the synthetizing enzyme leading to an increased synthesis of the actual transmitter. Other conditions influencing synthesis of TH are reported in Table 1.

B/ The conditions leading to a decreased TH activity in the sympathetic neuron and which may elicit a decrease in sympathetic activity are schematically shown below and reported in Table 1.

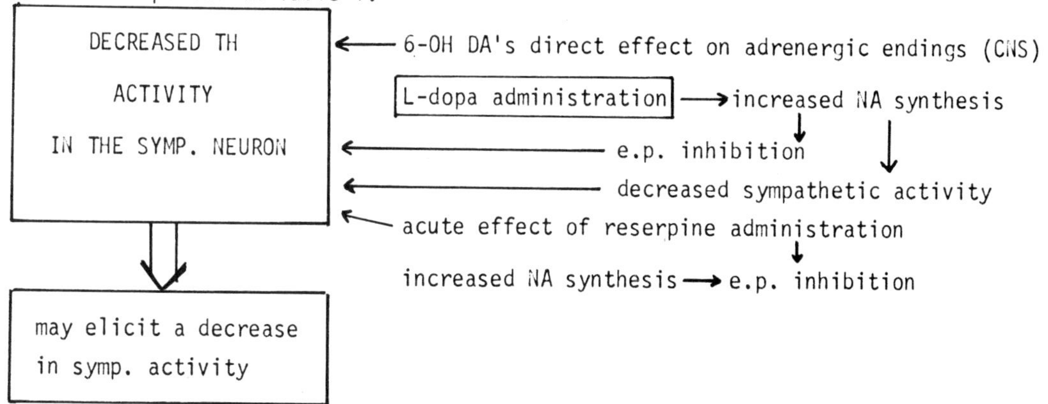

The experiments of Uretsky and Iversen (1970) demonstrate a fall in catecholamines, accompanied by a long lasting reduction in the activities of TH and DDC in brain areas which are rich in catecholamine containing nerve endings (hypothalamus and striatum), after administration of 6-OH DA (Table 1).

One can therefore assume that 6-OH DA, which causes a selective destruction of adrenergic nerve endings in the peripheral sympathetic nervous system (see diagram) exerts a similar effect on catecholaminergic neurons in the CNS.

A similar effect on TH activity of the rat adrenal gland is exerted by the administration of large doses (1 g/kg s.c. daily for 4-7 days) of the intermediate L-dopa (L-dihydroxyphenylalanine) (Dairman and Udenfriend, 1971).

The administration of L-dopa increases norepinephrine synthesis and produces a significant reduction in the activity of the "critical" biosynthetic enzyme TH, but not of L-ADC (Dairman and Udenfriend, 1971) (Table 1).

The reduction in enzyme activity is probably the result of a rapid mechanism of regulation which is considered to act through end-product inhibition. This represents a homeostatic mechanism capable of rapidly adjusting the rate of synthesis of the transmitter without modifying the total level of the synthetizing enzyme.

A decrease in the activity of the synthetizing mechanism is therefore an expression of the control mechanism compensating for the increase in transmitter synthesis produced by a physiological intermediate (in this case L-dopa).

In the adrenal gland a real decrease in enzyme level seems to be the consequence of L-dopa administration (Dairman and Udenfriend, 1971). This could be a result of either repressed synthesis or of an increase in the rate of enzyme degradation.

The increased transmitter synthesis produced by L-dopa could also result in a diminution of sympathetic nerve activity (postganglionically) and cause a more direct inhibition of the synthetizing enzyme (Fig. 4 and the previous diagram). The immediate effect of an acute administration of reserpine (depleting peripherally located catecholamines, e.g. in the heart) may actually be a decrease of TH activity in the heart (Thoenen et al. 1970) accompanied by a simultaneous increase in TH activity in the sympathetic (stellate) ganglia. The latter could be explained by a simultaneous compensatory increase of NA synthesis which may finally lead to a subsequent end-product inhibition (see Fig. 4 and previous diagram).

Fig. 4 illustrates schematically the situation in which (upper part of the figure) an accumulation of the transmitter in the cell body may be a consequence of a complete block of the impinging synaptic output, produced by preganglionic denervation (decentralization).

TABLE 1

Effect of various agents on transmitter-synthetizing enzymes

ENZYME	ORGAN	EFFECT	AGENT	REFERENCE
TH	adrenal cerv. sup. g.	increase	reserpine phenoxy- benzamine	Thoenen et al. 1969 a,b
TH	vas.deferens adrenal cerv. sup. g. brainstem	increase max after 3 days in ganglia	electr.stimul. reserpine In ganglia abol. by nicot.block.	Alousi, Weiner, 1966 Mueller et al. 1969 a, b,c Mueller et al. 1970
TH	stellate g. heart	increase	reserpine	Thoenen et al. 1970
TH DBH DDC	adrenal cerv. sup. g. hypothalamus	increase (not DDC)	cold-exposure	Thoenen, 1970
TH DDC	hypothalamus striatum	decrease	6-OH DA	Uretsky, Iversen, 1970
TH ChAc	adrenal	increase	amphetamine 6-OH DA reserpine	Mandell, Morgan, 1970
TH	brain	increase	amphetamine 6-OH DA reserpine	Mandell, Morgan, 1969 Mandell, Morgan, Oliver, 1970
TH	caudate n.	increase	cyclic AMP	Segal, Mandell, 1970
TPH	brain stem	increase	corticoste- rone reserpine	Azmitia, McEwen, 1969 Mandell, 1970
DBH PNMT	adrenal heart salivary g. stellate g.	increase	reserpine	Molinoff et al. 1970 Molinoff et al. 1971
TH	adrenal	increase	insulin	Weiner, Mosimann, 1970
TH	adrenal	decrease	L-dopa	Dairman, Udenfriend, 1971
TH	symp. g. vas.deferens	decrease after 24 hrs	6-OH DA	Cheah et al. 1971
TH	symp. g.	increase	reserpine	Cheah et al. 1971
L-ADC	adrenal brain	none	L-dopa	Dairman, Udenfriend, 1971
DDC	symp. g.	none	denervation	Giacobini, Noré, 1971

Explanation of symbols in Table 1. TH = tyrosine hydroxylase; DDC = dopa decarboxylase; L-ADC = aromatic L-aminoacid decarboxylase; DBH = dopamine -hydroxylase; PNMT = phenylethanolamine-N-methyltransferase; TPH = tryptophan-hydroxylase; ChAc = cholineacetylase.

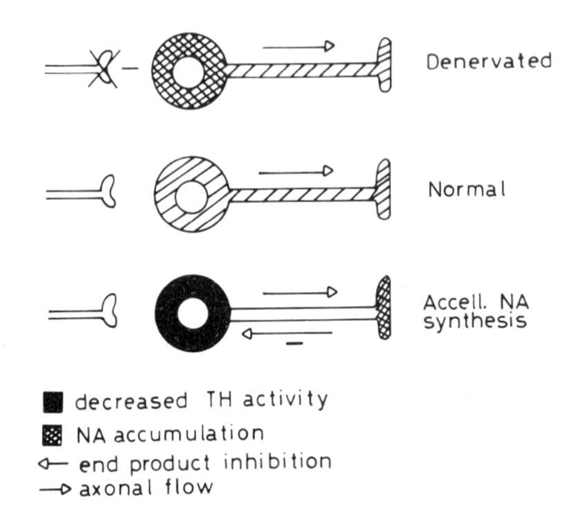

Fig. 4. Changes in tyrosine hydroxylase activity or NA related to denervation. See text.

Such an experiment was performed (Giacobini et al. 1970; Giacobini and Noré, 1971) by preganglionically denervating the lumbo-sacral ganglia of the cat and studying subsequently the variation of NA and of the intracellular inactivating enzyme (MAO) in total ganglion homogenates as well as in single nerve cells (cell bodies).

A small but significant increase (20-30%) in NA was found in both preparations (Giacobini et al. 1970; Giacobini and Noré, 1971) indicating an accumulation of the transmitter in the cell body.

This accumulation could depend on a) an increased rate of synthesis of the transmitter in the cell body, b) a decreased intracellular inactivation (MAO) or c) a decreased axoplasmic flow of the transmitter from the cell body to the peripheral endings (Fig. 4).

With respect to the first hypothesis, no change in TH activity has been found after decentralization in the superior cervical ganglion of the rat (Sedvall and Kopin, 1967 a, b and Sedvall et al. 1968); while treatment with 6-OH DA decreases

TH activity (together with MAO activity) by 30% during a period of 21 days (Cheah et al. 1971).

The second hypothesis could be ruled out on the basis of the results (Giacobini et al. 1970) showing an increase in MAO activity in denervated ganglion cells. Most likely, the axoplasmic flow was inhibited as a consequence of the decreased sympathetic activity (see previous diagram and part D) leading to an accumulation of the transmitter in the cell body.

C/ We shall now examine more closely the mechanisms of interaction by means of which an interference with postganglionic sympathetic function may elicit a reflex increase in preganglionic sympathetic activity.

Three conditions leading to interaction are listed in the diagram below:

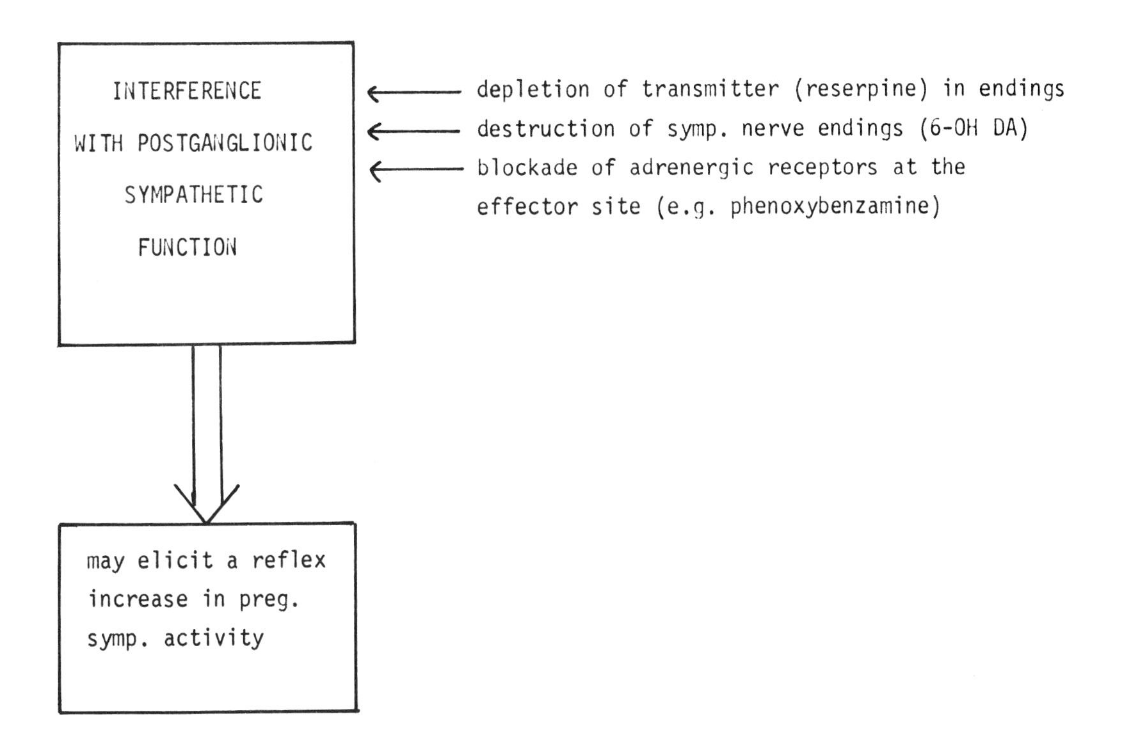

One of them should be analyzed more in detail since it may reveal more intimately the intracellular mechanisms of information and regulation in the nerve cell.

Fig. 5 represents sequentially the results of the experiments of Thoenen et al. (1970) and in particular illustrates the phase difference in the induction of transmitter synthesis in the cell body and nerve terminals respectively.

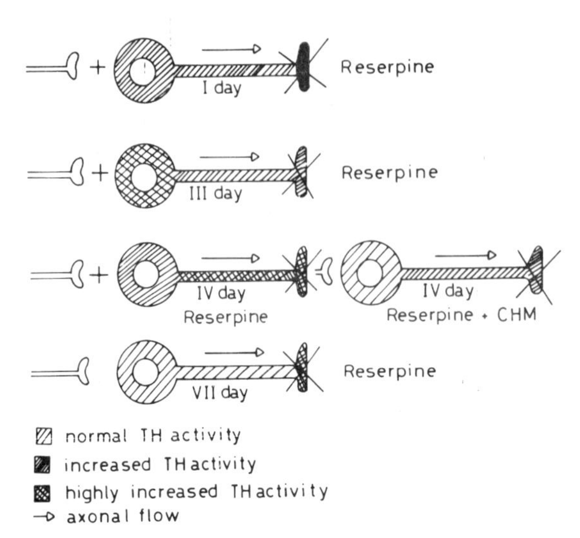

Fig. 5. Sequential changes in tyrosine hydroxylase activity related to the effect of reserpine. CHM = cycloheximide. See text.

The administration of reserpine depletes the transmitter in the peripheral synapse (heart), thus inducing an increase in perikarial TH which reaches a peak at the third day and then declines. At the 7th day, the TH activity in the cell body is back to its original value. In the terminals the increase in TH activity lags behind by two or three days reaching its highest level at 5 - 7 days, at which time the enzyme activity in the cell body has already begun to decline.

These differences in the time course of the rise and decline of TH after reserpine could be interpreted (Thoenen et al. 1970) as an initial increase of enzyme synthesis in the cell body followed by a subsequent transport of the enzyme to the distal part of the nerve cell. However, as discussed earlier, inhibiting the protein synthesis by means of cycloheximide decreases the enzyme activity in the heart terminals when the drug is given after TH has already begun to increase in the heart.

According to the authors (Thoenen et al. 1971) this would indicate that the induced enzyme is being principally synthetized in the nerve terminals and the lag in increased enzyme activity in the nerve terminals depends on the time necessary for the "regulatory information" to travel from the cell body to the endings.

D/ On the basis of what has been previously discussed we may be able to identify at least four different conditions (see diagram below), which theoretically could lead to an accumulation of the transmitter molecule in the sympathetic neuron (cell body):

a) decreased nerve stimulation (e.g. following denervation)

b) decreased intracellular inactivation (decreased MAO activity)

c) decreased axonal flow

d) increased rate of synthesis in the cell body, eventually combined with a, b or c.

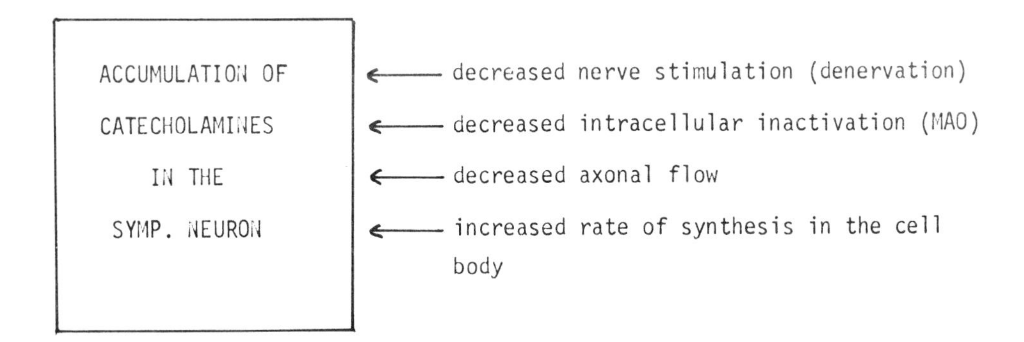

Our experiments (Giacobini et al. 1970) have shown that in the denervated (decentralized) ganglion the accumulation of the transmitter (NA) is accompanied by an increase of MAO activity and not by a decrease. Furthermore, in order to evaluate the role of the axoplasmic flow in the mechanism of regulation, we compared two different preparations (Giacobini, 1970 and unpublished results), one of them with ligated axons.

It is known that axonal ligation prevents the flow of cytoplasm and axonal material to the peripheral endings resulting into an accumulation of material at the place of ligation (Dahlström, 1969).

A marked decrease in the number of cells carrying ChAc, AChE as well as MAO indicated that the ligation affected the enzyme systems both in cholinergic and adrenergic cells (Fig. 6) (Giacobini, 1970).

On the other hand, in ligated cells which were no longer activated by synaptic input (decentralized cells) no changes were observed either in ChAc or AChE (as compared with non-ligated denervated cells). MAO activity, however, was still decreased in the cell body (Fig. 6). This suggests that incoming synaptic activity as well as axoplasmic flow may play a role in the regulation of intracellular inactivation mechanisms (MAO) in the cell body. Principally, the inhibition of axoplasmic transport for the inactivating enzyme (MAO) may reduce the synthesis of this enzyme in the cell body.

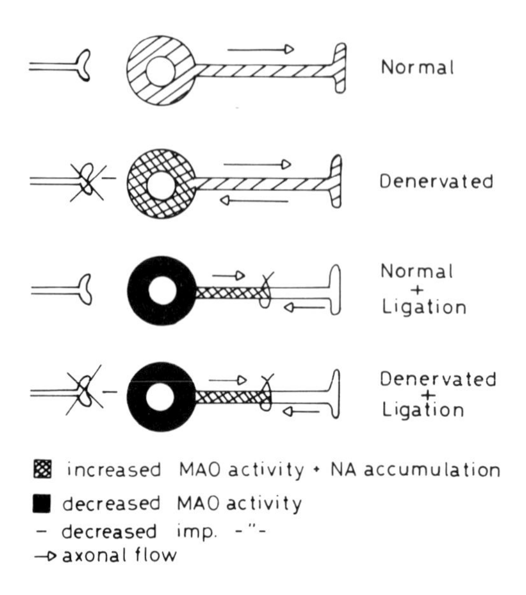

Fig. 6. Changes in MAO activity and NA related to denervation and ligation. See text.

The effect of various agents on transmitter-inactivating enzymes is reported in Table 2.

AChE activity is increased in the caudate nucleus by the action of cyclic AMP (Segal and Mandell, 1970). The decrease in MAO (and TH) activity in sympathetic ganglia and vas deferens following an early effect of 6-OH DA (Cheah et al. 1971) may be interpreted as a direct action of the drug on ganglion cells but it could also be secondary to its effect on the peripheral endings and to the following increased preganglionic sympathetic activity.

TABLE 2

Effect of various agents on transmitter-inactivating enzymes

ENZYME	ORGAN	EFFECT	AGENT	REFERENCE
COMT	symp. g.	none	denervation	Giacobini, Kerpel-Fronius, 1969
MAO	adrenal	increase	L-dopa	Tarver, Spector, 1970
MAO	adrenal heart stellate g.	none	reserpine	Molinoff et al. 1970
MAO	symp. g. cells	increase	denervation	Giacobini et al. 1970
MAO	caudate n.	none	cyclic AMP	Segal, Mandell, 1970
AChE	caudate n.	increase	cyclic AMP	Segal, Mandell, 1970
MAO	symp. g. vas.deferens	decrease after 24 hrs	6-OH DA	Cheah et al. 1971
MAO	symp. g.	none	reserpine	Cheah et al. 1971
MAO	optical pathways	increase	sensory deprivation (darkness)	Bigl et al. 1971
COMT	optical pathways	none	sensory deprivation (darkness)	Bigl et al. 1971
AChE	brain	decrease	undercutting	Chu et al. 1971
AChE	brain	none	undercutting + stim.	Chu et al. 1971

MAO = monoamineoxidase; COMT = catechol-O-methyltransferase; AChE = acetyl-cholinesterase.

Conclusions

The physiological mechanisms associated with changes in synthesis and inactivation of transmitters (e.g. catecholamines in sympathetic ganglia) clearly demonstrate a mutual interaction between a) the various parts of the nerve cell, principally between the two strategic sites (cell body and endings) involved in synthesis and degradation of the transmitter and b) the various cells involved in the same pathway. This interaction seems to be controlled by the level of synaptic activity. (Synapses A - B and B - C in Fig. 2) and be modulated by regulatory "chemical information" flowing in both directions (A ⟶ C and C ⟶ A in Fig. 2).

The molecular mechanisms involved in the regulation of catecholamine synthesis may be summarized as follows:

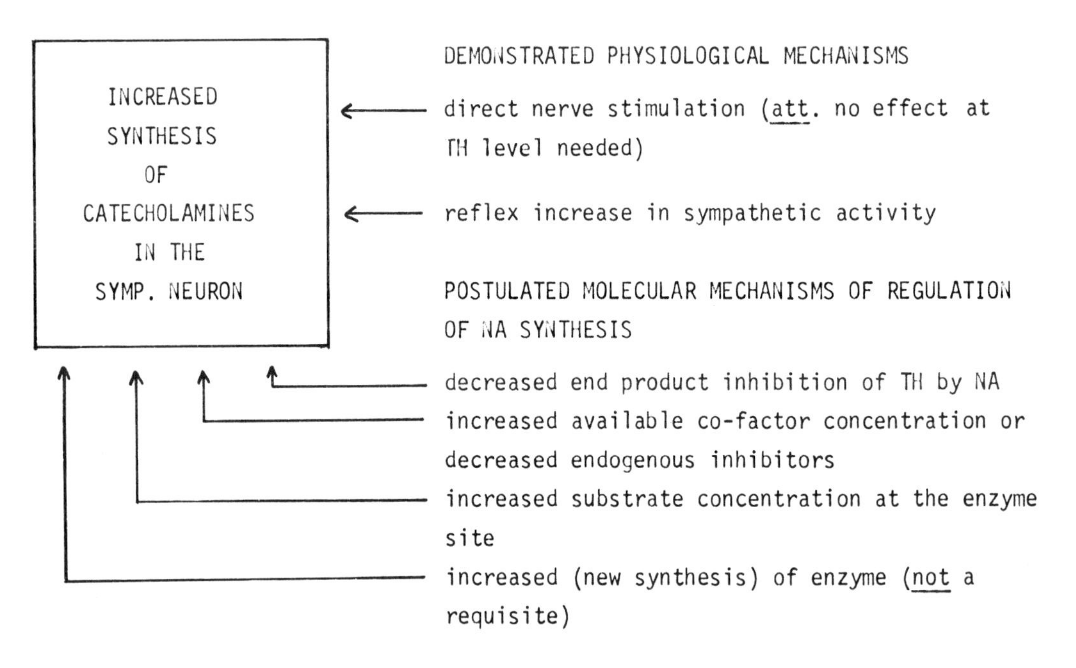

Udenfriend et al. (1965) have demonstrated that NA represents a negative feedback regulator of its own biosynthesis by competing for a pteridine cofactor at the active site of TH.

For a more detailed analysis of the molecular factors involved in regulation of NA biosynthesis we refer to the recent reviews of Weiner (1970) and of Axelrod (1971).

By means of the molecular mechanisms shown above it is possible for the nerve cell not only to control its own (cell B in Fig. 1) level of transmitter both in the cell body and at the synapse but also to regulate and adjust the level of transmitter in the connected neurons (cell A and C in Fig. 1).

A tentative model of neuronal interaction summarizing some of the conditions more extensively discussed in this article could therefore be schematically described as below. This model takes into consideration variations in impulse activity in both directions (presynaptically and postsynaptically), their effect on the transmitter level and the related enzymatic changes.

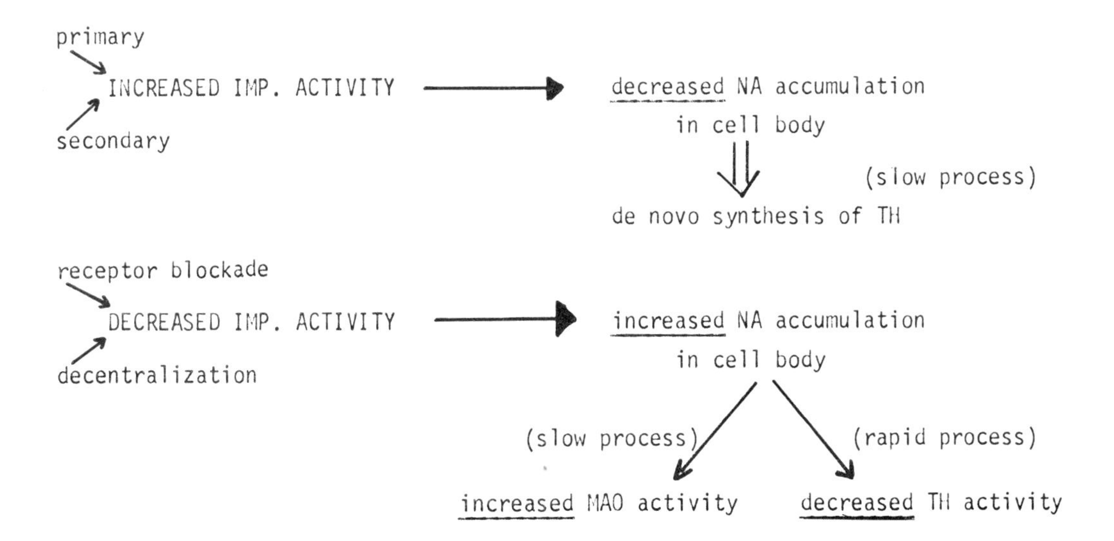

The first purpose of this discussion has been to critically analyse all the known factors which may lead to interaction in chemical neurotransmission.

The second purpose has been to emphasize the long term effects of several drugs on the enzymatic systems synthetizing and inactivating transmitters. In this connection it should be mentioned that other studies using another approach, namely isotopically labelled (*in vivo*) neurotransmitter pools, have clearly shown that factors like stress situations, electrochock treatment and centrally acting drugs modify significantly the turnover of neurotransmitters in the CNS (Costa, 1969; Costa, 1970).

Finally, with regard to intracellular mechanisms which enable neuronal activity to modify protein (enzyme) synthesis (Fig. 2), one should mention the recent work of Gisiger (1971) on RNA metabolism.

His results on stimulated rat sympathetic ganglia suggest that a) neuronal activity may cause an increase in RNA synthesis in ganglion cells, b) the interaction of the transmitter (acetylcholine) with its receptor is responsible for this metabolic stimulation and c) the signal for these changes arises from the activated postsynaptic membrane (Fig. 2).

ADDENDUM

After the completion of this manuscript, a paper by Brimijoin, S. and Molinoff, P.B. (J. Pharm. Expt. Therap. 178, 1971, 417-424) appeared on the effects of 6-OH DA on the activity of TH and DBH in sympathetic ganglia of the rat.

In both normally innervated and decentralized sympathetic ganglia, administration of 6-OH DA causes no changes in TH activity and a longlasting 50% decrease in DBH activity. Pretreatment with 6-OH DA prevents reserpine induced increases in the activity of these enzymes in ganglia. A similar effect is also produced by section of the postganglionic axons of the superior cervical ganglion. According to the authors, the causes that might have prevented 6-OH DA to induce the expected increase in TH or DBH activity in the ganglia may be:

a) 6-OH DA increases the rate of firing only in the nerves to the adrenal and not in the preganglionic sympathetic nerves,
b) 6-OH DA reduces the responsiveness of ganglia to stimuli that induce enzymes. Since pretreatment with 6-OH DA prevented reserpine from increasing the activity of TH and DBH in ganglia but not in adrenal glands, the second possibility seems to be more likely.

Alousi, A. and N. Weiner, (1966), The regulation of norepinephrine synthesis in sympathetic nerves: Effects of nerve stimulation, cocaine and catecholamine-releasing agents. Proceedings of the National Academy of Sciences, 56, 1491.

Azmitia, E.C. and B.S. McEwen, (1969), Corticosterone regulation of tryptophan hydroxylase in midbrain of the rat, Science 166, 1274.

Axelrod, J.B.,(1970), Synaptic activity, enzyme regulation, and protein synthesis in sympathetic neurons and medullary cells, Neurosciences Res. Prog. Bull. 8, 399.

Axelrod, J., R.A. Mueller and H. Thoenen, (1970), Neuronal and hormonal control of tyrosine hydroxylase and phenylethanolamine-N-methyltransferase activity, Bayer-Symposium II, 212.

Axelrod, J., (1971), Noradrenaline: Fate and control of its biosynthesis, Science 173, 598.

Bigl, V., L. Müller and D. Biesold, (1971), Developmental changes of monoamine-oxidase and catechol-O-methyltransferase in the structures of the optical pathway of the growing rat and the influence of functional alteration. Abst. Third International Meeting of the International Society for Neurochemistry, Budapest, p. 68.

Cheah, T.B., L.B. Geffen, B. Jarrott and A. Ostberg, (1971), Action of 6-hydroxy-dopamine on lamb sympathetic ganglia, vas deferens and adrenal medulla: a combined histochemical, ultrastructural and biochemical comparison with the effects of reserpine, Br. J. Pharmac. 42, 543.

Chu, Nai-Shin, L.T. Rutledge and O.Z. Sellinger, (1971), The effect of cortical undercutting and long-term electrical stimulation on synaptic acetylcholine-esterase, Brain Res. 29, 323.

Costa, E. (1969), Turnover rate of neuronal monoamines: pharmacological implications. In A. Cerletti and F.J. Bore (Eds.), The present status of psychotropic drugs, New York: Excerpta Medica Foundation, p. 11.

Costa, E. (1970), Simple neuronal models to estimate turnover rate of noradrenergic transmitters in vivo. Symposium on "Biochemistry of simple neuronal models". (Milano, 1969). Ed. E. Costa and E. Giacobini. Raven Press. p. 169.

Dahlström, A. (1969), Synthesis, transport and life-span of amine storage granules in sympathetic adrenergic neurons. Symp. of the Int. L. Society for Cell Biology, 8, 153.

Dairman, W. and S. Udenfriend (1971), Decrease in adrenal tyrosine hydroxylase and increase in norepinephrine synthesis in rats given L-dopa, Science 171, p. 1022.

Gisiger, V. (1971), Induction of RNA synthesis by acetylcholine stimulation of the postsynaptic membrane in a mammalian sympathetic ganglion. Abstr. 3rd Int. Meeting Int. Soc. Neurochem. Budapest. p. 256.

Giacobini, E. and S. Kerpel-Fronius, (1969), Catechol-O-methyltransferase in autonomic and sensitive ganglia of the cat. Acta physiol. scand. 75, 523.

Giacobini, E., K. Karjalainen, S. Kerpel-Fronius, M. Ritzén, (1970), Monoamines and monoamine oxidase in denervated sympathetic ganglia of the cat. Neuropharm. 9, 59.

Giacobini, E. (1971 a), Biochemistry of the developing autonomic neuron. In "Advances in Experimental Medicine and Biology. Ed. R. Paoletti and A.N. Davison, Plenum Press, New York, 13, 145.

Giacobini, E. (1971 b), Molecular mechanisms of nervous transmission and synaptic plasticity. Int. Symp. on the Histochemistry of Nervous Transmission (Helsinki, 1970). In "Progress in Brain Research", Elsevier & Co. Publ. Ed. O. Eränkö.

Giacobini, E. and B. Noré, (1971), Dopa-decarboxylase in autonomic and sensory ganglia of the cat. Acta physiol. scand. 82, 209.

Mandell, A.J. and M. Morgan. (1969), Increase in regional brain acetyltransferase activity induced with reserpine. Comm. Behav. Biol. 4, 247.

Mandell, A.J. (1970), Drug induced alterations in brain biosynthetic enzyme activity - - A model for adaptation to the environment by the central nervous system. Biochemistry of Brain and Behaviour, Plenum Press, p. 97.

Mandell, A.J. and M. Morgan, (1970), Amphetamine induced increase in tyrosine hydroxylase activity. Nature, 227, 75.

Mandell, A.J., M. Morgan and G.W. Oliver, (1970), The effects of in vivo administration of antidepressant and stimulant drugs on the specific activities of brain tyrosine hydroxylase and indoleamino-N-methyltransferase. In: M. Katz & T. William (Eds.). NIMH Workshop on the Biology of Depression, Washington, D.C. Government Printing Office.

Molinoff, P.B., S. Brimijoin, R. Weinshilboum and J. Axelrod, (1970), Neurally mediated increase in dopamine-β-hydroxylase activity. Proc. natn. Acad. Sci. 66, 453.

Molinoff, P.B., S. Brimijoin and J. Axelrod, (1971), Trans-synaptic induction of dopamine (3,4-dihydroxyphenylethylamine)-β-hydroxylase in adrenergic tissues of the rat. Biochem. J. 123, 32P.

Mueller, R.A., H. Thoenen and J. Axelrod, (1969 a), Inhibition of trans-synaptically increased tyrosine hydroxylase activity by cycloheximide and actinomycin D. Mol. Pharmacol. 5, 463.

Mueller, R.A., H. Thoenen and J. Axelrod, (1969 b), Increase in tyrosine hydroxylase activity after reserpine administration. J. Pharmacol. exp. Therap. 169, 74.

Mueller, R.A., H. Thoenen and J. Axelrod, (1969 c), Adrenal tyrosine hydroxylase: Compensatory increase in activity after chemical sympathectomy. Science, 158, 468.

Mueller, R.A., H. Thoenen and J. Axelrod, (1970), Inhibition of neuronally induced tyrosine hydroxylase by nicotinic receptor blockade. Europ. J. Pharmacol. 10, 51.

Patrick, R.L. and N. Kirshner, (1971), Effect of stimulation on the levels of tyrosine hydroxylase, dopamine-β-hydroxylase, and catecholamines in intact and denervated rat adrenal glands. Mol. Pharmacol. 7, 87.

Sedvall, G.C. and I.J. Kopin, (1967 a), Influence of sympathetic denervation and nerve impulse activity of tyrosine hydroxylase in the rat submaxillary gland. Biochem. Pharmacol. 16, 39.

Sedvall, G.C. and I.J. Kopin, (1967 b), Acceleration of norepinephrine synthesis in the rat submaxillary gland in vivo during sympathetic nerve stimulation. Life Sciences, 6, 45.

Sedvall, G.C., V.K. Weise and I.J. Kopin, (1968), The rate of norepinephrine synthesis measured in vivo during short intervals; influence of adrenergic nerve impulse activity. J. Pharmac. exp. Ther. 159, 274.

Segal, D.and A.J. Mandell, (1970), Behavioral activation of rats during intraventricular infusion of norepinephrine. Proceedings of the National Academy of Sciences. In press.

Sperry, R.W.. James Arthur Lecture. Publ. by The American Museum of Natural History, New York, 1964.

Tarver, J.H. and S. Spector, (1970), Catecholamine metabolic enzymes in the vasculature. Fed. Proc. 29, 278.

Thoenen, H., R.A. Mueller and J. Axelrod. (1969 a), Trans-synaptic induction of adrenal tyrosine hydroxylase. J. Pharmacol. exp. Therap. 169, 249.

Thoenen, H., R.A. Mueller and J. Axelrod, (1969 b), Increased tyrosine hydroxylase activity after drug-induced alteration of sympathetic transmission. Nature, 221, 1264.

Thoenen, H. (1970), Induction of tyrosine hydroxylase in peripheral and central adrenergic neurones by cold-exposure of rats. Nature, 228, 861.

Thoenen, H., R.A. Mueller and J. Axelrod, (1970), Phase difference in the induction of tyrosine hydroxylase in cell body and nerve terminals of sympathetic neurones. Proc. Natl. Acad. Sc. 65, 58.

Udenfriend, S.D., P. Zaltzman-Nirenberg and T. Nagatsu, (1965), Inhibitors of purified beef adrenal tyrosine hydroxylase. Biochem. Pharmacol. 14, 837.

Uretsky, N.J., L.L. Iversen, (1970), Effects of 6-hydroxydopamine on catecholamine containing neurones in the rat brain. J. Neurochem. 17, 269.

Weiner, N. (1970), Regulation of norepinephrine biosynthesis. Ann. Rev. Pharmacol. 10, 273.

Weiner, N., W.F. Mosimann. (1970), The effect of insulin on the catecholamine content and tyrosine hydroxylase activity of cat adrenal glands. Biochem. Pharmacol. 19, 1189.

THE ROLE OF NERVE STIMULATION IN THE REGULATION OF NEUROTRANSMITTER SYNTHESIS AND TURNOVER

NORMAN WEINER, JACK C. WAYMIRE AND KEDAR N. PRASAD
Departments of Pharmacology and Radiology
University of Colorado School of Medicine
Denver, Colorado 80220

ABSTRACT

In recent years a considerable amount of evidence has been accumulated to support the concept that increased adrenergic nervous activity is associated with enhanced synthesis of norepinephrine from tyrosine. In the isolated hypogastric nerve-vas deferens preparation of the guinea-pig, this enhanced synthesis can be partially or completely prevented by the addition of norepinephrine to the bath. Furthermore, in unstimulated preparations norepinephrine markedly inhibits its own synthesis from tyrosine. When dopa is used as precursor, synthesis of norepinephrine is not enhanced by nerve stimulation, suggesting that the effect of nerve stimulation is at the tyrosine hydroxylation step. Since the total amount of norepinephrine in the stimulated preparation is not different from that in the contralateral sham-stimulated organ, it is concluded that a small, chemically undetectable compartment of norepinephrine, presumably the free intraneuronal pool, is critical for this regulation and is depleted during nerve stimulation. Conversely, drugs such as tyramine, amphetamine and other indirectly acting sympathomimetic amines, which displace norepinephrine from storage sites; and reserpine, which blocks the uptake of catecholamines into storage vesicles, increase free intraneuronal norepinephrine and markedly inhibit tyrosine hydroxylation in intact tissue. Similarly, monoamine oxidase inhibitors inhibit tyrosine hydroxylase in the intact preparation, as does bretylium, which blocks the nerve stimulated release of norepinephrine. None of these substances inhibits isolated, partially purified tyrosine hydroxylase. The results with monoamine oxidase inhibitors suggest that intraneuronal monoamine oxidase may play a role in regulating norepinephrine synthesis by maintaining free intraneuronal norepinephrine in the axoplasm at low levels, thus minimizing end-product feedback inhibition of tyrosine hydroxylase.

Bretylium, in addition to indirectly inhibiting tyrosine hydroxylase by allowing the accumulation of free intraneuronal norepinphrine, markedly reduces the turnover of norepinephrine in the intact animal. Studies with bretylium indicate that catecholamine synthesis may be directly regulated by the frequency of impulse flow to adrenergic terminals, and that this effect may not be due entirely to the enhanced feedback inhibition of tyrosine hydroxylase by catecholamines.

Other mechanisms for the regulation of the biosynthesis of norepinephrine exist. In the hypogastric nerve-vas deferens preparation of the guinea pig, enhanced norepinephrine synthesis occurs during the period immediately following a one hour program of nerve stimulation. This post-stimulation increase in norepinephrine synthesis appears to be distinct from that which occurs during nerve stimulation, since the former is not abolished by norepinephrine. Although the enhanced synthesis in the post-stimulation period is inhibited by puromycin, there is no increase in the amount of tyrosine hydroxylase in homogenates of vasa deferentia prepared after one hour of intermittent nerve stimulation. Furthermore, the increase in norepinephrine synthesis from tyrosine in the post-stimulation period is not blocked by the addition of cycloheximide to the bath during the stimulation period. Since the concentrations of both puromycin and cycloheximide which are employed markedly inhibit overall protein synthesis in this tissue, it would appear that the effect of puromycin on norepinephrine synthesis in the vas deferens preparation may be unrelated to its ability to inhibit ribosomal protein synthesis.

Increased sympathetic nervous activity for prolonged periods is associated with an increase in the content of tyrosine hydroxylase in the adrenal medulla and in adrenergic neurons. A wide variety of pharmacological agents which either directly or indirectly bring about enhanced sympathetic nervous activity can produce an increase in tyrosine hydroxylase levels in the adrenal medulla and in adrenergically innervated tissues after one or more days. We have demonstrated increased tyrosine hydroxylase levels in the cat adrenal gland 12 hours after insulin hypoglycemic stress. The increases in tyrosine hydroxylase content can be blocked at least partially by inhibitors of protein synthesis. In mouse neuroblastoma cells grown in culture for prolonged periods, exposure of the cells to dibutyryl cyclic AMP or other stable analogs of cyclic AMP for 3 days is associated with maturation and differentiation of the cells and an increase in tyrosine hydroxylase levels. Cyclic AMP may be involved in the mechanism of enzyme induction in the adrenergic nervous system associated with prolonged increases in sympathetic nervous activity.

INTRODUCTION

Increased sympathetic nervous activity associated either with increased physical exercise, enhanced psychological stress, or a variety of pharmacological or pathological stresses, is associated with increased release of the mammalian sympathetic neurotransmitter, norepinephrine, increased metabolism of the released transmitter, and an increased output of norepinephrine and its metabolites in the urine (Euler et al, 1955; Weiner, 1970). In the absence of pharmacological intervention, such as simultaneous administration of substances which block either the synthesis or the storage of norepinephrine, tissue norepinephrine levels either remain relatively constant or fall transiently and fairly rapidly return to control levels (Luco and Goñi, 1948; Vogt, 1954). These observations strongly suggest that the synthesis of norepinephrine varies in a manner which is directly related to the activity of the sympathetic nervous system and must be in some way regulated by the events associated with nervous activity.

The biosynthetic pathway for norepinephrine synthesis has been well established and the enzymes involved in this pathway have been studied extensively (Udenfriend; 1966). Norepinephrine is formed largely from tyrosine in the intact organism. The first step involves aromatic hydroxylation to form dihydroxyphenylalanine (dopa) and is catalyzed by the enzyme tyrosine hydroxylase. This enzyme requires a reduced pteridine cofactor for activity (Nagatsu et al, 1964; Ikeda et al, 1966). Generally, studies have been performed with the synthetic cofactor, 6,7-dimethyltetrahydropterin. The natural cofactor involved in the aromatic hydroxylation of tyrosine appears to be tetrahydrobiopterin (Lloyd and Weiner, 1971). In addition to a reduced cofactor, tyrosine hydroxylase requires molecular oxygen and ferrous ion for activity (Ikeda et al, 1966). α,α-Dipyridyl, an iron chelator, is able to inhibit virtually completely the activity of tyrosine hydroxylase. The effect of this iron chelator cannot be overcome by addition of excess catalase, suggesting that the ferrous ion is not solely required for the destruction of peroxides formed during the reaction which may, in turn, inactivate the enzyme (Waymire et al, submitted for publication). Udenfriend and co-workers have shown that the level and activity of tyrosine hydroxylase in a variety of tissues determine the overall rate of norepinephrine synthesis and it is generally regarded that this first step in the biosynthetic pathway is the rate limiting reaction (Levitt et al, 1965).

The second step in the biosynthetic pathway involves decarboxylation of dopa to 3,4-dihydroxyphenylethylamine (dopamine). This reaction is catalyzed by L-aromatic amino acid decarboxylase, a typical decarboxylation enzyme with a rather broad substrate specificity for aromatic amino acids. The enzyme requires pyridoxal phosphate for activity (Lovenberg et al, 1962).

The final step in the synthesis of norepinephrine involves aliphatic hydroxylation catalyzed by the enzyme dopamine-β-hydroxylase. This enzyme, like tyrosine hydroxylase, is a mixed function oxidase requiring molecular oxygen and a reduced co-substrate, ascorbic acid. The enzyme appears to contain copper (Levin and Kaufman, 1961; Kaufman and Friedman, 1965). A variety of sulfhydryl reagents, including disulfuram, is able to inhibit dopamine-β-hydroxylase and consequently block the formation of norepinephrine, presumably by chelating the copper of the enzyme.

The enzymes involved in the biosynthesis of norepinephrine are present in all portions of the adrenergic neuron. Although the precise localization of the first enzyme in the biosynthetic pathway is not fully clarified, at least for all tissues, most investigators regard this enzyme as largely or exclusively present in the axoplasm (Musacchio, 1968; Weiner et al, 1971a; Musacchio et al, 1971). In some species, the enzyme has a marked propensity to bind to membranes. It appears certain that the enzyme is not present within the storage vesicles of the adrenergic neuron. L-Aromatic amino acid decarboxylase also appears to be a soluble enzyme localized exclusively in the axoplasm (Weiner, 1970). Dopamine-β-hydroxylase is a particulate enzyme which resides within the storage vesicle (Hagen and Welch, 1955; Kirshner, 1957). Thus, dopamine must be taken up from the axoplasm into the storage vesicle in order to reach the site of the enzyme dopamine-β-hydroxylase for subsequent conversion to norepinephrine.

Norepinephrine dynamics in the adrenergic neuron is extremely complex. In addition to the storage vesicles in which the bulk of norepinephrine appears to be bound to ATP and perhaps to macromolecular constituents of this organelle (Weiner and Jardetzky, 1964), both the neuronal plasma membrane and the storage vesicle membrane possess the ability to take up norepinephrine from their surroundings in a very efficient fashion (Axelrod, 1964; Weiner, 1964, 1970). The axonal membrane system seems to be common for a variety of phenylethylamine derivatives. It is sodium dependent and probably involves a carrier mediated transport process (Iversen and Kravitz, 1966; Bogdanski and Brodie, 1969). It is selectively blocked by cocaine and a variety of tricyclic antidepressants including imipramine and desmethylimipramine (Weiner, 1964, 1970). The storage vesicle uptake process is distinct from the axonal membrane uptake process in that it is not blocked by cocaine or the tricyclic antidepressants, but appears to be effectively blocked by reserpine and related alkaloids (Kirshner et al, 1965; Rutledge and Weiner, 1967). The vesicle uptake process requires ATP and magnesium for optimal activity and is temperature dependent (Kirshner, 1962). In addition to norepinephrine and dopamine, a variety of phenylethylamines are concentrated in the storage vesicle by this mechanism.

Enzymes for the catabolism of catecholamines are also present within the adrenergic neuron. The more important of these is monoamine oxidase, a mitochondrial enzyme capable of oxidatively deaminating norepinephrine, dopamine and phenylethylamines lacking an α-carbon substituent (Hagen and Weiner, 1959). Catechol-O-methyltransferase also appears to be present within the adrenergic neuron, but its role in the intraneuronal catabolism of norepinephrine and dopamine appears trivial (Axelrod, 1959; Jarrott and Iversen, 1971).

Thus the adrenergic neuron contains all of the enzymes both for the biosynthesis of norepinephrine and for norepinephrine catabolism, an axonal plasma membrane uptake mechanism and an intraneuronal system for the uptake and storage of norepinephrine in subcellular organelles. An appreciation of each of these processes and their interactions within the intact neuronal system is essential for an understanding of the intracellular dynamics, regulation of synthesis, and turnover of this neurotransmitter.

END-PRODUCT FEEDBACK REGULATION OF NOREPINEPHRINE SYNTHESIS DURING STIMULATION

Several years ago studies were initiated in our laboratories on the regulation of norepinephrine synthesis employing the isolated hypogastric nerve-vas deferens preparation of the guinea pig as described by Hucović (1961). In contrast to many *in vivo* systems in which norepinephrine synthesis and turnover have been examined, this isolated system has the advantage that precursor concentrations can be carefully regulated and monitored as can the total synthesis of catecholamines and their metabolites in both stimulated and unstimulated preparations. In all of these studies, both isolated hypogastric nerve-vas deferens preparations from a freshly killed guinea pig are set up in identical organ baths. One end of each vas deferens is attached to a glass rod and the other end is affixed to a strain gauge transducer. One hypogastric nerve is gently placed across a pair of platinum electrodes and the preparation is stimulated at supramaximal voltage. In all studies the stimulated preparation is compared with the sham stimulated contralateral control preparation from the same animal. Optimal, intermittent, sustained contractions are obtained with supramaximal voltage at a frequency of 25 pulses/sec for 5 sec out of each minute. In general, the preparation is stimulated for a period of one hour. At the end of the period of stimulation each tissue and its bathing solution are separately acidified with trichloroacetic acid, the tissues are homogenized, centrifuged and the pellets washed twice. When labeled tyrosine is used as precursor, the catecholamines are separated from tyrosine and tyrosine metabolites by successive chromatography on alumina and Dowex-50-Na^+ x 4 ion exchange columns. Catecholamines are eluted from alumina columns with 0.2 N acetic acid and norepinephrine and dopamine are successively eluted from the ion exchange column with 1N and 2N HCl. When labeled dopa is used as precursor, chromatography is performed only with the Dowex-50-Na^+ x 4 ion exchange column. The HCl eluates are either dried in a vacuum desiccator and counted by liquid scintillation spectrometry using a dioxane scintillation cocktail (Weiner and Rabadjija, 1968a), or an aliquot of the eluates is directly added to a Triton X-100 toluene scintillation cocktail for counting. In those studies in which labeled tyrosine is used as precursor, the alumina effluents containing tyrosine and non-catechol metabolites of tyrosine are passed through a Dowex-50-H^+ x 4 ion exchange column, washed with phosphate buffer and the tyrosine is eluted with ammonium hydroxide, counted by liquid scintillation spectrometry and assayed by the fluorescence procedure of Waalkes and Udenfriend (1957). All radioisotope results are corrected for quenching by use of an external standard and the catecholamine results are corrected for incomplete recoveries using an appropriate internal standard (Weiner and Rabadjija, 1968a,b).

When 3,5-^{3}H-tyrosine is added to the bath during the one hour period of intermittent nerve stimulation, there is a significant increase in catecholamine synthesis associated with nerve stimulation. The differences are not readily apparent when the average norepinephrine synthesis rate of a group of stimulated preparations is compared with the average rate of synthesis of sham stimulated controls, largely due to the variability in activity observed between animals (Fig. 1). However, when the stimulated control is compared with the contralateral sham stimulated organ from the same animal, consistent increases in catecholamine synthesis of about 45% are observed in the stimulated preparation. The bulk of the catecholamines formed from tyrosine is present as norepinephrine and is present almost entirely in the tissue. In contrast, when the tyrosine hydroxylase step is bypassed by using 2,4,6-^{3}H-dopa as catecholamine precursor, there is no significant increase in norepinephrine synthesis associated with nerve stimulation. Virtually identical results are obtained if monoamine oxidase is inhibited completely by the addition of 1.5×10^{-4} M pargyline to the medium (Fig. 1). This suggests that the enhanced norepinephrine synthesis from tyrosine associated with nerve stimulation is not related to an alteration in the catabolism of norepinephrine during the experiments. Catechol-O-methyltransferase apparently plays a minor or negligible role in the metabolism of catecholamines in the guinea pig vas deferens preparation (Weiner and Rabadjija, 1968a; Jarrott, 1971).

Thus, in these series of experiments we were able to demonstrate directly increased norepinephrine synthesis from tyrosine as a result of nerve stimulation. The altered rate of biosynthesis of norepinephrine appears to be localized at the tyrosine hydroxylase step. Udenfriend and coworkers, employing partially purified bovine adrenal tyrosine hydroxylase, demonstrated that catecholamines could inhibit the activity of this enzyme and the inhibition appears to be competitive with the reduced pteridine cofactor (Udenfriend et al, 1965; Ikeda et al, 1966). In view of these results, it seemed likely that the increase in norepinephrine synthesis observed with stimulation of the hypogastric nerve-vas deferens preparation of the guinea pig might be due to increased release of norepinephrine and

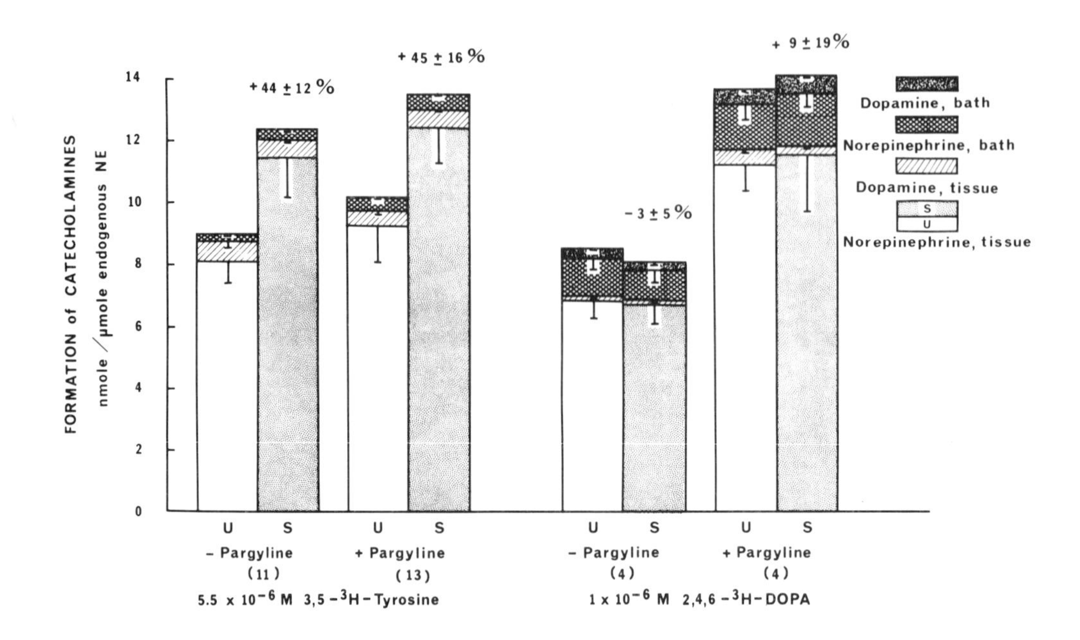

Fig. 1. The effect of nerve stimulation on catecholamine synthesis from 3,5,-^{3}H-tyrosine and from 2,4,6-^{3}H-dopa in vas deferens-hypogastric nerve preparations of the guinea pig. When stimulated preparations (S) are compared with contralateral unstimulated preparations (U) of the same animal, a significant increase ($p < .05$) of catecholamine synthesis from tyrosine, but not from dopa, is observed. Inhibition of monoamine oxidase with 1.5×10^{-4} M pargyline does not influence either the basal levels or the nerve stimulated increase of catecholamine synthesis from tyrosine. In the presence of the inhibitor of monoamine oxidase, labeled catecholamines from ^{3}H-dopa approximately double, presumably because newly synthesized amines are protected from degradation in this group (Modified from Weiner and Rabdjija, 1968a).

a reduction in the end-product feedback inhibition of the first step in the biosynthetic pathway. In an attempt to evaluate this, 6×10^{-6}M norepinephrine was added to the bath during the period of stimulation and the effect of this catecholamine on norepinephrine synthesis in the presence and absence of stimulation was evaluated. This concentration of norepinephrine inhibited basal levels of norepinephrine synthesis from tyrosine by 60 percent (Fig. 2). The much greater inhibitory effect of this concentration of norepinephrine on tyrosine hydroxylase in the intact tissue, as compared with soluble, fortified preparations of bovine adrenal enzyme (ED_{50} of approximately 1×10^{-3} M; Udenfriend et al, 1965) is probably attributable to at least two factors: First, although 6×10^{-6} M norepinephrine is added to the medium in our studies, as a consequence of the ability of the adrenergic neuron to take up and concentrate norepinephrine, the concentration of the added norepinephrine in the nerve terminals is probably many fold higher than that present in the medium. We have demonstrated the necessity of norepinephrine uptake into the adrenergic nerve terminal for this inhibition of norepinephrine synthesis, since the inhibitory effect of norepinephrine is virtually completely abolished if cocaine is simultaneously added to the bath. Secondly, Udenfriend and coworkers generally employed 1×10^{-3} M synthetic

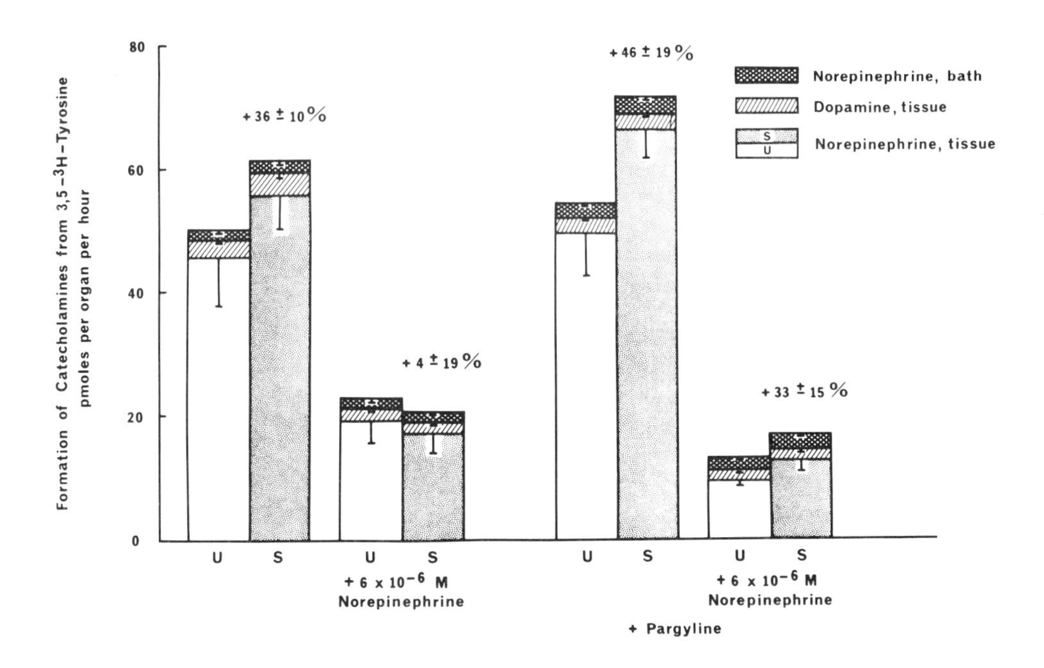

Fig. 2. Effect of norepinephrine on catecholamine synthesis from tyrosine in unstimulated (U) and stimulated (S) vas deferens-hypogastric nerve preparations of the guinea pig. In the absence of norepinephrine, nerve stimulation is associated with a significant increase in norepinephrine synthesis when each stimulated preparation is compared with the sham stimulated control of the same animal ($p < .05$). In the presence of 6×10^{-6} M norepinephrine, no significant difference in norepinephrine synthesis between the control and stimulated preparations is demonstrable. (Modified from Weiner and Rabdjija, 1968a).

pteridine cofactor in their studies of the kinetics of tyrosine hydroxylase of bovine adrenal medulla. The concentration of the cofactor present in the adrenergic nerve terminal is probably considerably less than 10^{-3} M, and since the inhibitory effect of norepinephrine is competitive with the pteridine cofactor, less norepinephrine will be required for equivalent inhibition of the enzyme in the non-fortified, intact preparation.

In addition to inhibiting the basal level of norepinephrine synthesis, exogenous norepinephrine virtually completely abolishes the increase in norepinephrine synthesis associated with nerve stimulation in the intact preparation. This effect also can be overcome by cocaine, suggesting that uptake of norepinephrine into the adrenergic nerve terminal is essential for this inhibition (Fig. 2; Alousi and Weiner, 1966; Weiner and Rabadjija, 1968a).

Since exogenous norepinephrine also is able to release norepinephrine from endogenous stores, the possibility that this effect of added norepinephrine was due to increased release and catabolism of newly synthesized norepinephrine was evaluated. In the presence of a monoamine oxidase inhibitor, the inhibitory effect of norepinephrine on the basal level of norepinephrine synthesis was reduced still further, presumably because that norepinephrine taken up from the medium was protected from oxidative deamination and higher levels of the catecholamines could accumulate in the nerve ending, resulting in an intensification of the inhibitory effect. Again, in the presence of norepinephrine and pargyline, the enhanced norepinephrine synthesis associated with nerve stimulation was reduced (Fig. 2). These results are consistent with the concept that the enhanced synthesis of norepinephrine associated with nerve stimulation is the result of reduced feedback inhibition of tyrosine hydroxylase because of reduced levels of a critical pool of intraneuronal norepinephrine associated with nerve stimulation (Alousi and Weiner, 1966).

In order to test further this hypothesis, the endogenous content of norepinephrine in stimulated hypogastric nerve-vas deferens preparations of guinea pig was compared with that of the contralateral sham stimulated preparations. The level of norepinephrine in these preparations was identical, suggesting that total norepinephrine in the adrenergic neuron is not the critical factor in regulating its own synthesis (Weiner and Rabadjija, 1968a). These results were not unexpected since tyrosine hydroxylase is believed to be a soluble enzyme and the bulk of the norepinephrine pool in adrenergic neurons is present within the storage vesicles. Thus, the major pool of norepinephrine would not be available to compete with pteridine cofactor for sites on the tyrosine hydroxylase enzyme (Weiner et al, 1971a). From a variety of pharmacological studies, we have concluded that the critical pool in the regulation of norepinephrine synthesis at the tyrosine hydroxylase step is a small, biochemically unmeasurable pool of free intraneuronal norepinephrine. Drugs which would be expected to increase the level of free intraneuronal norepinephrine and decrease the level of stored norepinephrine; e.g., indirect acting sympathomimetic amines (Weiner and Selvaratnam, 1968; Kopin et al, 1969), reserpine and monoamine oxidase inhibitors (Weiner and Bjur, in press), inhibit tyrosine hydroxylase in intact tissue but have negligible effects on isolated, purified tyrosine hydroxylase activity. These agents thus exert their effects on tyrosine hydroxylase indirectly and apparently do so by increasing the levels of free intraneuronal norepinephrine, either by shifting the norepinephrine stores from the bound to the free state or by protecting the free intraneuronal norepinephrine from catabolism (Fig. 3). Similarly, bretylium, which blocks the release of norepinephrine from electrically stimulated adrenergic neurons, presumably by inhibiting nerve terminal membrane depolarization, also inhibits norepinephrine synthesis by enhancing end-product feedback inhibition. The effects of bretylium appear to be multiple: (1) By preventing nerve stimulated release of norepinephrine, bretylium exposure would facilitate the loading of catecholamine stores to capacity and cause a proportional (or disproportionately greater) increase in free intraneuronal norepinephrine; (2) Free intraneuronal norepinephrine would be further increased by the ability of bretylium to inhibit intraneuronal monoamine oxidase (Kuntzman and Jacobson, 1963); (3) Bretylium causes the release of norepinephrine from storage sites (Hertting et al, 1962) (Fig. 3). Bretylium has no direct effect on isolated, partially purified tyrosine hydroxylase. The inhibitory effect of bretylium on tyrosine hydroxylase in intact tissue can be overcome at least partially by addition of excess pteridine cofactor to the medium, supporting the concept that the action of this drug involves enhancement of end-product feedback inhibition of tyrosine hydroxylase. Bretylium administration is associated with a marked reduction in the turnover of norepinephrine *in vivo* (Posiviata et al, 1971).

INCREASED NOREPINEPHRINE SYNTHESIS AFTER SHORT-TERM NERVE STIMULATION

In addition to the enhanced synthesis of norepinephrine which occurs during nerve stimulation, immediately following short-term stimulation of the hypogastric nerve-vas deferens preparation, there is an accelerated synthesis of norepinephrine from tyrosine which again appears to be primarily or solely due to increased activity of the enzyme tyrosine hydroxylase. In contrast to the enhanced synthesis during nerve stimulation, the post-stimulation increase in norepinephrine synthesis is not blocked by the addition of norepinephrine to the bath

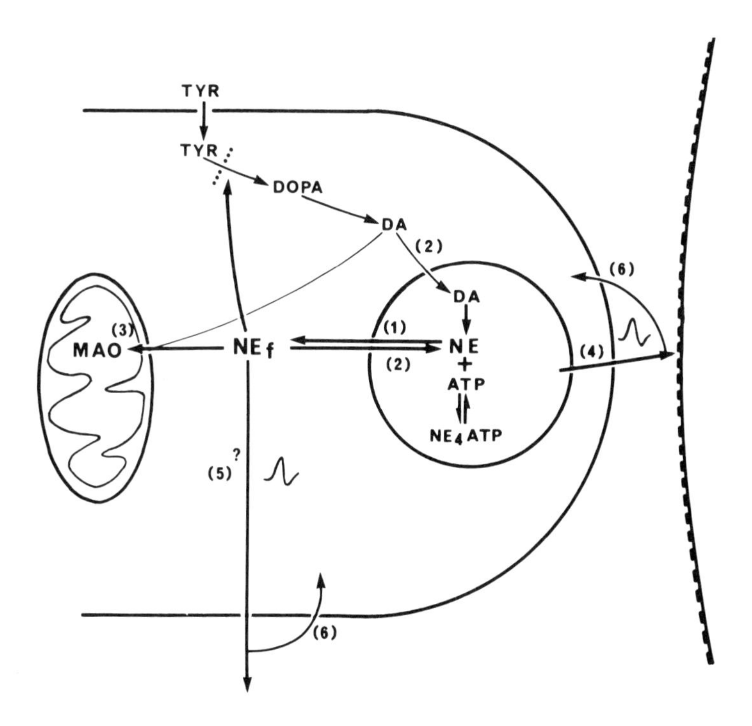

Fig. 3. Schematic drawing of the regulation of tyrosine hydroxylase by end-product feedback inhibition. Nerve stimulation would result in reduction of free intraneuronal norepinephrine (NE_f) and stimulation of tyrosine hydroxylase activity by (4), perhaps making more norepinephrine (NE) binding sites available, and perhaps by direct release of NE_f (5) from the axoplasm. Procedures which would lead to increased levels of free intraneuronal norepinephrine and enhanced inhibition of tyrosine hydroxylase activity include: (a) Inhibition of monoamine oxidase (MAO) (3); (b) Block of uptake of catecholamines into the vesicle, (2), e.g., by reserpine; (c) Release of norepinephrine from storage sites (1), e.g., by indirectly acting sympathomimetic amines; (d) Block of membrane depolarization and nerve stimulated release of norepinephrine (4) and perhaps (5), e.g., by bretylium. Bretylium also increases free intraneuronal norepinephrine by inhibition of MAO (3) and releases norepinephrine from storage sites (1). DA = dopamine; TYR = tyrosine.

during nerve stimulation but is markedly reduced by puromycin, a well known inhibitor of protein synthesis (Weiner and Rabadjija, 1968b). However, the post-stimulation increase in norepinephrine synthesis observed in the vas deferens-hypogastric nerve preparation of the guinea pig is not inhibited either by cycloheximide or actinomycin D. Furthermore, homogenates prepared from previously stimulated tissues do not contain amounts of tyrosine hydroxylase different from those in sham stimulated preparations. The addition of excess pteridine cofactor to the intact preparations does not abolish the enhanced synthesis of norepinephrine from tyrosine observed in the post-stimulation period (Thoa et al, 1971). Thus, the post-stimulation increase in norepinephrine synthesis from tyrosine does not appear to be the result of enhanced synthesis of either tyrosine hydroxylase or pteridine reductase and does not appear to be due to increased levels or activity of the reduced pteridine cofactor. Nerve terminal membrane depolarization is required for the post-stimulation increase in norepinephrine synthesis, since the effect is completely abolished by the addition of bretylium to the bath in the stimulation period in concentrations sufficient to inhibit contractile responses during nerve stimulation (Posiviata et al, 1971). Furthermore, calcium appears to be required for the post-stimulation increase in norepinephrine synthesis, since the post-stimulation effect is not observed in preparations exposed to calcium-free Krebs-Ringer bicarbonate solution. Under these circumstances, preganglionic nerve stimulation does not produce a contraction of the preparation.

Although puromycin and cycloheximide both inhibit ribosomal protein synthesis, the mechanisms by which these two agents act differ. Cycloheximide appears to inhibit either the attachment of transfer-RNA to the ribosome or the formation of peptide bonds. Thus, the utilization of amino acids is blocked by this agent. In contrast, puromycin does not prevent the formation of small peptides, but apparently prematurely terminates the elongation of the peptide chain by serving as an acceptor for the peptide chain. The oligopeptide to which puromycin is attached thus becomes the end-product of protein synthesis in the presence of this nucleoside analog (Franklin and Snow, 1971). Furthermore, chemically, puromycin may be considered both as a nucleoside analog and as a tyrosine analog. Thus, either by competing with tyrosine for tyrosine metabolism as a structural analog, or by using up the endogenous tyrosine in the formation of oligopeptides, puromycin may markedly affect tyrosine metabolism. The inhibitory effect of puromycin on the post-stimulation increase in norepinephrine synthesis thus may be related to its effects on tyrosine metabolism rather than to an inhibitory effect on protein synthesis.

The possible effects of puromycin on tyrosine metabolism suggested that the post-stimulation increase in norepinephrine synthesis might be an indirect consequence of altered tyrosine metabolism as a result of nerve stimulation. Thus, during the period of stimulation, tyrosine (and perhaps other amino acids) may be utilized at an accelerated rate, yielding a relative deficiency of this amino acid in the tissue. When labeled tyrosine is added to the bath in the post-stimulation period, the amino acid may be taken up at an accelerated rate and may accumulate in a pool deficient in endogenous tyrosine. Thus, accelerated formation of labeled norepinephrine from labeled tyrosine might result as a consequence of a reduced endogenous tyrosine pool. To assess this possibility, the stimulation experiments were performed with different concentrations of unlabeled tyrosine present in the medium during the period of stimulation. Subsequent to the period of stimulation, tissues were transferred to fresh physiological salt solution and labeled tyrosine was added in order to assess the formation of norepinephrine from this precursor. A significant increase in the synthesis of norepinephrine in the post-stimulation period was still demonstrable in those preparations previously incubated with unlabeled tyrosine during the period of stimulation. It thus appears that the post-stimulation increase in norepinephrine synthesis cannot be attributed to the relative lack of this amino acid at the time of addition of the labeled precursor, as a consequence of accelerated tyrosine utilization during the period of stimulation (Weiner et al, 1971b).

INCREASED LEVELS OF TYROSINE HYDROXYLASE AFTER PROLONGED SYMPATHETIC NERVOUS ACTIVITY

In contrast to the changes in the rate of norepinephrine synthesis during and after short-term adrenergic nervous system stimulation, increased sympathetic nervous activity for prolonged periods of time is associated with an increase in the content of tyrosine hydroxylase in the adrenal medulla and in adrenergic neurons (Dairman et al, 1968; Mueller et al, 1969a,b; Viveros et al, 1969; Thoenen et al, 1969). A wide variety of pharmacological agents which either directly or indirectly bring about enhanced sympathetic nervous activity can produce an increase in tyrosine hydroxylase levels in the adrenal medulla and in adrenergically innervated tissues. For example, the administration of agents which block the function of the sympathetic nervous system and presumably reflexly lead to increased adrenergic nerve stimulation, is associated with increases in tissue tyrosine hydroxylase levels after about one to two days. Viveros et al (1969) reported that increased levels of tyrosine hydroxylase are demonstrable in adrenal glands of rabbits subsequent to insulin stress. In a similar series of experiments, we examined the effects of insulin stress on catecholamine synthesis in cat adrenal glands. These experiments were performed on cats 6 to 16 days after unilateral denervation of one adrenal gland. Insulin, 4-5 units per kg, was administered intramuscularly to these animals and at varying times after insulin administration animals were killed and the innervated and denervated adrenal glands were examined for catecholamine content and tyrosine hydroxylase activity. Within 5 hours after the administration of insulin, the catecholamine content of the innervated adrenal gland is significantly lower than that of the contralateral denervated gland. At 8 hours after insulin, catecholamine content of the innervated gland is approximately 1/3 that present in the contralateral denervated organ. By 12-24 hours, some recovery of the catecholamine content of the innervated gland already has taken place, although the catecholamine content at 24 hours is still significantly lower in the innervated gland than in the denervated organ. Tyrosine hydroxylase activity in the innervated gland is significantly higher than that in the corresponding denervated organ 12 hours after insulin administration. The enzyme levels are still significantly elevated in the innervated gland 48 hours after insulin, although the levels appear to be returning toward normal at this time. At the 12-hour time period, when maximal changes in tyrosine hydroxylase levels are observed, there is a significant correlation between increases in tyrosine hydroxylase activity in the innervated gland and the extent of depletion of catecholamines in this organ. The increased adrenal tyrosine hydroxylase activity after insulin stress appears to be the consequence of synthesis of new enzyme since the effect can be at least partially blocked by administration of cycloheximide (Weiner and Mosimann, 1970).

It is quite likely that normal tonic impulses to the adrenal gland are sufficient to stimulate synthesis of tyrosine hydroxylase above basal levels observed in resting or non-innervated glands. Approximately one week after unilateral adrenal gland denervation, the catecholamine content of the denervated adrenal gland is significantly higher than the content of the contralateral innervated gland. In contrast, tyrosine hydroxylase content of the denervated gland is significantly lower than the tyrosine hydroxylase activity of the corresponding innervated organ (Weiner and Mosimann, 1970). Approximately 2 months after unilateral denervation of the adrenal gland, both the catecholamine content and the tyrosine hydroxylase activity of the innervated organ are significantly greater than that in the chronically denervated structure (Table 1). These results indicate that biochemical atrophy of the adrenal gland may occur after chronic denervation and imply that tonic nerve impulses to the adrenergic tissue are required to maintain the levels of both the biosynthetic enzymes and the catecholamines normally present in this organ.

INDUCTION OF TYROSINE HYDROXYLASE IN MOUSE NEUROBLASTOMA BY CYCLIC AMP ANALOGS

Attempts have been made to elucidate the mechanism by which the levels of tyrosine hydroxylase in sympathetic tissues are regulated. As a model system for these studies, we have initiated investigations on the enzyme tyrosine hydroxylase in mouse neuroblastoma cells in culture, employing a newly developed assay for the enzyme (Waymire et al, 1971). Mouse neuroblastoma cells have been maintained in culture for more than a year (Prasad, 1971). The original tumor of mouse neuroblastoma contains considerable amounts of tyrosine hydroxylase. However, when the mouse neuroblastoma cells are maintained in culture, mitosis takes place approximately every 24 hours and the tyrosine hydroxylase level in the cultured cells diminishes progressively and is barely detectable after a period

TABLE I

TYROSINE HYDROXYLASE AND CATECHOLAMINE CONTENT OF CAT ADRENAL GLANDS

Effect of Chronic Denervation

	Innervated Gland	Denervated Gland	Percent Difference
Catecholamine Content μmol per gland	0.73 ± 0.10	0.57 ± 0.08	21.8 ± 3.2*
Tyrosine Hydroxylase Activity μmole dopa formed per gland per hour	0.075 ± 0.001	0.044 ± 0.006	38.9 ± 5.9*

*Percent differences = $\frac{I-D}{I}$ x 100 $p < .001$

Adrenals were denervated (2R, 3L) two months prior to assay of glands.

Results are means ± S.E.M. of 5 cats.

of one week. Exposure of the cells in culture to 0.5 to 1.0 mM $N^6,O^{2'}$-dibutyryl cyclic 3'-5'-adenosine monophosphate (dibutyryl cyclic AMP) is associated with a progressive increase in the levels of tyrosine hydroxylase in the cells over several days. The rise in tyrosine hydroxylase persists for at least 2 days after removal of the cyclic nucleotide from the medium (Fig. 4). Preceding the rise in tyrosine hydroxylase is inhibition of cell division, increase in cell size, and cellular differentiation as indicated by formation of cytoplasmic processes greater than 50μ in length (Prasad and Hsie, 1971). Cessation of cell division and evidence of cellular differentiation are not invariably associated with enhanced tyrosine hydroxylase levels in mouse neuroblastoma cells, since x-irradiation can produce similar morphological changes without any increase in tyrosine hydroxylase (Prasad et al, submitted for publication). Conversely, exposure of the neuroblastoma cells to sodium butyrate 0.5 - 1 mM, is associated with increased levels of tyrosine hydroxylase without evidence of cellular differentiation (Table 2).

The ability of sodium butyrate to increase levels of tyrosine hydroxylase in mouse neuroblastoma cells was an unexpected complication of these studies. Although neither cyclic AMP nor 5'-AMP treatment is associated with increased levels of tyrosine hydroxylase in these cells, it is unlikely that the induction of tyrosine hydroxylase produced by dibutyryl cyclic AMP is mediated solely by the butyric acid formed by hydrolysis of this cyclic nucleotide. Exposure of the cells to the more stable N^6-monobutyryl 3'-5'-cyclic AMP also produces induction of tyrosine hydroxylase, as does treatment with other stable analogs of cyclic AMP, notably 8-aminomethyl cyclic AMP (Table 2). Furthermore, the effect of dibutyryl cyclic AMP on tyrosine hydroxylase activity proceeds at a more rapid rate and higher levels of tyrosine hydroxylase are ultimately achieved with this cyclic nucleotide than can be achieved with equivalent amounts of butyric acid. Finally, exposure of the cell culture to papaverine, a potent inhibitor of cyclic nucleotide phosphodiesterase, is associated with enhanced levels of tyrosine hydroxylase in mouse neuroblastoma cell cultures, suggesting that endogenously produced cyclic AMP, when allowed to accumulate in the cells by inhibition of its hydrolytic degradation, can lead to tyrosine hydroxylase induction (Table 2).

If the mouse neuroblastoma cell may be considered as a model system for adrenergic nervous tissue, these results suggest that the induction of tyrosine hydroxylase after prolonged adrenergic nervous system stimulation may be mediated

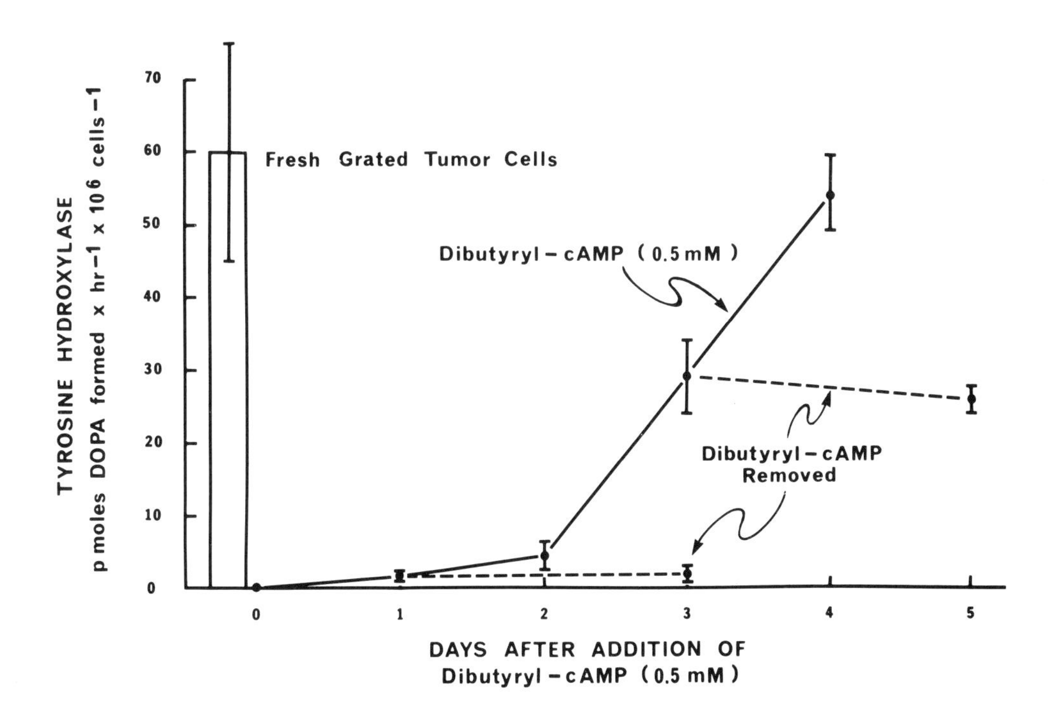

Fig. 4. Tyrosine hydroxylase levels in neuroblastoma tumor freshly removed from mice (bar on left) and in cultured mouse neuroblastoma cells after exposure to 0.5 mM N^6,$O^{2'}$-dibutyryl cyclic AMP for various periods of time (solid line). In some studies, after one or three days, the medium containing the cyclic nucleotide was replaced with fresh medium lacking this substance and the incubation was continued for two days (dashed lines). Tyrosine hydroxylase was assayed as described in Table 2.

by activation of adenyl cyclase and enhanced production of cyclic AMP. Addition to brain slices of any of several putative neurotransmitters, including norepinephrine and histamine, and membrane depolarization, produced either by electrical stimulation or by addition of potassium in high concentrations, are associated with activation of adenyl cyclase and elevated levels of cyclic AMP in neural tissue (Kakiuchi and Rall, 1968a,b; Kackiuchi et al, 1969). It is possible that one or more of these substances, or prostaglandins [some of which are known to be released by adrenergic nerve stimulation (Shaw and Ramwell, 1968), and which in some systems activate adenyl cyclase (Horton, 1969)] may activate adenyl cyclase in adrenergic neurons subjected to prolonged sympathetic nervous stimulation. The production of cyclic AMP in the latter cells may, in turn, initiate the series of events which ultimately lead to increased tyrosine hydroxylase levels and perhaps enhanced production of other enzymes related to the synthesis both of norepinephrine and of the synaptic vesicles in these cells.

TABLE 2

TYROSINE HYDROXYLASE LEVELS IN CULTURED MOUSE NEUROBLASTOMA CELLS

Effect of cyclic nucleotides, sodium butyrate and papaverine

Expt.	Addition*	Tyrosine Hydroxylase Levels pmoles $^{14}CO_2$ formed per 10^6 cells
I.	None	1.0
	3',5'-cAMP(2)**	0.7 (0.8; 0.5)
	N^6, $0^{2'}$-Dibutyryl-3',5'-cAMP	66.0
	Sodium butyrate	30.0
II.	None	1.9
	N^6,$0^{2'}$-Dibutyryl-3',5'-cAMP (3)	100.7 ± 1.6
	Sodium butyrate (3)	42.4 ± 1.0
	Papaverine, 0.13 mM, 1 day (2)	5.8 (6.6; 5.0)
III.	None	1.5
	Papaverine, 0.13 mM (2)	27.3 (28.6; 26.1)
IV.	None	2.3
	Sodium butyrate	91.4
	N^6-Monobutyryl-3',5'-cAMP (3)	70.2 ± 4.7
V.	Sodium butyrate	33.9
	8-Aminomethyl-3',5'-cAMP, 0.3 mM (2)	17.9 (16.0; 19.7)
	8-Furfurylamine-3',5'-cAMP, 0.3 mM (2)	3.8 (3.1; 4.6)

After incubation for the time specified in physiological salt solution containing agamma globulin newborn calf serum, cells were harvested, washed and lysed in 0.1% Triton-X-100 (Prasad et al, submitted). The lysed cells were assayed for tyrosine hydroxylase by the coupled decarboxylation method (Waymire et al, 1971) in 200 mM sodium acetate, pH 6.1; 1 mM $FeSO_4$, 2 mM $DMPH_4$, 100 units hog kidney aromatic amino acid decarboxylase and 0.1 mM L-tyrosine-1-^{14}COOH. The $^{14}CO_2$ was collected in NCS solubilizer after terminating the reaction with trichloroacetic acid and the $^{14}CO_2$ was counted by liquid scintillation spectrometry.

Results are presented as individual experiments (I-V) because the magnitude of the response of the cells to treatment with the several substances varied among the different harvested cell populations.

*Unless otherwise specified, additions were made at a final concentration of 0.5 mM and exposure was for 3 days.

**Numbers in parentheses represent number of individual determinations. If unspecified, single experimental determinations are presented.

REFERENCES

Alousi, A., and N. Weiner, (1966), The regulation of norepinephrine synthesis in sympathetic nerves. Effect of nerve stimulation, cocaine and catecholamine releasing agents. Proc. Natl. Acad. Sci. U. S. 56, 1491.

Axelrod, J., (1959), Metabolism of epinephrine and other sympathomimetic amines. Physiol. Rev. 39, 751.

Axelrod, J., (1964), The uptake and release of catecholamines and the effect of drugs. In: Progress in Brain Research, Vol. 8, Biogenic Amines, (Himwich, H. E., and Himwich, W. A., Eds., Elsevier Publishing Co., Amsterdam, The Netherlands) p. 81.

Bogdanski, D. F. and B. B. Brodie, (1969), The effects of inorganic ions on the storage and uptake of ^{3}H-norepinephrine by rat heart slices. J. Pharmacol. Exp. Therap. 165, 181.

Dairman, W., R. Gordon, S. Spector, A. Sjoerdsma and S. Udenfriend, (1968), Increased synthesis of catecholamines in the intact rat following administration of α-adrenergic blocking agents. Mol. Pharmacol. 4, 457.

Euler, U. S. von, R. Luft, and T. Sundin, (1955), The urinary excretion of noradrenaline and adrenaline in healthy subjects during recumbancy and standing. Acta Physiol. Scand. 34, 169.

Franklin, T. J. and G. A. Snow, (1971), Biochemistry of antimicrobial action. (Academic Press, New York), p. 87.

Hagen, P. and N. Weiner, (1959), Enzymic oxidation of pharmacologically active amines. Fed. Proc. 18, 1005.

Hagen, P. and A. D. Welch, (1956), The adrenal medulla and the biosynthesis of pressor amines. Rec. Progr. Hormone Res. 12, 27.

Hertting, G., J. Axelrod, and R. W. Patrick, (1962), Actions of bretylium and guanethidine on the uptake and release of ^{3}H-noradrenaline. Brit. J. Pharmacol. 18, 161.

Horton, E. W., (1969), Hypotheses on physiological roles of prostaglandins. Physiol. Rev. 49, 122.

Huković, S., (1961), Responses of the isolated sympathetic nerve-ductus deferens preparation of the guinea pig. Brit. J. Pharmacol. Chemotherap. 16, 188.

Ikeda, M., L. A. Fahien, and S. Udenfriend, (1966), A kinetic study of bovine adrenal tyrosine hydroxylase. J. Biol. Chem. 241, 4452.

Iversen, L. L. and E. A. Kravitz, (1966), Sodium dependence of transmitter uptake at adrenergic nerve terminals. Mol. Pharmacol. 2, 360.

Jarrott, B., (1971), Occurrence and properties of catechol-O-methyl transferase in adrenergic neurons. J. Neurochem. 18, 17.

Jarrott, B. and L. L. Iversen, (1971), Noradrenaline metabolizing enzymes in normal and sympathetically denervated vas deferens. J. Neurochem. 18, 1.

Kakiuchi, S. and T. W. Rall, (1968a), The influence of chemical agents on the accumulation of adenosine 3',5'-phosphate in slices of rabbit cerebellum. Mol. Pharmacol. 4, 367.

Kakiuchi, S. and T. W. Rall, (1968b), Studies on adenosine 3',5'-phosphate in rabbit cerebral cortex. Mol. Pharmacol. 4, 379.

Kakiuchi, S., T. W. Rall, and H. McIlwain, (1969), The effect of electrical stimulation upon the accumulation of adenosine 3',5'-phosphate in isolated cerebral tissue. J. Neurochem. 16, 485.

Kaufman, S. and S. Friedman, (1965), Dopamine-β-hydroxylase. Pharmacol. Rev. 17, 71.

Kirshner, N., (1957), Pathway of noradrenaline formation from dopa. J. Biol. Chem. 226, 821.

Kirshner, N., (1962), Uptake of catecholamines by a particulate fraction of the adrenal medulla. J. Biol. Chem. 237, 2311.

Kirshner, N., M. Rorie, and D. L. Kamin, (1965), Inhibition of dopamine uptake in vitro by reserpine administered in vivo. J. Pharmacol. Exp. Therap. 141, 285.

Kopin, I. J., V. K. Weise, and G. C. Sedvall, (1969), Effect of false transmitters on norepinephrine synthesis. J. Pharmacol. Exp. Therap. 170, 246.

Kuntzman, R. and M. M. Jacobson, (1963), The inhibition of monoamine oxidase by benzyl and phenethylguanidines related to bretylium and guanethidine. Annals New York Acad. of Sci. 107, 945.

Levine, E. Y. and S. Kaufman, (1961), Studies on the enzyme catalyzing the conversion of 3,4-dihydroxyphenylethylamine to norepinephrine. J. Biol. Chem. 236, 2043.

Levitt, M., S. Spector, A. Sjoerdsma, and S. Udenfriend, (1965), Elucidation of the rate-limiting step in norepinephrine biosynthesis in the perfused guinea-pig heart. J. Pharmacol. Exp. Therap. 148, 1.

Lloyd, T. and N. Weiner, (In Press), Isolation and characterization of a tyrosine hydroxylase cofactor from bovine adrenal medulla. Mol. Pharmacol.

Lovenberg, W., H. Weissbach, and S. Udenfriend, (1962), Aromatic L-amino acid decarboxylase. J. Biol. Chem. 237, 89.

Luco, J. V. and F. Goñi, (1948), Synaptic fatigue and chemical mediators of postganglionic fibers. J. Neurophys. 11, 497.

Mueller, R. A., H. Thoenen, and J. Axelrod, (1969a), Increase in tyrosine hydroxylase activity after reserpine administration. J. Pharmacol. Exp. Therap. 196, 74.

Mueller, R. A., H. Thoenen and J. Axelrod, (1969b), Adrenal tyrosine hydroxylase: Compensatory increase in activity after chemical sympathectomy. Science 163, 468.

Musacchio, J. M., (1968), Subcellular distribution of adrenal tyrosine hydroxylase. Biochem. Pharmacol. 17, 1470.

Musacchio, J. M., J. J. Wurzburger, and G. L. D'Angelo, (1971), Different molecular forms of bovine adrenal tyrosine hydroxylase. Mol. Pharmacol. 7, 136.

Nagatsu, T., M. Levitt and S. Udenfriend, (1964), Tyrosine hydroxylase: The initial step in norepinephrine biosynthesis. J. Biol. Chem. 238, 2190.

Posiviata, M., G. Becker, and N. Weiner, (1971), The effect of bretylium tosylate on catecholamine synthesis in the guinea-pig isoated vas deferens preparation. The Pharmacologist 13, 253.

Prasad, K. N., (1971), Effect of dopamine and 6-hydroxydopamine on mouse neuroblastoma cells in vitro. Cancer Research 31, 1457.

Prasad, K. N. and A. W. Hsie, (1971), Dibutyryl adenosine 3'-5'-cyclic monophosphate-induced morphologic differentiation of mouse neuroblastoma cells in vitro. Nature New Biol. 233, 141.

Prasad, K. N., J. C. Waymire and N. Weiner, (Submitted for publication), Tyrosine hydroxylase: Relationship between differentiation and enzyme activity in cultured mouse neuroblastoma cells exposed to dibutyryl cyclic AMP or x-irradiation.

Rutledge, C. O., and N. Weiner, (1967), The effect of reserpine upon the synthesis of norepinephrine in the isolated rabbit heart. J. Pharmacol. Exp. Therap. 157, 290.

Shaw, J. E. and P. W. Ramwell, (1968), Release of prostaglandin from rat epididymal fat pad on nervous and hormonal stimulation. J. Biol. Chem. 243, 1498.

Thoa, N. B., D. G. Johnson, I. J. Kopin and N. Weiner, (1971), Acceleration of catecholamine formation in the guinea pig-vas deferens following hypogastric nerve stimulation: roles of tyrosine hydroxylase and new protein synthesis. J. Pharmacol. Exp. Ther. 178, 442.

Thoenen, H., R. A. Mueller and J. Axelrod, (1969), Trans-synaptic induction of adrenal tyrosine hydroxylase. J. Pharmacol. Exp. Therap. 169, 249.

Udenfriend, S., (1966), Biosynthesis of the sympathetic neurotransmitter, norepinephrine. The Harvey Lectures 60, 57.

Udenfriend, S., P. Zaltzman-Nirenberg and T. Nagatsu, (1965), Inhibitors of purified beef adrenal tyrosine hydroxylase. Biochem. Pharmacol. 14, 837.

Viveros, O. H., L. Arqueros, R. J. Connett, and N. Kirshner, (1969), Mechanism of secretion from the adrenal medulla. IV. The fate of storage vesicles following insulin and reserpine administration. Mol. Pharmacol. 5, 69.

Vogt, M., (1954), The concentration of sympathin in different parts of the central nervous system under normal conditions and after the administration of drugs. J. Physiol. (London) 123, 451.

Waalkes, T. P. and S. Udenfriend, (1957), A fluorometric method for the estimation of tyrosine in plasma and tissues. J. Lab. Clin. Med. 50, 733.

Waymire, J. C., R. Bjur and N. Weiner, (1971), Assay of tyrosine hydroxylase by coupled decarboxylation of dopa formed from 1-^{14}C-L-tyrosine. Anal. Biochem. 43, 588.

Waymire, J. C., N. Weiner, F. H. Schneider, M. Goldstein, and L. S. Freedman, (Submitted for publication), Tyrosine hydroxylase in human adrenal and pheochromocytoma: localization, kinetics and catecholamine inhibition.

Weiner, N., (1964), The catecholamines; biosynthesis, storage and release, metabolism and metabolic effects. In: The Hormones, Vol. 4, (ed. by G. Pincus, K. V. Thimann and E. B. Astwood, Academic Press, N. Y.) p. 403.

Weiner, N., (1970), Regulation of norepinephrine biosynthesis. Ann. Rev. Pharmacol. 10, 273.

Weiner, N. and R. Bjur (In press). The role of intraneuronal monoamine oxidase in the regulation of norepinephrine synthesis. In: Monoamine oxidases: New vistas. (E. Costa and M. Sandler, eds.) Vol. 5, "Advances in Biochemical Psychopharmacology", Raven Press.

Weiner, N. and O. Jardetzky, (1964), A study of catecholamine nucleotide complexes by nuclear magnetic resonance spectroscopy. Arch. Pathol. Pharmakol. 248, 308.

Weiner, N., and W. F. Mosimann, (1970), The effect of insulin on the catecholamine content of cat adrenal glands. Biochem. Pharmacol. 19, 1189.

Weiner, N., M. Posiviata and G. Becker, (1971b), Further studies on the post-stimulation increase in catecholamine synthesis in the isolated vas deferens. The Pharmacologist 13, 253.

Weiner, N. and M. Rabadjija, (1968a), The effect of nerve stimulation on the synthesis and metabolism of norepinephrine in the isolated guinea-pig hypogastric nerve-vas deferens preparation. J. Pharmacol. Exp. Ther. 160, 61.

Weiner, N. and M. Rabadjija, (1968b), The regulation of norepinephrine synthesis. Effect of puromycin on the accelerated synthesis of norepinephrine associated with nerve stimulation. J. Pharmacol. Exp. Ther. 164, 103.

Weiner, N. and I. Selvaratnam, (1968), The effect of tyramine on the synthesis of norepinephrine. J. Pharmacol. Exp. Therap. 161, 21.

Weiner, N., J. C. Waymire and F. H. Schneider, (1971a), The localization and kinetics of tyrosine hydroxylase of the adrenals of several species and of human chromaffin tumors. Acta Cientifica Venezolana. 22, Suppl. 2, 179.

Acknowledgment: The authors are grateful to Dr. R. K. Robins, ICN Research Institute, Irvine, California, for generously providing the stable cyclic nucleotide analogs 8-aminomethyl cyclic AMP (YH-13) and 8-furfurylamino-cyclic AMP (YH-10).

This work was supported by USPHS Grants NS07642, NS07927, NS09230 and RR05357.

IMMUNOLOGY

VARIABILITY, SYMMETRY AND PERIODICITY IN THE STRUCTURE OF IMMUNOGLOBULINS

GERALD M. EDELMAN
The Rockefeller University
New York, New York, USA

Introduction of an antigen into vertebrate organisms provokes the synthesis of antibodies that can bind specifically to that antigen or to molecules with structures similar to it (Landsteiner, 1945). The capacity of the immune system to respond to an enormous range of stereochemically distinct antigenic molecules is as remarkable as the specificity with which each antigen is bound. Moreover, upon a second encounter with the same antigen, there is immunological memory, i.e. the specific response is more rapid and larger amounts of specific antibodies are synthesized. Lymphoid cells and the antibodies synthesized by them constitute a molecular recognition system which is remarkable for its specificity, range of response and control (Killander, 1967; Cairns, 1967).

Studies of the structure of antibody molecules (Edelman, 1959; Porter, 1959; Edelman and Poulik, 1961; Edelman and Gall, 1969) and of the cellular dynamics of the immune response (Cairns, 1967) lend strong support to selective theories of immunity (Jerne, 1955; Burnet, 1959) which assume that the information required for the specificity and range of the immune response resides in the organism _before_ exposure to the antigen. Antigens serve to stimulate (or select) only those lymphoid cells that synthesize complementary antibodies. This cellular system is clonal, i.e. each cell synthesizes a single variety of antibody and only after encounter with the appropriate complementary antigen is it stimulated to form progeny cells synthesizing the identical type of antibody at greatly increased rates. Clonal selective theories require a minimum of three conditions: 1) the organism must contain information for the synthesis of an _enormous_ repertoire of different antigen-binding sites most of which may never be used, 2) there must be a mechanism for antigen trapping to favor encounter with the appropriate lymphoid cells, 3) lymphoid cells must contain an amplifier of high gain which is triggered by the antigen, so that after selection of the appropriate cells, a significant number of antibody molecules of the correct specificity are produced. This last condition is, in fact, realized by cell division and increased antibody synthesis (Cairns, 1967).

In view of the special requirements of clonal selection, it is perhaps not surprising that immunoglobulins have unusual structures. It is now clear, for example, that the basis for the variety of binding sites required by selective theories is variation in the amino acid sequences of the first 110-120 residues of immunoglobulin light and heavy polypeptide chains. Moreover, structural work on antibodies has indicated that extensive gene duplication is a basic mechanism in the evolution of the immune system. The intrinsic genetic interest of these observations is obvious. Less obvious, but of equal importance is that they provide a unique opportunity to analyze the relationship between the primary and tertiary structures of a group of proteins that have been selected during evolution for extensive variations in amino acid sequence as well as for a remarkable regional differentiation of function. Primary structural analyses suggest that the tertiary structure of immunoglobulin molecules is organized in a series of domains having different functions.

Although there are at least five classes and several subclasses of immunoglobulins (World Health Organization, 1964; World Health Organization, 1968), the detailed discussion in this paper will be restricted to the γG immunoglobulins, for there are good reasons to believe that this most prevalent class is representative of the major features of antibodies. We now know the complete covalent structure of a γG immunoglobulin molecule. Although the three-dimensional structure at atomic resolution and the details of the antigen-binding sites have not been determined, a good deal can be said about the relation of antibody structure to function by comparing the entire amino acid sequence with sequences of portions of different immunoglobulin molecules analyzed by other workers (Killander, 1967; Cairns, 1967).

Diversity and Evolution by Gene Duplication

A detailed comparison of the exact amino acid sequences of various immunoglobulin polypeptide chains has yielded some major surprises. Immunoglobulin chains have been found to contain a region in which all the heterogeneity of the subclass is located; these variable or V regions comprise the amino terminal sequence of 120 or so residues of the polypeptide chains (figure 1). *A priori*, one would expect V regions to be concerned with antigen binding; this has been experimentally confirmed by affinity-labelling techniques (Singer and Doolittle, 1966).

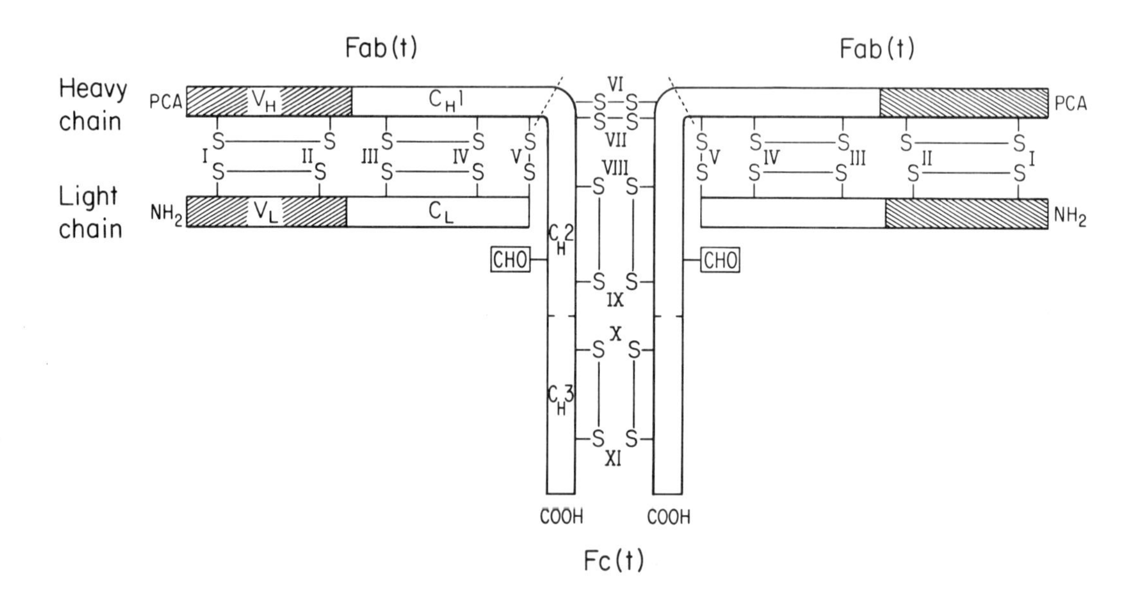

Fig. 1. Overall arrangement of chains and disulfide bonds of a human γG1 immunoglobulin (myeloma protein Eu). Half cystinyl residues are numbered I-XI beginning from the NH_2 terminus. Numbers I-V designate corresponding residues in light and heavy chains. PCA - pyrollidonecarboxylic acid. CHO - carbohydrate. "Fab(t) and "Fc(t) refer to fragments produced by trypsin, which cleaves the heavy chain as indicated by dashed lines above half-cystinyl residues VI. V_H, V_L - homologous variable regions of heavy and light chains; C_H1, C_H2, C_H3 - homology regions comprising C_H or constant region of heavy chain; C_L - constant region of light chain which is homologous to C_H1, C_H2, and C_H3.

As predicted from early experiments on the chain structure (Edelman, 1963), the different immunoglobulin classes are differentiated from each other according to the structure of their heavy chains. The essential class specificity can be localized even more specifically in the carboxyl terminal portions of the heavy chains in the constant or C regions. These portions of the molecule mediate different effector functions such as complement fixation and cell fixation which occur after binding of the antigen by the V regions (see Edelman and Gall, 1969). There is now evidence that some of the immunoglobulin classes (World Health Organization 1968; Ishizaka *et al*., 1966) carry out quite special effector functions.

Additional hints about the evolutionary origin and functional differentiation of immunoglobulins emerge from analyses of their primary structure (Edelman *et al*., 1969; Edelman, 1970a). A consideration of the amino acid sequences of γG immunoglobulin (figure 2, figure 3) supports the following conclusions:

1. V_H and V_L are homologous to each other but are not obviously homologous

to C_H or C_L. V regions from the same molecule appear to be no more closely related than V regions from different molecules.

2. $C_{\gamma 1}$ consists of three homology regions, C_H1, C_H2, and C_H3, each of which is closely homologous to the others and to C_L.

3. Each variable region and each constant homology region contains one disulfide bond. This indicates that the intrachain disulfide bonds are linearly and periodically distributed in the structure.

4. The region which contains all of the interchain disulfide bonds is in the middle of the linear sequence of the heavy chain and has no homologous counterpart in other portions of the heavy or light chains.

SEQUENCE HOMOLOGY IN EU VARIABLE REGIONS

										1									10
EU V_L	(RESIDUES 1-108)									ASP	ILE	**GLN**	MET	THR	**GLN**	**SER**	PRO	SER	THR
EU V_H	(RESIDUES 1-114)									PCA	VAL	**GLN**	LEU	VAL	**GLN**	**SER**	GLY	–	ALA
									20										30
LEU	SER	ALA	SER	VAL	**GLY**	ASP	ARG	**VAL**	THR	ILE	THR	**CYS**	ARG	**ALA**	**SER**	GLN	SER	ILE	ASN
GLU	VAL	LYS	LYS	PRO	**GLY**	SER	SER	**VAL**	LYS	VAL	SER	**CYS**	LYS	**ALA**	**SER**	GLY	GLY	THR	PHE
											40								
THR	–	–	TRP	LEU	ALA	**TRP**	TYR	GLN	**GLN**	LYS	**PRO**	**GLY**	LYS	ALA	PRO	LYS	LEU	LEU	MET
SER	ARG	SER	ALA	ILE	ILE	**TRP**	VAL	ARG	**GLN**	ALA	**PRO**	**GLY**	GLN	GLY	LEU	GLU	TRP	MET	GLY
	50											60							
TYR	LYS	ALA	SER	SER	–	LEU	GLU	SER	GLY	VAL	PRO	SER	ARG	**PHE**	ILE	**GLY**	SER	GLY	SER
GLY	ILE	VAL	PRO	MET	PHE	GLY	PRO	PRO	ASN	TYR	ALA	GLN	LYS	**PHE**	GLN	**GLY**	–	ARG	VAL
		70																	80
GLY	THR	GLU	PHE	THR	–	–	–	–	–	–	–	LEU	THR	ILE	**SER**	**SER**	**LEU**	GLN	PRO
THR	ILE	THR	ALA	ASP	GLU	SER	THR	ASN	THR	ALA	TYR	MET	GLU	LEU	**SER**	**SER**	**LEU**	ARG	SER
									90										
ASP	**ASP**	PHE	**ALA**	THR	**TYR**	TYR	**CYS**	GLN	GLN	–	**TYR**	ASN	SER	ASP	**SER**	LYS	MET	PHE	GLY
GLU	**ASP**	THR	**ALA**	PHE	**TYR**	PHE	**CYS**	ALA	GLY	GLY	**TYR**	GLY	ILE	TYR	**SER**	PRO	GLU	GLU	TYR
100																			
GLN	**GLY**	THR	LYS	**VAL**	GLU	VAL	LYS	GLY											
ASN	**GLY**	GLY	LEU	**VAL**	THR														

Fig. 2. Comparison of the amino acid sequences of the variable regions V_L and V_H. Deletions indicated by dashes have been introduced to maximize the homology. Identical residues are shaded.

The results are consistent with the hypothesis (Singer and Doolittle, 1966; Hill *et al.*, 1966) that immunoglobulin chains arose by duplication of a precursor gene of about 330 nucleotides in length. Because there is no clear cut evidence of homology between V and C regions, it is somewhat harder to decide whether V and C regions evolved from a single gene. The alternative is that there were two precursor genes for V and C which were originally unrelated but were brought together because of selective advantages of combining their functions in a single product molecule (Edelman, 1970a).

SEQUENCE HOMOLOGY IN EU CONSTANT REGIONS

		110									120	
EU C_L (RESIDUES 109-214)	THR	VAL	ALA	ALA	PRO	SER	VAL	PHE	ILE	PHE	PRO	PRO	SER
EU C_H1 (RESIDUES 119-220)	SER	THR	LYS	GLY	PRO	SER	VAL	PHE	PRO	LEU	ALA	PRO	SER
EU C_H2 (RESIDUES 234-341)	LEU	LEU	GLY	GLY	PRO	SER	VAL	PHE	LEU	PHE	PRO	PRO	LYS
EU C_H3 (RESIDUES 342-446)	GLN	PRO	ARG	GLU	PRO	GLN	VAL	TYR	THR	LEU	PRO	PRO	SER

									130										
ASP	GLU	GLN	–	–	LEU	LYS	SER	GLY	THR	ALA	SER	VAL	VAL	CYS	LEU	LEU	ASN	ASN	PHE
SER	LYS	SER	–	–	THR	SER	GLY	GLY	THR	ALA	ALA	LEU	GLY	CYS	LEU	VAL	LYS	ASP	TYR
PRO	LYS	ASP	THR	LEU	MET	ILE	SER	ARG	THR	PRO	GLU	VAL	THR	CYS	VAL	VAL	VAL	ASP	VAL
ARG	GLU	GLU	–	–	MET	THR	LYS	ASN	GLN	VAL	SER	LEU	THR	CYS	LEU	VAL	LYS	GLY	PHE

140												150							
TYR	PRO	ARG	GLU	ALA	LYS	VAL	–	–	GLN	TRP	LYS	VAL	ASP	ASN	ALA	LEU	GLN	SER	GLY
PHE	PRO	GLU	PRO	VAL	THR	VAL	–	–	SER	TRP	ASN	SER	–	GLY	ALA	LEU	THR	SER	GLY
SER	HIS	GLU	ASP	PRO	GLN	VAL	LYS	PHE	ASN	TRP	TYR	VAL	ASP	GLY	–	VAL	GLN	VAL	HIS
TYR	PRO	SER	ASP	ILE	ALA	VAL	–	–	GLU	TRP	GLU	SER	ASN	ASP	–	GLY	GLU	PRO	GLU

		160										170							
ASN	SER	GLN	GLU	SER	VAL	THR	GLU	GLN	ASP	SER	LYS	ASP	SER	THR	TYR	SER	LEU	SER	SER
–	VAL	HIS	THR	PHE	PRO	ALA	VAL	LEU	GLN	SER	–	SER	GLY	LEU	TYR	SER	LEU	SER	SER
ASN	ALA	LYS	THR	LYS	PRO	ARG	GLU	GLN	GLN	TYR	–	ASP	SER	THR	TYR	ARG	VAL	VAL	SER
ASN	TYR	LYS	THR	THR	PRO	PRO	VAL	LEU	ASP	SER	–	ASP	GLY	SER	PHE	PHE	LEU	TYR	SER

		180										190							
THR	LEU	THR	LEU	SER	LYS	ALA	ASP	TYR	GLU	LYS	HIS	LYS	VAL	TYR	ALA	CYS	GLU	VAL	THR
VAL	VAL	THR	VAL	PRO	SER	SER	SER	LEU	GLY	THR	GLN	–	THR	TYR	ILE	CYS	ASN	VAL	ASN
VAL	LEU	THR	VAL	LEU	HIS	GLN	ASN	TRP	LEU	ASP	GLY	LYS	GLU	TYR	LYS	CYS	LYS	VAL	SER
LYS	LEU	THR	VAL	ASP	LYS	SER	ARG	TRP	GLN	GLU	GLY	ASN	VAL	PHE	SER	CYS	SER	VAL	MET

		200													210				
HIS	GLN	GLY	LEU	SER	SER	PRO	VAL	THR	–	LYS	SER	PHE	–	–	ASN	ARG	GLY	GLU	CYS
HIS	LYS	PRO	SER	ASN	THR	LYS	VAL	–	ASP	LYS	ARG	VAL	–	–	GLU	PRO	LYS	SER	CYS
ASN	LYS	ALA	LEU	PRO	ALA	PRO	ILE	–	GLU	LYS	THR	ILE	SER	LYS	ALA	LYS	GLY		
HIS	GLU	ALA	LEU	HIS	ASN	HIS	TYR	THR	GLN	LYS	SER	LEU	SER	LEU	SER	PRO	GLY		

Fig. 3. Comparison of the amino acid sequence of constant homology regions C_L, C_H1, C_H2 and C_H3. Deletions indicated by dashes have been introduced to maximize the homology. Identical residues are darkly shaded; both dark and light shading are used to indicate identities which occur in pairs in the same position.

Comparison of the primary structures of a great variety of immunoglobulins shows that the diversity of V regions has several distinct features (figures 4 and 5). Although the figures show only heavy chains, both light and heavy chain V regions fall into subgroups of related sequences. The sequences of different subgroups are sufficiently different to warrant the conclusion that subgroups are specified by different genes. Within a subgroup, the variation may be accounted for by single base substitutions in the genetic code. Moreover, certain positions show many substitutions (the so-called hypervariable regions) whereas others, such as the 1/2 cystine residues at positions 23 and 88 of κ chains, have never been observed to vary at all.

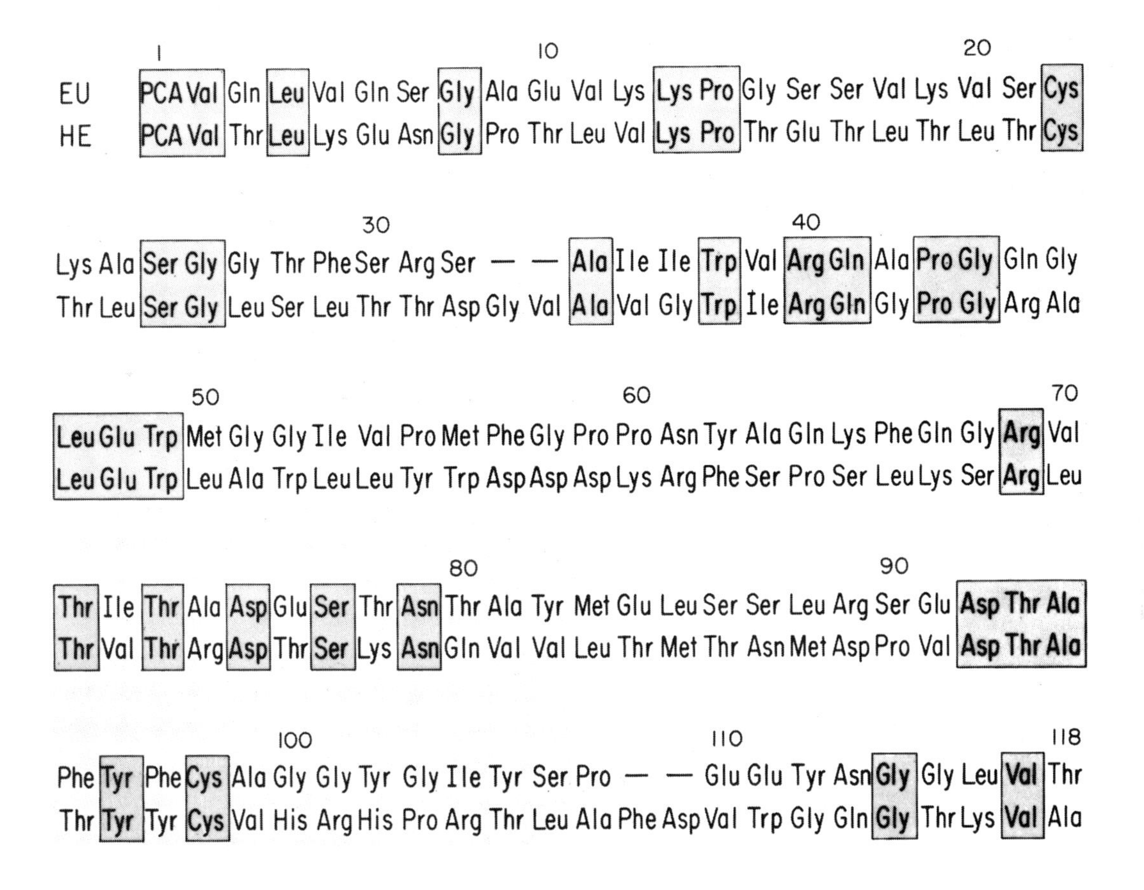

Fig. 4. Comparison of V_H regions of protein Eu (V_{HI}) and protein He (V_{HIII}). V_H regions within each subgroup are more closely similar in sequence than are these two proteins.

This set of observations indicates that both mutation and selection have operated to yield variable region structure and that intra-subgroup variation is the major source of immunoglobulin diversity. Although the origin of intrasubgroup diversity remains unknown, several other sources of antibody diversity are now understood, and they shall be discussed in some detail below.

	1	10
HE (γG)	PCA Val Thr Leu Lys Glu Asn Gly Pro Thr Leu Val Lys Pro Thr Glu Thr Leu Thr	
DAW (γG)	PCA Val Thr Leu Arg Glu Ser Gly Pro Ala Leu Val Arg Pro Thr Gln Thr Leu Thr	
COR (γG)	PCA Val Thr	
OU (γM)	PCA Val Thr Leu Thr Glu Ser Gly Pro Ala Leu Val Lys Pro Lys Gln Pro Leu Thr	

20 30 40

Leu Thr Cys Thr Leu Ser Gly Leu Ser Leu Thr Thr Asp Gly Val Ala Val Gly Trp Ile Arg Gln Gly Pro Gly
Leu Thr Cys Thr Phe Ser Gly Phe Ser Leu Ser Gly Glu Thr Met Cys Val Ala Trp Ile Arg Gln Pro Pro Gly
Met Cys Val Gly Trp Ile Arg Gln Pro Pro Gly
Leu Thr Cys Thr Phe Ser Gly Phe Ser Leu Ser Thr Ser Arg Met Arg Val Ser Trp Ile Arg Arg Pro Pro Gly

50 60

Arg Ala Leu Glu Trp Leu Ala Trp Leu Leu Tyr Trp Asp Asp Asp Lys Arg Phe — Ser Pro Ser Leu Lys Ser
Glu Ala Leu Glu Trp Leu Ala Trp Asp Ile Leu — Asn Asp Asp Lys Tyr Tyr — Gly Ala Ser Leu Glu Thr
Lys Gly Leu Glu Trp Leu Ala — Arg Ile Asx Trp Asp Asp Asp Lys Tyr Tyr — Asn Thr Ser Leu Glu Thr
Lys Ala Leu Glu Trp Leu Ala — Arg Ile Asx — Asx Asx Asp Lys Phe Tyr Trp Ser Thr Ser Leu Arg Thr

70 80 90

Arg Leu Thr Val Thr Arg Asp Thr Ser Lys Asn Gln Val Val Leu Thr Met Thr Asn Met Asp Pro Val Asp Thr
Arg Leu Ala Val Ser Lys Asp Thr Ser Lys Asn Gln Val Val Leu Ser Met Asn Thr Val Gly Pro Gly Asp Thr
Arg Leu Thr **Ile** Ser Lys Asp Thr Ser Arg Asn Gln Val Val Leu Thr Met — — — Asp Pro Val Asp Thr
Arg Leu Ser **Ile** Ser Lys Asn Asp Ser Lys Asn Gln Val Val Leu Ile Met Ile Asn Val Asn Pro Val Asp Thr

100 110

Ala Thr Tyr Tyr Cys Val His Arg His Pro Arg Thr Leu Ala — — — Phe Asp Val Trp Gly Gln Gly Thr
Ala Thr Tyr Tyr Cys Ala Arg Ser Cys Gly Ser Gln — — — — Tyr Phe Asp Tyr Trp Gly Gln Gly Ile
Ala Thr Tyr Tyr Cys Ala Arg Ile Thr Val Ile Pro Ala Pro Ala Gly Tyr Met Asp Val Trp Gly Arg Gly Thr
Ala Thr Tyr Tyr Cys Ala Arg Val Val Asn Ser Val Met

118

Lys Val Ala
Leu Val Thr
Pro Val Thr

Fig. 5. Comparison of V_{HIII} regions showing variations within a single heavy chain subgroup. Proteins Daw and Cor are described by Press and Hogg (1969) and protein Ou is from μ chains (Wikler et al., 1969).

Sites and Molecular Arrangement: The Domain Hypothesis

A consideration of the disulfide bond arrangements (Gall and Edelman, 1970) and complete primary structure of γG1 immunoglobulin (Edelman *et al.*, 1969; Edelman, 1970a) suggests a number of general conclusions about the tertiary and quarternary structure that could not be drawn in primary structural analyses of other proteins. The bases for this difference reside in the repetition of homologous regions, the linear periodic arrangement of the intrachain disulfide bonds, and the unique location of the interchain bonds. For example, given the identity of both light chains and both heavy chains it was obvious even before there was direct crystallographic evidence (Terry *et al.*, 1968), that a two-fold axis must pass through the heavy chain disulfide bonds.

A striking illustration of the periodicity of the molecule is conveyed by a poppit bead model (Edelman, 1970b; Edelman, 1970c) of the γG1 immunoglobulin Eu (figure 6). If one considers the homologies shown in figures 2 and 3, one plausible arrangement of the chains is a T-shaped congeries of compact domains each formed by a separate V homology region or C homology region (Edelman, 1970a). In such an arrangement, each domain is stabilized by a single, intrachain disulfide bond and is linked to neighboring domains by less tightly folded stretches of the polypeptide chain probably with very close apposition of neighboring domains. A two-fold pseudosymmetry axis relates the V_LC_L to the V_HC_H1 domains and a true dyad axis going through the inter-heavy chain disulfide bonds relates the C_H2-C_H3 domains of each heavy chain. An attractive additional hypothesis is that the tertiary structures of each of the homologous domains are similar; each such domain would contribute to at least one active site mediating a function of the class of immunoglobulin to which the heavy chain belongs. The function of V region domains is obviously antigen binding. Although the function of each constant region domain has not been specified, there is some evidence (Kehoe and Fougereau, 1969) that C_H2 may play a role in complement fixation.

Support for the domain hypothesis has come from electron microscopic observations of γA immunoglobulin (Green *et al.*, 1971; Dourmashkin *et al.*, 1971) in which globular structures corresponding to the domains in the Fab region were observed. Moreover, cleavage of the molecule by a variety of proteolytic enzymes at stretches of chains between or neighboring each homology region have been observed by several workers (Solomon and McLaughlin, 1969; Karlsson *et al.*, 1969; Turner and Bennich, 1968). ORD and CD studies of separately cleaved V and C regions (Bjork *et al.*, 1971) and experiments on the reconstitution of heavy chain dimers (Bjork and Tanford, 1971) lend additional credence to the idea that homology regions are folded in domains.

The most interesting of the stretches between domains is the so-called hinge region (figure 7) for it has been shown that the molecule can be bent in this region. The existence of such a region strongly influences the overall shape of γG immunoglobulin. Valentine and Green (1967) have shown that immunoglobulins linked in rings to divalent haptens could take on a Y shape but the evidence was less convincing that individual molecules had this shape. Any interpretation of the shape of the immunoglobulin must consider the proposal first raised by Feinstein (Feinstein and Rowe, 1965) that the molecule might be bent or "click open" and the proposal (Noelken *et al.*, 1965) that regions corresponding to Fab and Fc are *freely* flexible and connected like balls on a string.

A CPK model of the covalent structure around the interchain disulfide bonds of γG1 immunoglobulin Eu (figure 7) indicates that some rotation would be permitted around lysyl residue 222 of the heavy chain, although in the proline-rich region around the inter-heavy chain bonds rotation is relatively restricted. Evidence from rotational relaxation studies (Wahl and Weber, 1967; Weltman and Edelman, 1967; Yguerabide *et al.*, 1970) is consistent with the presence of limited flexibility in the hinge region of single γG immunoglobulin molecules in solution. The relaxation times are compatible with a relatively extended structure but give no clues as to the shape of the molecule. Measurements of low angle X-ray scattering and ultracentrifugal studies (Charlwood and Utsumi, 1969) have suggested that in solution the molecule has larger dimensions than those measured in the electron microscope. In addition, analysis of anomalies in the cross section

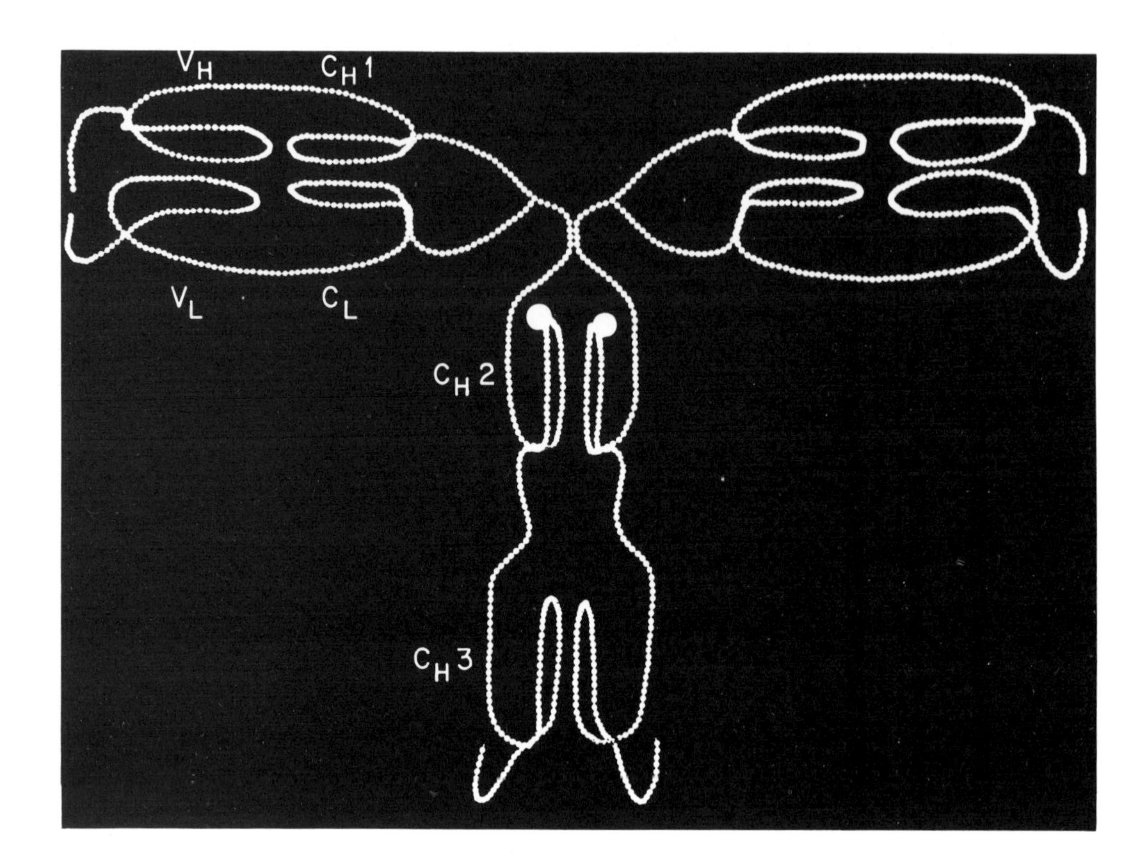

Fig. 6. Poppit bead model showing corresponding alignment of the intrachain disulfide loops within domains. This is not intended as a representation of the detailed three-dimensional structure. Each bead represents a single amino acid residue and the large balls represent the carbohydrate portions of the heavy chains. The homology regions which constitute each domain are indicated. V_L, V_H - domains made up of variable homology regions. C_L, C_H1, C_H2, C_H3 - domains made up of constant homology regions.

curves of the X-ray scattering data (Pilz *et al.*, 1970) suggested that γG immunoglobulin unbound to antigen was T-shaped (figure 8). The other models including that proposing completely free rotation (Noelken *et al.*,1965) did not appear to fit the data.

Recently, an X-ray crystallographic study of γG1 immunoglobulin at 6 Å resolution has been carried out by Sarma *et al.* (1971). The electron density map suggests a T-shaped structure similar to that proposed on the basis of the low angle X-ray studies in solution. The dimensions were larger than those shown in the electron micrographs but smaller than those calculated from the experiments on low angle X-ray scattering. There was no evidence of very low electron density or of extended polypeptide chain in the hinge region as suggested by Noelken *et al.* (1965). Clearly, the X-ray crystallographic evidence is the most reliable indication of the molecular shape of at least one form of γG immunoglobulin. Because the molecule can have different gross shapes in solution and in the crystal, however, solution studies continue to be of great interest. For example, studies of time-dependent fluorescent depolarization (Yguerabide *et al.*, 1970)

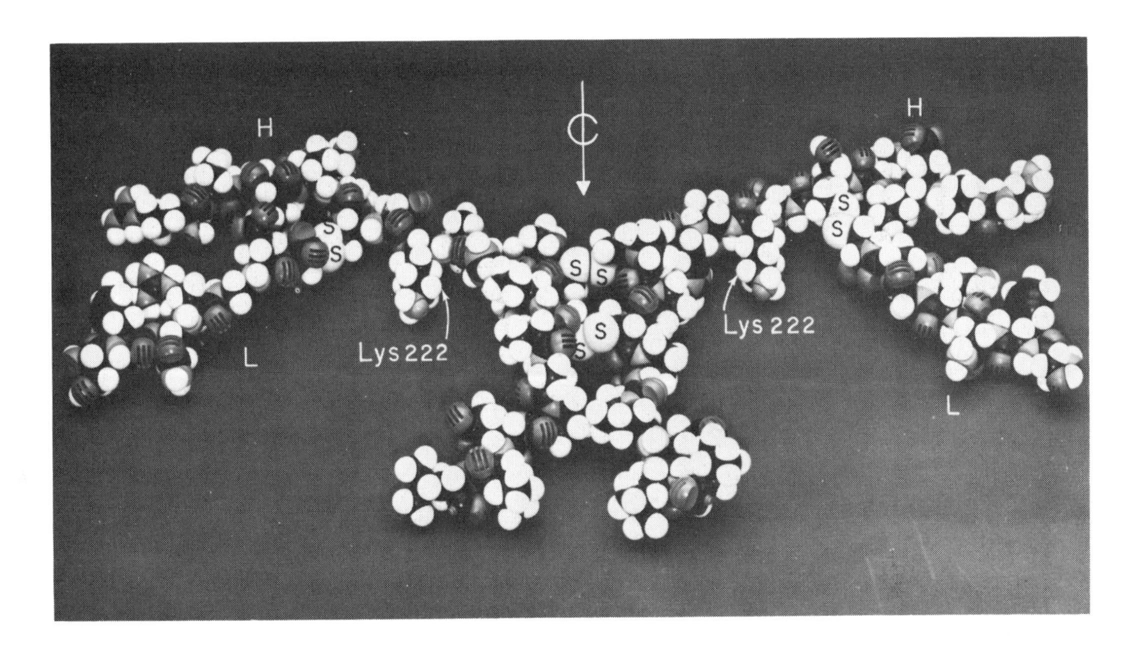

Fig. 7. CPK space-filling model of the covalent structure around the interchain disulfide bonds and hinge region of γG1 immunoglobulin. The lysyl residue at which cleavage by trypsin occurs to form Fab(t) and Fc(t) fragments is shown. H - heavy chains; L - light chains. The arrow shows the dyad axis.

suggest that the Fab arms can rotate through a restricted arc of about 30° from their rest position probably without hindrance. These studies together with the electron microscopic results suggest that only after attachment of large antigens can the molecule assume a more acute Y shape. It is a reasonable hypothesis that the Fab arms can be stabilized in a Y shape after attachment to polyvalent antigens (figure 9). This may expose buried regions such as complement binding sites in the C_H2 domain, and serve to modulate the effector functions of the molecule.

It is evident that the antigen-combining site must have a special structure for although the site is likely to be in a fixed position at the ends of the Fab arms (Valentine and Green, 1967) the basic structure of the V regions must allow large variations in shape. The arrangement of the V region and the homologies pointed out above suggest the hypothesis that disulfide bonds play a key role in stabilizing the combining site. Whatever the details of the folding, the homology between V_L and V_H regions is consistent with a coordinate location of their intrachain bonds. Examination of the portions of V regions which appear to be hypervariable and the location of residues by affinity labelling suggests the hypothesis that the V region disulfide bond plays a pivotal role in stabilizing the site (figure 10). A great deal of variation occurs in residues of the light chain near the intrachain disulfide bonds of the V region and this variation may directly alter the shape of the site. Changes in shape of the site may also arise by modulation of the folding around the V region disulfide bond. This would bring different portions of the loop together with this bond as a pivot (figure 10). By this means, the site could remain fixed but at the same time its local topography could vary without major rearrangements in the neighboring structure.

This model is in accord with the fact that amino acid sequence variations are distributed among the first 108 to 120 residues in the whole length of the V regions. It allows for the presence (Edelman and Gall, 1969) of alterations in contact residues, modulating residues (those that cause a change in the shape of the site even though they are distant from the site) and compensatory residues

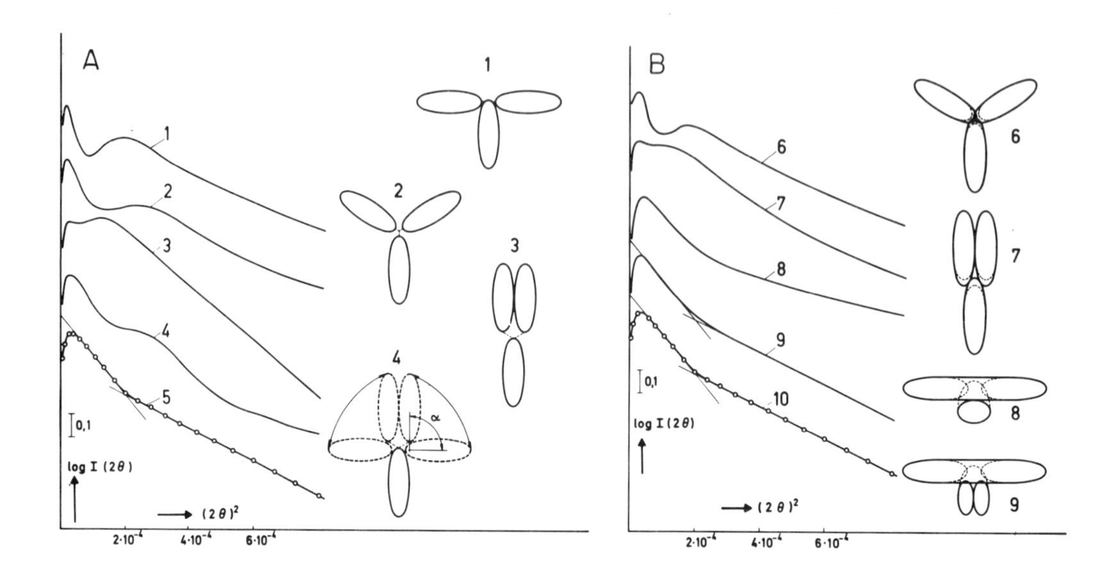

Fig. 8. Low angle X-ray scattering of γG immunoglobulin Eu (Pilz et al., 1970). A. Comparison of the calculated cross section curves for various models of compact ellipsoids which do not touch each other (1-4) with the experimental cross section curve (5). The scattering curves of shapes obtained by variation of angle α have been averaged so that deviation from the experimental curve is smallest. B. Comparison of the calculated cross section curves for various models in which ellipsoids touch or overlap each other (6-9) with the experimental cross section (10). Curve 9 represents the best approximation to the experimental curve obtained with these models. The dashed lines correspond to the ellipsoids equivalent to the individual fragments. I is the intensity of the scattered beam and 2θ is the scattering angle.

(those that must be altered to accommodate a change in one of the other two types). Perhaps the most attractive feature of this model is that it can explain the occurrence of hypervariable and variable positions. The evolutionary cost of variation may be less near the invariant disulfide bond because variations so near the site would require less adjustment elsewhere. Away from this region, more radical changes in the three-dimensional structure might be required to accomodate a mutational change and therefore such mutations might not be as strongly selected for. This seems more attractive than the notion (Wu and Kabat, 1970) that only the hypervariable regions are concerned with the site, relegating all other amino acid variations to the category of unexplained "noise."

Immunoglobulins as Models for Analysis of the Relationship of Primary to Tertiary Structure

Unlike enzymes, the light and heavy chains of immunoglobulins appear to have been selected so that their sequences vary, giving enormous diversity in the three-dimensional structure of antigen-combining sites. This suggests that immunoglobulins may be excellent models for analysis of the fundamental problem of the determination of tertiary by primary structure (Edelman, 1970a). Bence Jones proteins have been shown to be light chains (Edelman and Gally, 1962; Schwartz and Edelman, 1963) and are ideal for this purpose for they have a number

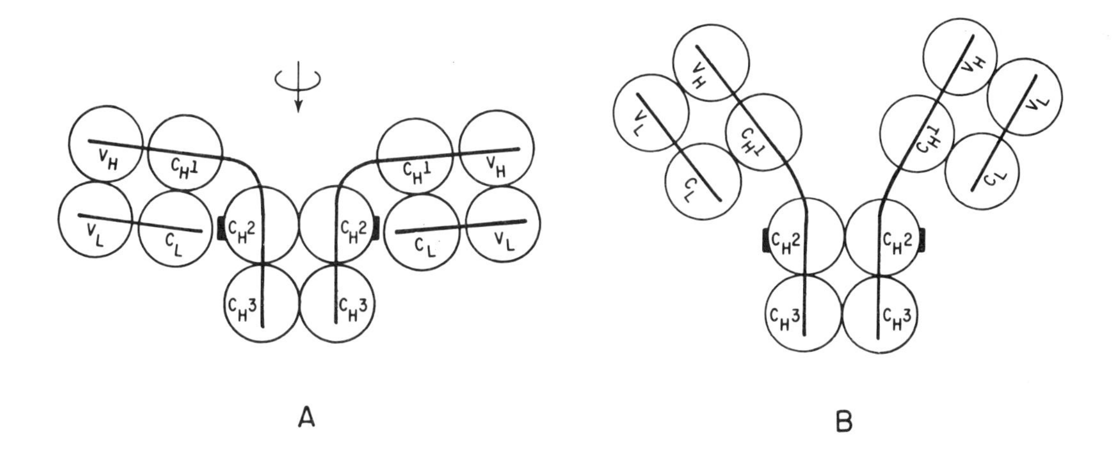

Fig. 9. The domain hypothesis showing pivoting in the hinge region. Homology regions (see figures 2 and 3) which constitute each domain are indicated. V_L, V_H- domains made up of variable homology regions. C_L, C_H1, C_H2, C_H3 - domains made up of constant homology regions. Within each of these groups of homology regions, domains are assumed to have similar three-dimensional structures and each is assumed to contribute to an active site. The V domain sites contribute to antigen-binding functions and the C domain sites to effector functions. (A) Hypothetical arrangement of domains in the free immunoglobulin molecule. The arrow refers to a dyad axis of symmetry. (B) Suggested rearrangement after antigen binding. (▮) Complement binding site.

of desirable properties: 1) they represent an enormously diverse set of protein sequences, 2) the sequences are disposed around several basic patterns, namely κ and λ chains and their subgroups. Each pattern has presumably evolved to serve different functions, the V regions for variation in shape, and the C region for conservation of shape and interchain binding. 3) There is evidence to suggest that each sequence has a different tertiary structure (Edelman and Gally, 1962). 4) Bence Jones proteins may be crystallized (Solomon and McLaughlin, 1969; Karlsson _et al._, 1969). 5) The molecular weights of the chains are relatively low.

Sequence analysis of a suitably chosen set of crystalline Bence Jones proteins in the same subgroup can be correlated with analysis of their three-dimensional structures. The correlation would be particularly favored if the structures were isomorphous. Even if they were not, the presence of identical C_L domains may simplify the comparison of two different Bence Jones proteins. Additional comparisons among Bence Jones proteins, Fab fragments and whole molecules, would be quite valuable in determining the basis for particular conformation as well as in analyzing the nature of the antigen-combining sites.

Interactions between Light and Heavy Chains: the p X q Hypothesis

An examination of the amino acid sequences indicates that the variation in V regions of light and heavy chains of all immunoglobulin classes is similar and that it arises by the same process. It is therefore useful to reconsider the once debated question of the relationship of the two kinds of chains to the antigen-combining site. Early studies (Edelman _et al._, 1961; Edelman _et al._, 1963a) of the chain structure of purified antibodies of different specificities revealed clearly for the first time that their heterogeneity was restricted in comparison with normal immunoglobulins. Because these differences among antibodies were

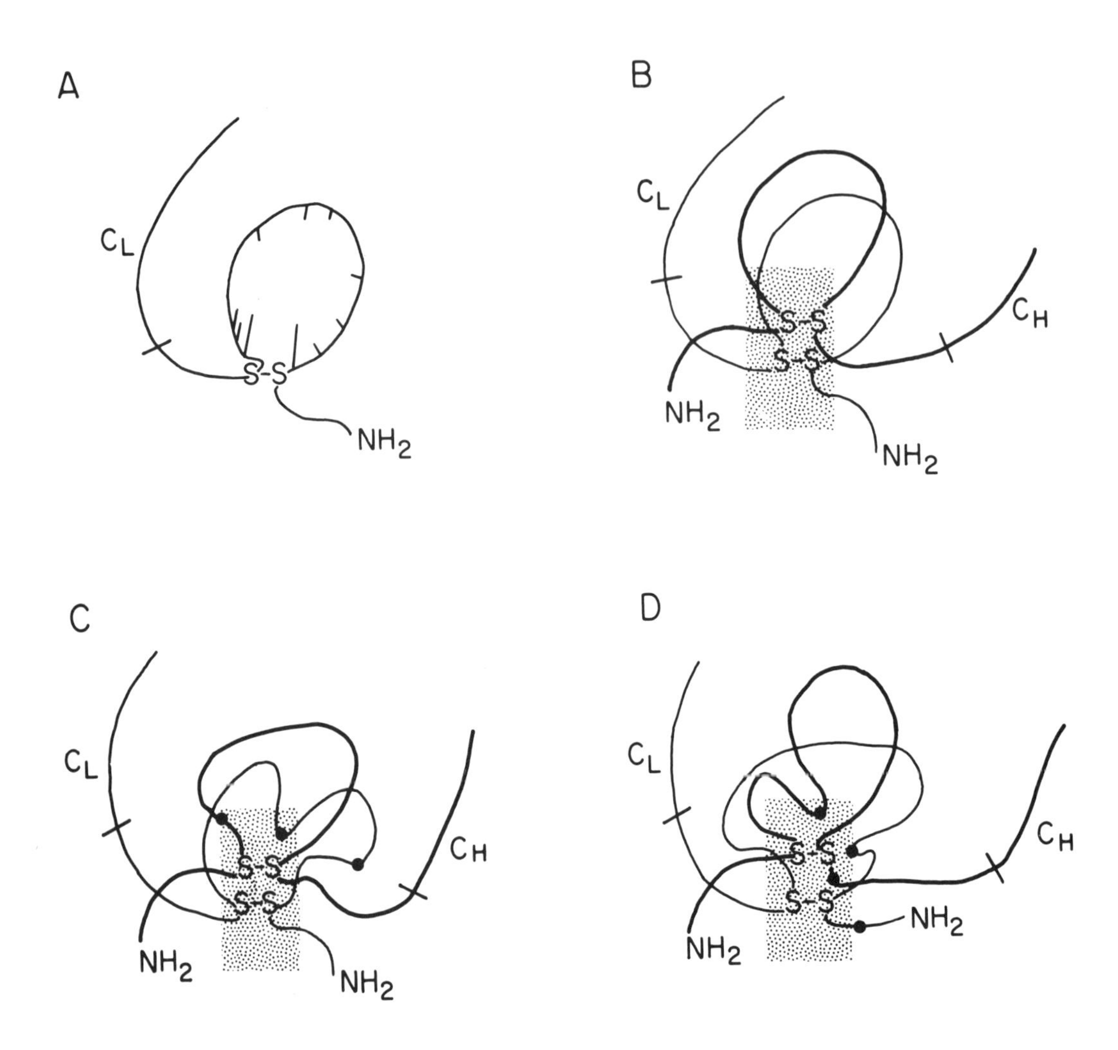

Fig. 10. Schematic diagrams to suggest the possible role of V region disulfide bonds in stabilizing antigen-binding sites. A) ϰ chain disulfide loop. Positions where variations in sequence occur are indicated by bars. B) Coordinate loops of heavy and light chains. Site region is indicated by shading. C) and D) Effect of amino acid replacements: changes in shape of the site region by alteration of the structure around the disulfide bond by substitution of contact, modulating and compensatory residues. Shading indicates position of the site region and dots indicate amino acid replacements. The configurations of the chains are arbitrary and are not intended to represent the actual folding.

observed in denaturing solvents after extensive reduction and alkylation, it was surmised that polypeptide chains from antibodies of different specificities differed in their amino acid compositions and sequences (Edelman, 1963). The studies of Koshland (1967) on the amino acid composition of purified antibodies completely supported this view. It was also clear that antibodies of the same specificity must be degenerate for chains of antibody molecules made in a single individual against a single antigen do not all have the same primary structure.

Two questions emerged from these observations: 1) Do light and heavy chains both contribute to the specificity of the active site? 2) Is there any special relationship between the light and heavy chains of a given antibody i.e. can each light chain form an antibody molecule with one and only one heavy chain? From the results of chain reconstitution experiments (Edelman *et al.*, 1963b; Olins and Edelman, 1964; Fougereau and Edelman, 1964), the answer to the first question was yes and the answer to the second was no. Although this interpretation has been contested (Porter and Weir, 1966), the main body of evidence now supports the conclusion that both light and heavy chains contribute to the active site. The evidence derives from the reconstitution experiments, affinity labelling experiments (Singer and Doolittle, 1966), and perhaps most convincingly, the general resemblance of the structure of the variable regions of light and heavy chains of myeloma proteins (Edelman *et al.*, 1969; Edelman, 1970a) to those of normal immunoglobulins (Milstein *et al.*, 1969; Fruchter *et al.*, 1970).

The importance of the conclusion that both chains contribute to the specificity is that it greatly reduces the number of genes required by *any* theory of antibody variability, i.e.. if there are p heavy chains and q light chains, as many as p X q different immunoglobulins can be formed. This proposal has also been challenged (Tanford, 1968) and evidence has been reported which has placed it in doubt. The contrary proposal is that each light chain will interact to give a native structure with only one heavy chain (Dorrington *et al.*, 1967). This would require a very special mechanism for genetic complementation among genes at completely unlinked loci (i.e. those for heavy chain and light chain V regions) and for this reason it seems highly unlikely (Edelman and Gall, 1969).

Moreover, the observations (Tanford, 1968) leading to the conclusion that a native structure can only be achieved by specific pairing have failed to be confirmed (Dorrington and Tanford, 1970).

There remains the observation that reconstitution of myeloma proteins from light and heavy chains appears to favor specific association (Metzger and Mannik, 1964). This result requires further experimental extension: the most urgently needed experiment is to attempt to reconstitute a large number of different myeloma proteins with V regions belonging to the same and different subgroups. Whatever the result, the data can be consistently explained on the basis of clonal selection of those molecules with stronger V_H-V_L interactions rather on the basis of any prior special relationship between light and heavy chain genes. So far, there is no unequivocal *in vitro* evidence to contradict the p X q hypothesis or the random assortment of chains.

The Origin of Diversity and Evidence for Somatic Translocation of Genetic Information

The most fundamental problem in the study of antibody structure is the origin of antibody diversity. Arguments for various theories have been elaborated elsewhere (Killander, 1967; Cairns, 1967); here, it might be useful to discuss only those features of the theories that might be of interest in relating primary to tertiary structure. It is worth noting that each theory requires special assumptions and there is no "minimal" or "simplest" theory among those extant.

Comparison of the various theories is made easier by considering those aspects of diversity that can already be satisfactorily explained. It is clear, for example, that V region subgroups occur in light chains and that they appear to be specified by non-allelic genes (Milstein and Munroe, 1970; Hood and Talmage, 1970). It is also clear that subgroups occur in heavy chains (see figures 4 and 5). The evidence indicates that regardless of the number of genes specifying each subgroup, the subgroups themselves must have arisen by gene duplication during evolution and therefore there must have been a selective advantage in preserving the different subgroups as bases for variation. Divergence to yield a subgroup might have provided a particularly valuable basis for variation for a certain group of antigens. This does not imply, however, that there is a one-to-one selection of each antibody by each antigen during evolution. It has been suggested elsewhere (Gally and Edelman, 1970) that V region allotypes in the rabbit may have served a similar purpose and that evolution of κ and λ chains may be explained in the same way.

The existence of subgroups and the finding of simple Mendelian markers in C regions both support the remarkable conclusion (Milstein and Munroe, 1970; Hood and Talmage, 1970; Gally and Edelman, 1970) that information from two genes is required to specify a single immunoglobulin chain. Analysis of linkage relationships in the rabbit indicates that κ, λ, and heavy chains must be specified by at least three unlinked gene clusters (Gally and Edelman, 1970). The appearance of the same kind of V region in association with different C regions requires a special somatic mechanism (Dreyer and Bennett, 1965; Gally and Edelman, 1970) for constructing a structural VC gene for each chain.

It has been proposed (Gally and Edelman, 1970) that translocation of a V to a C gene would be a satisfactory mechanism. The genetic evidence indicates that the information for immunoglobulin polypeptide chains is specified in a gene cluster consisting of V and C genes and there is strong support for the idea that this cluster functions to link information from V and C genes. It has been found (Oudin and Michel, 1969) that the same idiotypic markers can occur in V regions of antibodies of γM and γG classes. Moreover, a myeloma tumor has been described which secretes γG and γM immunoglobulins having identical light chains and apparently identical V_H regions but different C_H regions (Wang _et al_., 1970). Most recently, Pernis (Pernis _et al_., 1971) has observed γM immunoglobulins on the surface of lymphocytes which contain intracellular γG immunoglobulin. Consistent with the interpretation that their V regions were identical, both γG and γM proteins had the same V region allotypes.

It is reasonable to suggest that the immunoglobulin gene cluster or translocon is the basic unit of evolution in the antibody system. Whatever the mechanism of translocation of gene information turns out to be, translocation can account nicely for clonal commitment of lymphoid cells and allelic exclusion. Moreover, a mechanism for linking V to C at the polynucleotide level would be likely to increase the diversity. On the basis of random chain assortment, the known number of variant positions in V region subgroups and the number of subgroups, one would expect at least 200 different kinds of antibody V regions and 1000 different molecules _before introducing any variation within each subgroup_. All of these variants can be explained by the evolution of different subgroup genes, the occurrence of translocation within gene clusters and the random assortment of heavy and light chains.

The fundamental problem of the origin of diversity therefore reduces to the question: How many genes are there _within_ each subgroup? The germ line theory states that in the genome there is a V gene for each V region. This theory must explain how antibody V regions within a subgroup are kept alike during evolution and, above all, how Mendelian genetic markers can occur in V regions. The existence of rabbit allotypes and strain-specific V region markers in certain mouse κ chain V regions (Edelman and Gottlieb, 1970) has not been adequately explained by this theory. Indeed, enormous degrees of natural selection would be required to maintain such markers, for recombination and mutation during evolution would be expected to destroy them or prevent them from being present on all members of a large multigene family. The gene-expansion model of Hood and Talmage (1970), which attempts to account for the origin of diversity completely by evolution in the germ line, requires special assumptions. In particular, it requires that an adequate and different set of V region genes must be generated frequently by gene duplication from one or a few genes. These events are followed by accumulation of mutations and appropriate selection during speciation and previous adaptations are not utilized. It is difficult to see what combination of selective forces could achieve this result. In addition, this theory makes the _ad hoc_ assumption that unequal crossing over (required in the germ line to generate new V genes) is prevented from occurring in somatic cells.

The somatic mutation theory (Cohn, 1968) presumably requires only one gene in each V region subgroup. Although this theory has no problem in accounting for allotypes or the evolution of V genes, it requires an ad hoc mechanism of somatic hypermutation as well as a very specific selection mechanism to produce properly folded molecules in a very short time. A satisfactory basis for this selective process has not been detailed. Clonal selection would not appear to be adequate, for it is clearly not equivalent to natural selection during evolution.

The somatic recombination theory, which assumes that there is a small number of V genes in each subgroup each carrying mutations selected during evolution, can explain the presence of genetic markers in V regions by a process of gene conversion during evolution (Gally and Edelman. 1970; Edelman and Gally, 1968). This theory requires that the V genes recombine frequently in the germ line and the soma and that the recombination process takes place via hybrid DNA formation and repair synthesis.

Obviously, no theory of diversity is free of special assumptions. The somatic recombination theory does, however, attempt to explain both variation and the fusion of information from V and C genes in one mechanism (figure 11).

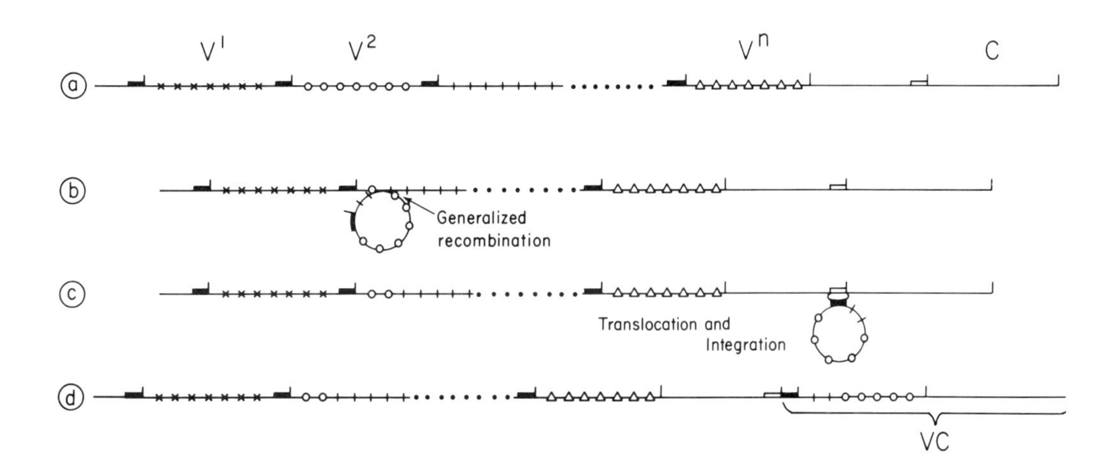

Fig. 11. Hypothetical mechanism for V gene translocation. Shown are V genes of only one of the subgroups of a gene cluster or translocon. (a) Base substitutions which result in changes in sequence from that of the basic sequence of the subgroup are indicated along each V gene by ×, O, |, and Δ. Intrachromosomal recombination between adjacent V genes would lead to the formation of a V gene episome (b). This process would be analogous to generalized recombination seen in bacterial systems, i.e. recombination would occur wherever homologous DNA strands pair to form hybrids but not necessarily at any specific nucleotide sequences. The episome formed could therefore contain new sequence variants composed of sequences from two adjacent V genes. As a first step in the formation of a complete VC gene, integration of the episome (c) is assumed to require an enzyme capable of recognizing specific nucleotide sequences in both the episome and the C gene (d). The hypothetical enzyme is shown bound to the attachment regions of the episome (▬) and the C gene (▭). This scheme is intended to illustrate the idea of translocation; obviously special mechanisms to accomplish this may exist in chromosomes.

In contrast, both germ line theories and somatic mutation theories assume that an entire V or C gene must be translocated somatically without recombination and therefore must specify some means for excising or reading the V gene at its exact beginning and end. In any case, the existence of heavy chain disease variants (Terry and Ohms, 1970) is consistent with the idea that V-C fusion takes place at the DNA level and light chain V region variants with deletions of almost an entire V region (Smithies et al.. 1971) can also be interpreted as anomalies of translocation. Data on these deletion variants are among the most useful kinds of information that can be obtained from further structural analysis of antibodies for it does not seem likely that more normal subgroup variants can prove that a theory of diversity is correct. The answer to the question of the origin of diversity may have to await the design of innovative experiments on chromosomes, DNA and cell variation.

REFERENCES

Bjork, I., F.A. Karlsson and I. Berggård. (1971), Independent folding of the variable and constant halves of a λ immunoglobulin light chain. Proc. Natl. Acad. Sci. U.S.. 68, 1707.

Bjork, I. and Tanford, C. (1971), Gross conformation of free polypeptide chains from rabbit immunoglobulin G. I. Heavy chain. Biochemistry, 10, 1271.

Burnet, F.M. The Clonal Selection Theory of Acquired Immunity, (1959), Vanderbilt University Press, Nashville, Tennessee.

Cairns, J. (1967), Cold Spring Harbor Symp. Quant. Biol., 32.

Charlwood, P.A. and S. Utsumi, (1969). Conformation changes and dissociation of Fc fragments of rabbit immunoglobulin G as a function of pH. Biochem. J., 112, 357.

Cohn, M. (1968), The molecular biology of expectation. in Nucleic Acids in Immunology. (O.J. Plescia and W. Braun, eds.), Springer-Verlag, New York, p. 671.

Dorrington. K.J., and C. Tanford, (1970), Molecular size and conformation of immunoglobulins. Adv. Immunol. 12, 333.

Dorrington. K.J., M.H. Zarlengo, and C. Tanford. (1967), Conformational change in complementarity in the combination of H and L chains of immunoglobulin G. Proc. Natl. Acad. Sci. U.S., 58, 991.

Dourmashkin, R.R., G. Virella, and R.M.E. Parkhouse (1971), Electon microscopy of human and mouse myeloma serum IgA, J. Mol. Biol., 56, 207.

Dreyer, W.J. and J.C. Bennett. (1965), The molecular basis of antibody formation: a paradox. Proc. Natl. Acad. Sci. U.S., 54, 864.

Edelman, G.M. (1959), Dissociation of γ-globulin. J. Am. Chem. Soc., 81, 3155.

Edelman, G.M. (1963), Unresolved problems in the analysis of the structure of antibodies. Immunopathology, IIIrd International Symposium, La Jolla, Calif., Schwabe and Co., Publishers. Basel, p. 57.

Edelman, G.M. (1970a), The covalent structure of a human γG immunoglobulin. XI. Functional implications. Biochemistry 9, 3197.

Edelman, G.M. (1970b), Antibody structure: a molecular basis for specificity and control in the immune response, in Control Processes in Multicellular Organisms (G.E. Wolstenholme and J. Knight, eds.), J.and A. Churchill, London, p. 304.

Edelman, G.M. (1970c), The complete amino acid sequence and arrangement of disulfide bonds of a γG immunoglobulin molecule. in Developmental Aspects of Antibody Formation and Structure (J. Šterzl and I. Řiha, eds.), Academia Publishing House, Prague, p. 381.

Edelman, G.M., B. Benacerraf and Z. Ovary, (1963a), Structure and specificity of guinea pig 7S antibodies, J. Exp. Med., 118, 229.

Edelman, G.M.. B. Benacerraf, Z. Ovary and M.D. Poulik. (1961), Structural differences among antibodies of different specificities, Proc. Natl. Acad. Sci. U.S., 47, 1751.

Edelman, G.M., B.A. Cunningham, W.E. Gall, P.D. Gottlieb, U. Rutishauser, and M.J. Waxdal (1969), The covalent structure of an entire γG immunoglobulin molecule. Proc. Natl. Acad. Sci. U.S., 63, 78.

Edelman. G.M. and W.E. Gall (1969), The antibody problem, Ann. Rev. Biochem., 38, 415.
Edelman, G.M.and J.A Gally (1968), Antibody structure, diversity,and specificitv, Brookhaven Symposium, 21, 328.
Edelman, G.M. and J.A. Gally (1962), The nature of Bence Jones proteins, J. Exp. Med. 116, 207.
Edelman, G.M. and P.D. Gottlieb (1970), A genetic marker in the variable region of light chains of mouse immunoglobulins. Proc. Natl. Acad. Sci. U.S., 67, 1192.
Edelman, G.M., D.E. Olins, J.A. Gally, and N.D. Zinder. (1963b), Reconstitution of immunologic activity by interaction of polypeptide chains of antibodies. Proc. Natl. Acad. Sci. U.S., 51, 753.
Edelman, G.M. and M.D. Poulik (1961), Studies on structural units of the γ-globulins. J. Exp. Med. 113, 861.
Feinstein, A., and A.J. Rowe (1965), Molecular mechanism of formation of an antigen-antibody complex. Nature 205, 147.
Fougereau, M., and G.M. Edelman (1964), Resemblance of the gross arrangement of polypeptide chains in reconstituted and native γ-globulins. Biochemistry, 3, 1120.
Fruchter, R.G., S.P. Jackson, L.E. Mole, and R.R. Porter (1970), Sequence studies of the Fd section of the heavy chain of rabbit immunoglobulin G. Biochem. J., 116, 249.
Gall, W.E. and G.M. Edelman (1970), The covalent structure of a human γG immunoglobulin. X. The intrachain disulfide bonds. Biochemistry, 9, 3188.
Gally, J.A. and G.M. Edelman (1970), Somatic translocation of antibody genes. Nature, 227, 341.
Green, N.M., Dourmashkin, R.R. and R.M.E. Parkhouse (1971), Electron microscopy of complexes between IgA (MOPC 315) and a bifunctional hapten., J. Mol. Biol., 56, 203.
Hill, R.L., R. Delaney, R.E. Fellows, Jr., and H.E. Lebovitz (1966), The evolutionary origins of the immunoglobulins. Proc. Natl. Acad. Sci. U.S., 56, 1762.
Hood, L.and D.W. Talmage (1970), Mechanism of antibody diversity: germ line basis for variability. Science 168, 325.
Ishizaka, K., T. Ishizaka, and M.M. Hornbrook (1966), Physico-chemical properties of reaginic antibody V. Correlation of reaginic activity with γE-globulin antibody. J. Immunol. 97, 840.
Jerne, N.K. (1955), The natural selection theory of antibody formation, Proc. Natl. Acad. Sci. U.S., 41, 849.
Karlsson, F.A., P.A. Peterson, and I. Berggård (1969), Properties of halves of immunoglobulin light chains, Proc. Natl. Acad. Sci. U.S., 64, 1257.
Kehoe, J.M. and M. Fougereau(1969), Immunoglobulin peptide with complement fixing activity. Nature 224, 1212.
Killander, J. ed. (1967), Gamma Globulins, Structure and Control of Biosynthesis, Nobel Symposium, 3. Almqvist and Wiksell, Stockholm.
Koshland, M.E. (1967), Location of specificity and allotypic amino acid residues in antibody Fd fragments. Cold Spring Harbor Symp. Quant. Biol., 32, 119.
Landsteiner, K. (1945), The Specificity of Serological Reactions, 2nd ed., Harvard University Press, Cambridge, Mass.
Metzger, H. and M. Mannik (1964), Recombination of antibody polypeptide chains in the presence of antigen. J. Exp. Med., 120, 765.
Milstein, C., C.P. Milstein, and A. Feinstein (1969), Non-allelic nature of the basic sequences of normal immunoglobulin κ chains. Nature 221, 151.
Milstein, C. and A.J. Munroe (1970), The genetic basis of antibody specificity. Annual Review of Microbiology, 24, 335.
Noelken, M.E., C.A. Nelson, C.E. Buckley III, and C. Tanford (1965), Gross conformation of rabbit 7S γ-immunoglobulin and its papain-cleaved fragments. J. Biol. Chem., 240, 218.
Olins, D.E. and G.M. Edelman (1964), Reconstitution of 7S molecules from L and H polypeptide chains of antibodies of γ-globulins. J. Exp. Med., 119, 789.
Oudin, J. and M. Michel (1969), Idiotypy of rabbit antibodies. II. Comparison

of idiotypy of various kinds of antibodies formed in the same rabbits against Salmonella typhi. J. Exp. Med., 130, 619.

Pernis, B., L. Forni, and L. Amante (1971), Ann. N.Y. Acad. Sci., in press.

Pilz, I., Puchwein, G., Kratky, O., Herbst, M., Haager, O., Gall, W.E. and G.M. Edelman (1970). Small angle X-ray scattering of a homogeneous γG1 immunoglobulin. Biochemistry, 9, 211.

Porter, R.R. (1959), The hydrolysis of rabbit γ-globulin and antibodies with crystalline papain. Biochem. J., 73, 119.

Porter, R.R. and R.C. Weir (1966), Subunits of immunoglobulins and their relationship to antibody specificity. J. Cell Physiol. 67, Suppl. 1: 51-64.

Press, E.M. and N.M. Hogg (1969), Comparative study of two immunoglobulin G Fd fragments. Nature 223, 807.

Sarma, V.R., E.W. Silverton, D.R. Davies and W.D. Terry (1971), The three-dimensional structure at 6 Å resolution of a human γG1 immunoglobulin molecule. J. Biol. Chem., 246, 3753.

Schwartz, J.H., and G.M. Edelman (1963), Comparisons of Bence Jones proteins and L polypeptide chains of myeloma globulins after hydrolysis with trypsin. J. Exp. Med., 118, 41.

Singer, S.J. and R.E. Doolittle (1966), Antibody active sites and immunoglobulin molecules, Science, 153, 13.

Smithies, O., D.M. Gibson, E.M. Fanning, M.E. Percy, D.M. Parr and G.E. Connell (1971), Deletions in immunoglobulin peptide chains as evidence for breakage and repair in DNA. Science, 172, 574.

Solomon, A. and C.L. McLaughlin (1969), Bence Jones proteins and light chains of immunoglobulins. I. Formation and characterization of amino-terminal (variant) and carboxyl-terminal (constant) halves. J. Biol. Chem., 244, 3393.

Tanford, C. (1968) Chemical basis for antibody diversity and specificity. Accts. Chem. Research, 1, 161.

Terry, W.D., B.W. Matthews, and D.R. Davies (1968) Crystallographic studies of a human immunoglobulin. Nature 220, 239.

Terry, W.D. and J. Ohms, (1970) Implications of heavy chain disease protein sequences for multiple gene theories of immunoglobulin synthesis. Proc. Natl. Acad. Sci. U.S., 66, 558.

Turner, M.W. and H. Bennich (1968) Subfragments from the Fc fragment of human immunoglobulin G. Biochem. J., 107, 171.

Valentine, R.C. and N.M. Green (1967) Electron microscopy of an antibody hapten complex, J. Mol. Biol., 27, 615.

Wahl, P., and G. Weber (1967), Fluorescence depolarization of rabbit gamma globulin conjugates. J. Mol. Biol., 30, 371.

Wang, A.C., S.K. Wilson, J.E. Hopper, H.H. Fudenberg, and A. Nisonoff, (1970), Proc. Natl. Acad. Sci. U.S., 66, 337.

Weltman, J.K. and G.M. Edelman (1967), Fluorescence polarization of γG immunoglobulin. Biochemistry 6, 1437.

Wikler, M., H. Kohler, T. Shinoda, and F.W. Putnam (1969), Macroglobulin structure: Homology of Mu and Gamma heavy chains of human immunoglobulins. Science, 163, 75.

World Health Organization, (1964), Bull. World Health Organ. 30, 447.

World Health Organization, (1968), Bull. World Health Organ. 38, 151.

Wu, I.I. and E.A. Kabat (1970), An analysis of the sequences of the variable regions of Bence Jones proteins and myeloma light chains and their implications for antibody complementarity. J. Exp. Med., 132, 211.

Yguerabide, J., H.F. Epstein, and L. Stryer (1970), Segmental flexibility in an antibody molecule. J. Mol. Biol., 51, 573.

THE LYMPHOCYTE SURFACE AND ITS REACTIONS TO ANTIGEN

G.J.V. NOSSAL
The Walter and Eliza Hall Institute of Medical Research
Melbourne, Victoria 3050, Australia

Abstract: This brief review deals with cell interactions at various levels of the immune response. 1) The interaction between uncommitted haematogenous stem cell and inducers of lymphopoiesis. 2) The reaction between T and B cells and vascular and other cells in lymphoid tissue which dictate lymphocyte migration patterns. 3) Preparatory to a discussion of lymphocyte antigen interactions, the Ig receptors on lymphocytes are considered, from the viewpoint of heterogeneity amongst lymphocytes, and of arrangement and molecular characteristics of the receptors. Evidence for a surprising degree of metabolic turnover of receptors is presented. 4) The importance of multi-point binding of antigenic determinants to lymphocytes is discussed, particularly in relation to collaboration between T and B lymphocytes. 5) Some speculations are presented on the mechanisms by which lymphocytes make the decision between immunity and tolerance.

1. INTRODUCTION

Cellular interactions are important at many different points in a consideration of immunology. They are involved in the decision taken by a multipotential stem cell to differentiate towards a lymphocyte. They clearly influence lymphocyte traffic patterns <u>in vivo</u>. Most important of all, they are crucially involved in the response of lymphocytes to antigens. In this brief review, I will attempt to summarize current work in my Unit directed at gaining a better understanding of this vital confrontation between lymphocyte surface and antigenic determinant, and of the factors which decide whether the result will be the induction of immunity or the induction of immunological tolerance, a shutting off of the cell's immune capacity. My coworkers in this project are too numerous to list as co-authors of this review, but each has contributed in an important way. They include J.L. Atwell, R.E. Cone, M. Feldmann, R. Huchet, H. Lewis, T.E. Mandel, J.J. Marchalonis, J.M.C. Mitchell, R.T. Rolley, V. Santer, J.W. Schrader, H. Wagner and J.D. Wilson. Much background information involving other Units at the Walter and Eliza Hall Institute and many other investigators will be included, frequently without full citation, because of limitations of space. The non-immunological reader is referred to two recent works for fuller documentation (Nossal and Ada, 1971; Miller, 1972).

2. DIFFERENTIATION OF T AND B LYMPHOCYTES

Before embarking on the main topic, brief reference should be made to two levels of cellular interaction that will not be extensively discussed in this symposium. The first is the interaction involved in the commitment of a cell to the lymphoid pathway of differentiation. Models involving embryological development give us the clearest clues, but the process undoubtedly continues also in adult life. There are blood-forming progenitor or stem cells, which in all probability are multipotential, first noted in the embryonic yolk sac. These migrate into the anlagen of various organs such as the fetal liver, the thymus and the spleen at defined stages of embryonic life (Moore and Owen, 1969). Let us take the thymus as an example. The epithelial rudiment of this organ is invaded by blood-borne stem cells as early as $11\,^1/_2$ days of embryonic life in the mouse, very shortly after the rudiment has descended from the pharyngeal pouch. For about 24 hr the stem cells to be found there display continued multipotency, that is they can, on transfer to an appropriate environment, engender erythrocytes and polymorphonuclear leukocytes as well as lymphocytes (Metcalf and Moore, 1971). After that, they become committed to lymphoid development. It appears probable that some cell constituent of the epithelial-reticular framework of the thymus makes a factor that acts as a lymphocyte differentiating stimulus, i.e. "tells" the stem cell to engage in lymphopoiesis. Once a cell emerges from the thymus committed to immunological function, it cannot revert to any other activity. The inductive role of the framework of a lymphoid organ embodies certain general principles. Just as a thymus depopulated of lymphocytes and stem cells can teach entering stem cells to remake a proper thymus, so a spleen rudiment can be grafted into a host animal, and will develop into a properly-functioning spleen, using host stem cells to do the job. The molecular basis of this inductive function is still unknown. So far, none of the many claimed "thymic hormones" stand up to rigid scrutiny.

The thymus forms and exports one type of lymphocyte, called a T lymphocyte. This cell type is responsible for so-called cellular immune phenomena, such as the delayed hypersensitivity reaction (e.g. the tuberculin reaction), cell-mediated, non-complement-dependent killing of antigenic cells (organ transplants, tumour grafts), or certain "helper" functions in antibody formation that will be more extensively dealt with by Dr. N.A. Mitchison later in this Symposium. There is a second type of lymphocyte, the B lymphocyte, which is the cell type responsible for actual antibody production. There is good reason to believe that B lymphocytes, too, are generated from stem cells through the action of a differentiating inducer. This comes largely from work in the chicken, where there exists a thymus-like organ known as the bursa of Fabricius which accepts stem cells and exports B lymphocytes. In mammals, there is no organ corresponding to the bursa, and it is not certain in what organ B cell differentiation occurs. The bone marrow certainly contains both stem cells and B cells, and may have assumed a bursa-like function. Alternatively, the cells making the inducer molecule that acts on stem cells to convert them into B cells may be widely deployed throughout the lymphoid system in mammals.

A big part of modern academic immunology concerns the differences between T and B lymphocytes. However, both types of cell also have important characteristics in common. Both have surface receptors for antigens. In both cases, the cell initially formed is inactive without antigenic stimulation. When antigen activates the cell, both T and B lymphocytes enter rapid, repeated mitotic cycles and produce a crop of differentiated effector cells which actually perform the immunological task, be this cell-mediated immunity or antibody secretion. In both systems, appropriate pre-exposure to antigen can alter the subsequent response to antigenic re-challenge. Thus, the lymphocyte population can be rendered more responsive, a state termed immunological memory; or less reactive, resulting in partial or even complete immunological tolerance. Moreover, in both systems, as we shall see below, the individual cells are heterogeneous with respect to the serologic specificity of their surface receptors.

3. MIGRATION PATTERNS OF T AND B LYMPHOCYTES

Lymphocytes are dynamic in their migratory behaviour, moving from lymph node to blood via the lymphatic circulation, and thence back to lymph node or spleen. This remigration back into lymphatic tissue is an essential feature of T lymphocyte behaviour; if it is prevented by draining off all the lymphocytes that enter the circulation from the thoracic duct, the centres of new T cell formation cannot make up for the loss and the total number of T lymphocytes in the animal falls drastically (Gowans and McGregor, 1965). Lymphocyte migration involves a number of types of cell interaction. First, lymphocytes escape from the circulation into lymph nodes via a peculiar kind of venule, with an unusual cellular lining. They actually enter the lining endothelial cell and emerge on its other side. Treatment of lymphocytes with certain glycolytic enzymes can impair this homing capacity (Gesner and Ginsburg, 1964). Secondly, once inside a lymph node or spleen, T and B lymphocytes behave very differently from each other, each population seeking out its own microenvironment, which can be strictly delineated anatomically. This must involve unknown interactions with other cellular elements in the lymph node. Finally, within the so-called B area, cells move from one sub-section to another at differing stages of their life cycle, driven by forces which are still totally obscure.

4. HETEROGENEITY AMONGST LYMPHOCYTES IN RECEPTOR SPECIFICITY AND ANTIGEN RESPONSE

At several points in the immunological part of this symposium, it will become clear that almost universal agreement has been reached on the validity of Burnet's (1957) clonal selection hypothesis. Stated in its simplest form, this says that through some somatic genetic process of diversification, different lymphocytes come to have surface receptors of different specificities. Each lymphocyte displays on its surface immunoglobulin molecules of one specificity. When an antigen sufficiently complimentary to that Ig arrives, the cell will be stimulated both to divide and to differentiate into an actively-secreting cell. As there is great variation in the affinity of antibodies for antigens, the concept of complimentarity is clearly not an all-or-none thing, and will depend, *inter alia*, on antigen concentration. High doses of antigen select for stimulation a greater total number of cells, including many of low affinity, while small antigen doses favour the stimulation of few, high affinity cells (Nussenzweig and Benacerraf, 1967). In all cases, the combining portion (V gene product) of the receptor immunoglobulin is an accurate sample of the antibody finally made by the progeny of the stimulated cell.

How good is the evidence for clonal selection ? Our early demonstration that one cell could only make one antibody (Nossal and Lederberg, 1958) was merely consistent with it. Very persuasive evidence has since been built up for clonal selection amongst B cells (Nossal and Ada, 1971), but the case for T cells is not quite as strong. In the case of B cells, it has been shown that lymphocytes from unimmunized animals show heterogeneity in their capacity to bind antigen, a small sub-set only being capable of binding any particular antigen. As might be expected, the proportion of cells showing labelling in radioautographs varies widely with the molar concentration and specific radioactivity of the isotopically marked antigen. Usually, investigators set up experiments in such a way that one cell in 1,000 or 10,000 is seen to bind antigen, though it is possible that the proportion of cells capable of being activated by that antigen is considerably smaller. Secondly, when lymphocyte populations are held with ^{125}I-antigen for some hours, a specific "suicide" of antigen-binding lymphocytes can be induced, such that the lymphocyte population, on appropriate transfer to a host animal, cannot respond to the antigen concerned, while retaining its capacity to respond to all other antigens (Ada and Byrt, 1969). Thirdly, a similar depletion of capacity to respond to an antigen can be achieved by filtering cells through antigen-coated columns (Wigzell and Anderson, 1969). For a variety of technical reasons, we are still some way from totally validating clonal selection through procedures which enrich a population for cells with a given receptor, but rapid progress is being made in this area. Moreover, detailed studies of specificity and affinity of cell surface receptors support the concept of identity between them and the antibody finally produced.

The total homogeneity of antibody actually secreted by an activated cell has now been documented very fully (see Marchalonis and Nossal, 1968, for references). The same formal documentation of homogeneity of receptor immunoglobulin on the surface of unstimulated B cells has not been achieved. In fact, there is a serious embarrassment, in that many B cells seem, by apparently reliable serologic criteria, to have two or even three classes of immunoglobulin chain on their surface. This may not mean that they display receptor of more than one specificity, for several reasons. First, the identical variable portion of the heavy chain, which (together with the variable portion of the light chain) confers antibody specificity may be attached to constant regions of different classes. Secondly, not all Ig at the lymphocyte surface may be receptor material. Some may be passively absorbed from the serum, though this is unlikely to be the sole source of the problem. It is possible that individual B cell clones switch from IgM to IgG production as they develop, without a change in serologic specificity (Nossal et al, 1964).

For T lymphocytes, our knowledge is much less complete. There is argument about whether the size of an antigenic determinant recognized by a T cell is different from that of three to six amino acids or sugars, the known size of the antibody combining site. Many of the more recent experiments, particularly those on the inhibition of union between T cells and antigen-coated erythrocytes (rosettes), and also on the specificity of T cell helper function, suggest that the T cell receptor must also be an immunoglobulin, but final concensus has not been reached on this point. We do know that T cells are heterogeneous in their antigen-binding capacity, and are subject to radioactive antigen "suicide". There are some lines of work suggesting that IgM molecules constitute the antigen receptors for both T and B lymphocytes. If so, there are clearly important differences in the arrangement and display of these molecules on the lymphocyte surface.

5. MOLECULAR AND METABOLIC CHARACTERISTICS OF LYMPHOCYTE SURFACE Ig

In the mouse, it is easy to demonstrate both Ig light chains and μ heavy chains on the surface of B cells. This can be accomplished by immunofluorescent or radioautographic techniques, using labelled, specific antiglobulin antibodies, e.g. antibody to mouse Ig made in a rabbit (Raff et al, 1970; Nossal et al, 1972). Alternatively, it is possible to use the gentle lactoperoxidase method of Marchalonis to radioiodinate accessible tyrosine residues on the surface of an intact living lymphocyte, and then to isolate labelled Ig from the cell (Marchalonis et al, 1971; Bauer et al, 1971) Both types of approach also reveal some γ-type heavy chains on B cell surfaces. When applied to T cells, however, the two methods, in our hands, give paradoxical results. Using the serologic approach of attaching antiglobulin antibodies there is a 100 to 200-fold difference in labelling intensity, i.e. under equivalent conditions, B cells bind much more antiglobulin than do T cells. However, when both types of cells are radioiodinated at the surface, with the lactoperoxidase technique, Marchalonis and colleagues find readily identifiable labelling of Ig material on both types of cell, with a relatively minor difference in apparent yield of numbers of labelled Ig molecules per cell, though T cells show only μ and light chains, and no γ chains.

When antigen binding is used to label T and B cells, it is again found that under equivalent conditions, B cells label more heavily. Moreover, they require a much greater excess of non-radioactive antigen to inhibit radioactive antigen binding than is the case for T cells. This supports the apparent conclusion from the antiglobulin experiments and argues for a higher density of Ig receptors on B than on T cells. The relative difficulty of removing T cells of a given specificity through use of antigen-coated columns points in the same direction

There are two lines of reasoning, either or both of which might contribute to a solution of the dilemma. The first line relates to the detailed arrangement of Ig receptors at the lymphocyte surface. If this is relatively dense, one 7S Ig molecule acting as a labelling reagent might stretch between two adjacent Ig receptors on the surface, making the Ig-Ig union a much stronger one. If the spacing were sparser by a factor of only two, it is still possible that the degree of binding of a labelled antireceptor Ig, now depending on solely a univalent bond, would be lower by a much higher factor. Moreover, most of the antibody activity of rabbit anti-mouse Ig sera is directed against determinants on the Fc portion of mouse Ig. If this is buried in the resting T cell, as has been suggested (Greaves, 1970), or if it is sterically blocked by an adjacent protein (e.g. the θ antigen), anti-Ig sera might bind much less efficiently on that account. The second line of reasoning extends the theme of accessibility. We simply do not know what proportion of the total Ig present on a cell surface is labelled by each technique. The free radicals generated when lactoperoxidase, hydrogen peroxide and ^{125}I react near the cell surface may have different penetration characteristics than whole, labelled Ig molecules. It must not be forgotten that lymphocytes are surrounded by a glycocalix of, as yet, poorly understood character.

Dr. Mitchison will develop one aspect of the dynamic character of Ig receptors on lymphocyte surfaces as a salient feature of his article. I wish to focus on another but probably related one, namely the metabolic turnover of Ig receptors on T and B cells. We have studied this by two techniques, neither of them totally immune to criticism. Cells are prelabelled with either ^{125}I-antiglobulin sera or the surface radioiodination technique. Then metabolism at 37° is allowed to proceed. The amount of radioactive Ig released into the medium is monitored in each case. With antiglobulin sera as the labelling reagent, the potential exists for altering the innate behaviour of receptors, although a number of control studies suggest that the very low concentrations of antiglobulin required for this work do not cause obvious rearrangement of cell surface Ig as a whole. With the radioiodination method, which is probably the more satisfactory of the techniques, one cannot totally deny the possibility of some damage to the cell surface, although viability tests fail to show such damage. Both tests suggest surprisingly rapid receptor turnover. The half-life of B cell surface Ig is around two hours and that of T cell Ig two to three times greater. Antiglobulins do cause some pinocytosis of receptor-antireceptor complexes, but rapid regeneration follows.

If lymphocyte receptors are indeed constantly shed from the surface, one might regard the resting lymphocyte as a cell slowly secreting antibody. A first approximation to the rate of Ig synthesis is 50,000 molecules per hour, or about 100 times slower than a fully-differentiated antibody-forming cell. Moreover, one's concept of antigen-lymphocyte reactions is drastically altered.

6. ANTIGEN MATRIX FORMATION

If Ig molecules are constantly being pushed from the lymphocyte surface, it is easy to see that antigens with only one antigenic determinant would be swept from the lymphocyte surface and would not remain there unless the antigen concentration was high. In fact, we and others have shown that such monovalent antigens are poor initiators of immune responses and poor inhibitors of antigen-lymphocyte interactions. On the other hand, if antigen is presented to the lymphocyte as a matrix of spaced, linked determinants, it will remain tethered to the lymphocyte surface even if one antigenic determinant is pushed off by metabolism. As the shed receptor is replaced by regeneration, it may well find a free antigenic valency to which it can bind, thus restoring the status quo. We find that the hapten DNP bound to bacterial flagella (a polymerized protein with repeating flagellin monomer units) inhibits binding of activated lymphocytes to DNP-coated erythrocytes much better than does DNP bound to human globulin (DNP-HG). More importantly, when cells that have bound either inhibitor are brought to 37° and allowed to metabolize, DNP-HG rosette inhibition is almost entirely abrogated within two hours but DNP-Fla inhibition is intact after four hours. This observation, when combined with the observations that Dr. Mitchison will present later in this Symposium, illustrates the great importance of multi-point binding of antigen to lymphocytes for effective interaction.

There are several important _in vivo_ mechanisms for causing formation of antigenic matrices. For example, natural or acquired IgM antibody to an antigen can accelerate antibody formation. This, most likely, is due to antigen-antibody complexes in the zone of antigen excess being deposited on surfaces of reticulo-endothelial cells in the lymphoid organs. Injection of antigen coated on to alum particles or to the surface of lipid oil droplets aids antibody formation. In the latter case, we have noted lymphocytes immediately adjacent to the oil droplet turning into activated, rapidly dividing cells. Antigen may become attached to the surface of macrophages, and an antigen matrix may be created. Perhaps the most interesting system involves the collaboration between T and B cells, and before discussing this in vivo, we must first draw attention to some _in vitro_ experiments.

7. IN VITRO COLLABORATION BETWEEN T AND B LYMPHOCYTES

Many immune responses require the collaboration of T and B lymphocytes. Mitchison (1971) has devised model systems which have thrown much light on the basis of this collaboration. He has employed hapten-protein conjugates as antigens, and has shown that B cells can make antibody to the hapten only if T cells activated against the carrier protein are present. Mitchison's hypothesis is that T cells pick up the conjugate by the carrier portion, creating a spaced antigenic matrix at the surface of the T cell, with the hapten portion jutting out free, for presentation to the B cell. This is believed to cause B cell activation.

Feldmann and Basten (1972) in my laboratory have recently shown that this explanation cannot be the sole one to explain T-B collaboration. They have devised a system for antibody formation in which T cells are physically separated from B cells by a nucleopore membrane of 1 μpore diameter. A number of convincing control experiments showed that this was not able to be penetrated by detectable numbers of cells. When the T cells were appropriately activated by hapten-protein conjugates, a factor passed through the nucleopore membrane and stimulated the B cells to form antibody, the efficiency being almost as good as when T and B cells were mixed. When the B cell population was treated to remove all macrophages, collaboration no longer worked.

Our present working hypothesis to explain these results is that activated T cells release their surface receptor, and that receptor-antigen complexes pass through the nucleopore membrane, attach to macrophage surfaces and the resulting antigen matrix activates B cells.

8. IN VIVO COUNTERPART OF IN VITRO T-B COLLABORATION

I should now like to link several observations together to present a speculative view of how T-B collaboration works in vivo. A little extra background information is needed. First, IgG antibody formation, which usually follows a preceding phase of IgM antibody formation, is usually more dependent on T-cell help than is the IgM response. Secondly, congenitally athymic mice, lacking T cells and therefore generally poor formers of IgG antibody, are very deficient in a special type of lymphocyte aggregation called a germinal centre. Thirdly, germinal centres form in specific areas of the lymph node cortex and spleen, called primary lymphoid follicles, as a direct result of antigenic stimulation. Fourthly, antigen can be shown to localize in lymphoid follicels on the surface of specialized dendritic follicle cells. Fifthly, while primary follicles consist of B cells, germinal centres are formed through an invasion of this area by T cells (Gutman and Weissman, 1971). It is tempting to agree with the suggestion of Gutman and Weissman that T-B collaboration takes place in the germinal centre. The dendritic follicle cell could then pick up the antigen-matrix generating Ig made by the germinal centre T cells and present it to B cells at the edge of the germinal centre. After collaboration and B cell stimulation, the B cell may move from the germinal centre to the medulla of the lymph node or the red pulp of the spleen, and complete its differentiation to full antibody-secreting status there.

9. TOLERANCE-IMMUNITY SIGNAL DISCRIMINATION

Antigen may so affect a lymphocyte population as to render it hypo- or unresponsive to subsequent antigenic challenge. This we term immunological tolerance. In vivo, T cells can be rendered tolerant with a lower antigen dose and in a shorter time than can B cells (Chiller et al, 1971). Also, injection of deaggregated, monomeric antigens tends to cause tolerance while polymeric, aggregated or particulate antigens tend to cause immunity (Dresser, 1962). However, in vivo or in vitro, remarkably small doses of antigen can cause tolerance if complexed with 7S antibody but in the zone of antigen excess (Feldmann and Diener, 1971). In fact, one paradox of in vitro experimentation is that tolerization is surprisingly not achieved by monomeric antigens, but can readily be achieved for both T and B cells by supraimmunogenic concentrations of polymerized antigens.

In seeking a unifying concept to explain immunity and tolerance, we have come to believe in the importance of antigen concentration and arrangement at a very localized patch on the lymphocyte surface. We have already stressed that activation of the cell requires multi-point binding of antigenic determinants to the lymphocyte surface. Our hypothesis is that this localized binding can exceed an optimal value, with a cluster of Ig receptors binding too many antigenic determinants. Immunogenesis could result from 1) the lymphocyte encountering natural polymers with appropriately spaced antigenic determinants; 2) the lymphocyte meeting antigen stuck on to a reticulo-endothelial cell (macrophage or dendritic follicle cell), the intermediary antigen glue being either T cell product or IgM antibody, and the spacing being designed to be optimal. Tolerogenesis could result from antigen overloading, so that the cell, over all its surface, has too many receptors occupied. It could arise from a cell meeting a polymer very highly substituted with antigen, so that localized excessive receptor saturation occurs (Feldmann, 1972). Finally, 7S antibody could act as a focussing device plastering more antigen again on to a localized cell surface patch.

The static picture just presented is a conscious oversimplification. Time parameters are obviously at least as important as space parameters in immune induction. It takes five days or more for a single cell stimulated by antigen to produce its clone of executive cells, and by serial transfer techniques such a clone can be kept going much longer. There is some evidence that tolerance induction at the level of a single cell takes around one day, and is reversible by trypsinization before then (Diener, 1971). It may well be that there are initial, common steps in immunogenesis and tolerogenesis, and that the decision hangs in the balance for some hours with respect to a given cell, with localized overstimulation by antigen being the destructive force leading to tolerance.

REFERENCES

Ada, G.L. and P. Byrt, (1969), Specific inactivation of antigen reactive cells with ^{125}I-labelled antigen, Nature (London) 222, 1291.

Baur, S., E.S. Vitetta, C.J. Sherr, I. Schenkein, and J.W. Uhr, (1971), Isolation of heavy and light chains of immunoglobulin from the surfaces of lymphoid cells, J. Immunol. 106, 1133.

Burnet, F.M., (1957), A modification of Jerne's theory of antibody production using the concept of clonal selection, Aust. J. Sci. 20, 67.

Chiller, J.M., G.S. Habicht, and W.O. Weigle, (1971), Kinetic differences in unresponsiveness of thymus and bone marrow cells. Science, 171, 813.

Diener, E., (1971), Symposium on Regulation of the Immune Response "Progress in Immunology", (Academic Press) in the press.

Dresser, D.W., (1962), Specific inhibition of antibody production. II. Paralysis induced in adult mice by small quantities of protein antigen, Immunology, 5, 378.

Feldmann, M., (1972), Induction of immunity and tolerance in vitro by hapten-protein conjugates. I. The relationship between the degree of hapten conjugation and the immunogenicity of DNP-polymerized flagellin, J.exp.Med., in the press.

Feldmann, M. and A. Basten, (1972), Manuscript in preparation, to be submitted to Nature New Biology.

Feldmann, M., and E. Diener, (1971), Antibody-mediated suppression of the immune response in vitro. III. Low zone tolerance in vitro, Immunol. 21, 387.
Gesner, B.M. and V. Ginsburg, (1964), Effect of glycosidases on the fate of transfused lymphocytes, Proc.Nat.Acad.Sci. U.S. 52,750.
Gowans, J.L. and D.D. McGregor, (1965), The immunological activities of lymphocytes, Progr. Allergy 9, 1.
Greaves, M.F., (1970), Biological effects of anti-immunoglobulins. Evidence for immunoglobulin receptors on "T" and "B" lymphocytes, Transplant Rev. 5, 45.
Gutman, G. and I.L. Weissman, (1971), The bone marrow origin of lymphoid primary follicle small lymphocytes. in "Morphological and Fundamental Aspects of Immunity", ed. Lindahl-Kiesling, Alm and Hanna (Plenum Press) 595.
Marchalonis, J.J. and G.J.V. Nossal, (1968), Electrophoretic analysis of antibody produced by single cells. Proc.Nat.Acad.Sci.U.S. 61, 860.
Marchalonis, J.J., R.E. Cone and V. Santer, (1971), Enzymic iodination: a probe for accessible surface proteins of normal and neoplastic lymphocytes, Biochem.J. 124, 921.
Metcalf, D. and M.A.S Moore, (1971), "Hemopoietic Cells" (North Holland, Amsterdam, Frontiers of Biology Series), in the press.
Miller, J.F.A.P., (1972), Collaboration between lymphocytes in the immune response, International Review of Cytology, in the press.
Mitchison, N.A., (1971), The carrier effect in the secondary response to hapten protein conjugates. I. Measurement of the effect with transferred cells and objections to the local environment hypothesis, European J. Immunol. 1, 10.
Moore, M.A.S. and J.T. Owen, (1967), Experimental studies on the development of the thymus, J.exp.Med. 126, 715.
Nossal, G.J.V. and G.L. Ada, (1971), Antigens Lymphoid Cells and the Immune Response, (Academic Press, New York), 324 pp.
Nossal, G.J.V. and J. Lederberg, (1958), Antibody production by single cells. Nature, 181, 1419.
Nossal, G.J.V., A. Abbot, J. Mitchell and Z. Lummus, (1968), Antigens in immunity. XV. Ultrastructural features of antigen capture in primary and secondary lymphoid follicles. J.Exp.Med. 127, 277.
Nossal, G.J.V., A. Szenberg, G.L. Ada and C.M. Austin, (1964), Single cell studies on 19 S antibody formation, J.Exp.Med. 119, 485.
Nossal, G.J.V., N.L. Warner, H. Lewis, and J. Sprent, (1972), Quantitative features of a sandwich radioimmunolabelling technique for lymphocyte surface receptors, J.exp.Med., in the press.
Nussenzweig, V and B. Benacerraf, (1967), Synthesis, structure and specificity of 7 S guinea pig immunoglobulins. in "Nobel Symposium 3 on Gamma Globulins" (ed. J. Killander), p. 233, Almquist and Wiksell, Stockholm.
Raff, M.C., M. Sternberg and R.B. Taylor, (1970), Immunoglobulin determinants on the surface of mouse lymphoid cells, Nature (London) 225, 553.
Wigzell, H. and B. Andersson (1969), Cell separation on antigen coated columns. Elimination of high rate antibody-forming cells and immunological memory cells, J.Exp.Med. 129, 23.

CELLULAR SELECTION REGULATING ANTIBODY AFFINITY DURING THE IMMUNE RESPONSE

Birger Andersson
Department of Tumor Biology, Karolinska Institutet,
Stockholm, Sweden

Abstract: Antibody affinity, a marker of the variable region of immunoglobulin has been studied. Antibody is the humoral product of the B-cell line of immunocompetent cells. The sensitivity of antibody plaque forming cells (PFC:s) to inhibition by free antigen was used as a measure for antibody affinity. Selection of cells producing antibody of different affinity was followed during different experimental conditions where the immunizing antigen concentration was varied. It was found that a big antigen dose stimulated antibody formation of an average low affinity as compared to the antibody formed after a small antigen dose at corresponding times after immunization. The affinity increased with time after immunization. Inhibition of the immune response by administered "passive" antibody was followed at the cellular level. High affinity antibody formation was more resistant to such inhibition, and thus they were selected for during inhibition by administered antibody. Induction of partial paralysis by a big dose of soluble antigen resulted in a small number of cells forming antibody of a relatively low affinity for the antigen. It is concluded that high affinity antibody formation is relatively sensitive to induction of paralysis and low affinity antibody formation is relatively resistant. Induction of paralysis by the use of small doses of soluble antigen - low zone tolerance - resulted only in small changes in affinity and it is thus concluded that the B-cell line is not the immediate target cell for low zone tolerance. Suppression of the immunity against the carrier protein (T-cell immunity), did not affect antibody affinity (B-cell function). Unspecific suppression by X-irradiation did not change the affinity. The findings described are in agreement with a cellular selection theory, postulating a similar binding constant for the cell-associated antigen specific receptor and the product of the cell released after stimulation. It is also concluded that the affinity is regulated mainly by cellular selection within the B-cell line, the precursors of humoral antibody formation. Direct proof for these postulates is obtained from experiments where immunological memory cells were fractionated on antigen coated immunoadsorbent column. High affinity memory cells showed a higher tendency to specifically be retained in such columns. Thus, a cell predetermined to release high affinity antibody is characterized also by a high affinity receptor for the antigen. Furthermore, since in these fractionation experiments an excess of anti-carrier immune cells (T-cell immunity) was ascertained by the addition of such cells, it is concluded that the specialization with regard to affinity takes place within the B-cell line. The specialization within the B-cell line is further demonstrated in limiting dilution experiments, where hapten specific cells (B-cells) were diluted in an excess of carrier specific (T-cells). Clones of anti-hapten antibody forming cells with a marked degree of homogeneity with regard to affinity were obtained.

1. INTRODUCTION

An early step in the induction of immunity or immunological tolerance seems to be an interaction between the antigen and specific receptors on the surface of the antibody forming cell precursors. If a certain critical number of antigen molecules must be bound to the cell surface in order to trigger the cell, then both the antigen concentration around the cell and the affinity of the receptors on the cell will determine whether a cell will produce antibody or not. Presumably also a certain, still higher number of antigen molecules on the surface will render the cell tolerant, and also in this case both antigen concentration and receptor affinity will determine whether the cell will become tolerant or not. The cellular selection theory for antibody affinity postulates identity of affinity between receptor and the antibody released from the same cell after triggering. It follows e.g. that a low antigen concentration will trigger only cells with high affinity receptors and the antibody formed will be of a relatively high affinity.

Table 1
Factors affecting antibody affinity

Experimental procedure	Mode of action	Effect on affinity
Time after immunization	Gradual decrease in antigen concentration	Gradual increase
High antigen dose	Initial high antigen concentration	Initially low
Low antigen dose	Initial low antigen concentration	Initially high
Administered antibody	Reduction in antigen concentration	Increase
High zone paralysis	Increase in antigen concentration	Decrease
Paralysis against carrier protein	T-cell immunity decreased	No effect
Unspecific suppression by X-irradiation	Cell division impaired	No effect
Low zone paralysis	T-cell immunity decreased	No effect

There is experimental evidence for this outlined hypothesis. Antigen dose, time after immunization, feedback inhibition with passive antibody and the induction ofpartial tolerance are factors which have been shown to influence the affinity of serum antibody (Siskind 1969), and the results are in concordance with the theoretical thermodynamic considerations mentioned above.

In Table 1 the effects of variations in antigen concentration on the distribution of cells producing antibody of varying affinity is summarized.

The present communication describes experiments performed at the cellular level in order to further elucidate and more directly illustrate selection leading to antibody of varying affinity. The affinity of membrane receptors have been analyzed by the use of fractionation of cells on immunoadsorbent columns. Some experiments will also be described indicating that the cellular selection takes place within the B-cell line of immunocompetent cells.

2.METHODOLOGY

Detailed description of the above mentioned techniques have been published (Andersson 1970), (Wigzell 1969).

The experimental animals were adult inbred mice of $(C3H \times C57Bl)F_1$ genotype.

Antigens were different heterologous albumins either in their native form or substituted with haptens such as n-iodo-phenyl-acetic acid (NIP) or dinitro-phenyl (DNP).

Antibody forming cells (PFC:s) were enumerated by the hemolytic plaque assay (Jerne 1963) using indicator erythrocytes labelled with the respective albumin or hapten. Antibody affinity was determined by an inhibition assay where free antigen was added to the hemolytic plaque forming system and the concentration of antigen giving 50% inhibition (I-50) was used as a measure for affinity (Andersson 1970).

Cell fractionation on immunoadsorbent columns was performed with Degalan polymeta-acrylic plastic beads labelled with the respective hapten-protein conjugate (Wigzell 1970).

3. REGULATION OF AFFINITY AT THE CELLULAR LEVEL

3:1 Effects of antigen dose and time after immunization.- Earlier studies of serum antibodies have indicated that there is a gradual increase in affinity during the immune response. High antigen doses seem to induce an antibody response which is initially of a lower affinity as compared to the response to a small dose of antigen. In fig. 1 an experiment is shown that demonstrates directly at the cellular level that a selection of cells forming antibody of varying affinity can be obtained in the anti-NIP response of mice. The PFC:s occurring as a result of immunization with 500 μg of antigen released antibody of a lower affinity than PFC:s occurring at corresponding times after immunization with 5 μg of antigen. It can also be seen that with time after immunization the affinity increased gradually in both experimental groups.

3:2 Effect of high dose immunological paralysis.- According to the concept of cellular selection the antigen concentration around the cell and the binding affinity of the cellular receptors for antigen will determine if the cell will remain unstimulated, be triggered into antibody formation or become paralyzed. The high affinity cells will become paralyzed more easily than the low affinity cells. In fig. 2 an experiment is shown that provides experimental evidence for this postulate. Mice immune to BSA were injected with a paralyzing dose of BSA. It can be seen that a substantial fraction of the PFC:s disappeared in the paralyzed animals as compared to the controls. Furthermore, affinity studies showed that it was the high affinity PFC:s which were selectively eliminated as the result of immunological paralysis.

3:3 Effect of administered antibody.- Feed-back inhibition of antibody synthesis by administration of "passive" antibody presumably acts by neutralizing the antigen. Since the effective immunogen concentration thereby will be reduced, changes in antibody affinity can be expected. Fig. 3 shows the effect of an injection anti-BSA antibody on the number of BSA specific PFC:s in animals earlier immunized with BSA. The number of PFC:s is reduced in the antiserum treated animals. It is also evident that the low affinity PFC:s were selectively inhibited, since the remaining PFC:s released antibody of a relatively high affinity as compared to the controls.

3:4 Evidence that affinity selection takes place within the B-cell line.- The humoral antibody formation in mice against the antigens used in the work described in this communication is dependent upon a cooperation between T-cells and B-cells. The experiments shown in table 2 illustrate this. A cooperation between hapten specific (B-cells) and carrier protein specific (T-cells) is illustrated in the DNP-system used by the author in affinity studies. Reports from several laboratories show that the B-cells are the actual antibody forming cells and the T-cells act as specific helpers but do not secrete humoral antibody (Davies 1969, Mitchison 1971). Bearing this in mind the interpretation of the findings presented in table 3 is facilitated. It can be seen that if mice immune to DNP-BSA are given a paralyzing injec-

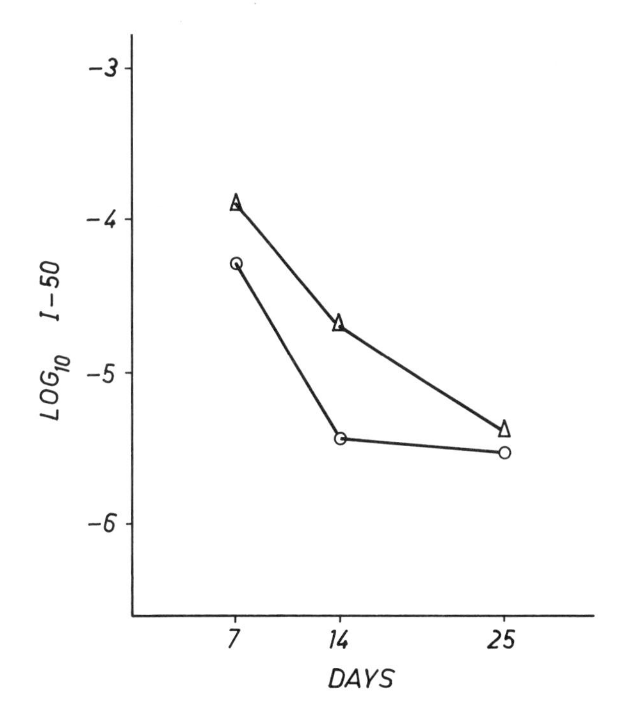

Fig. 1. Affinity of the antibody response of mice immunized with $NIP_{10}OA$. Effect at the cellular level of antigen dose and time after immunization. I-50 is the Molar concentration of NIP hapten inhibiting 50% of the PFC:s.

△——△ Immunized with 500 μg
○——○ Immunized with 5 μg

tion of BSA the number of DNP-PFC will be decreased, thus being due to a paralysis of the BSA-specific T-cells. The absence of an effect on the affinity of the anti-DNP response indicated that no cellular selection within the B-cell line was taking place during this type of immunosuppression. This of course strongly indicates that the cellular selection with regard to affinity is taking place within the hapten specific B-cell line in this system. In this connection can also be shown the effect of nonspecific suppresssion on antibody affinity. Mice undergoing a primary response to BSA were given 200R whole body irradiation. It can be seen in table 4 that the suppression induced in this way did not change the affinity of the antibody production. This is to be expected, since presumably B-cells of different affinity show the same sensitivity to X-irradiation, and the T-cell suppression occurring in this experiment will as has been illustrated above not affect the affinity.

3:5 <u>Effect of low dose immunological paralysis</u>.- If mice are given repeatedly injections of soluble BSA paralysis will be induced both at a high dosage level and at a low dosage level as well (Mitchison 1964). The high dose paralysis presumably acts on the B-cells, i.e. the cells directly involved in humoral antibody formation. This is indicated by the fact that a cellular selection with regard to affinity can be observed, and this is hard to explain in any other way than a direct action on the B-cells with their surface receptors expressing the affinity for the antigen. The nature of low dose paralysis is less well understood, but there is some evidence that the target cells involved here are the T-cells, i.e. the helper cells that not

Fig. 2. Variation in the antibody response of mice immunized with 100 /ug of BSA: The effect of 10 mg BSA i.v.

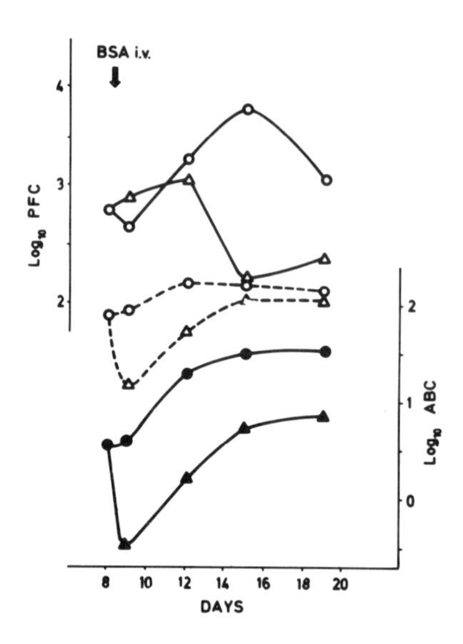

Fig. 2a. Quantitative changes in number of PFC:s in the regional lymph node and in the serum antibody levels.
○———○ PFC:s in control animals
△———△ PFC:s in animals given 10 mg BSA i.v.
○- - - -○ ABC at 100 /ug in control animals
△- - - -△ ABC at 100 /ug in animals given 10 mg BSA i.v.
●———● ABC at 1 /ug in control animals
▲———▲ ABC at 1 /ug in animals given 10 mg BSA i.v.

by themselves are forming humoral antibody (Mitchison 1971). An experiment further supporting this is shown in table 5. It can be seen that in low dose paralysed mice the antibody affinity is not significantly reduced, indicating that a cellular selection within the B-cells had not taken place.

3:6 PFC:s induced by high and low doses of antigen show different sensitivity to induction of paralysis and antibody induced suppression.- An immune response against BSA was evoked by a high dose of BSA in order to stimulate cells with an average low affinity. Another group of animals were immunized with a low dose of BSA in order to stimulate mainly high affinity cells. The high and low affinity systems were then compared with regard to their susceptibility to paralysis and antiserum suppression. The results are listed in table 6. The low affinity system was less sensitive to paralysis but more sensitive to antiserum suppression. The findings are in perfect agreement with the concept of an identity between cellular receptors responsible for induction of immunity and paralysis. Furthermore, cells present in an immune system obviously express a similar affinity of their humoral antibody and of their receptors operating in induction of immunity and paralysis.

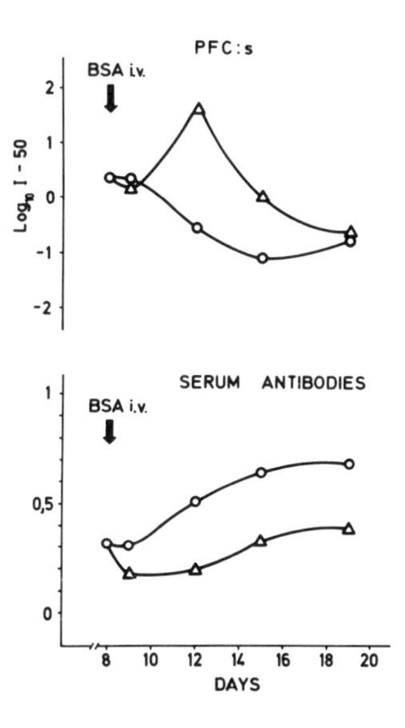

Fig. 2b. Qualitative changes in antibodies released by PFC:s and in serum antibody. I-50 is the concentration of free BSA giving 50% inhibition of PFC:s. S is the factor of avidity of serum antibodies. ABC is the antigen binding capacity. (From Andersson 1971).

○ ———— ○ control animals
△ ———— △ animals given 10 mg BSA i.v.

Table 2
Cellular cooperation in the anti-hapten response of mice.

Hapten specific cells	Carrier specific cells	Immunogen	Antibody response[a)] NIP-PFC	DNP-PFC
Anti-NIP-OA + anti DNP-OA 10^6	————	10 µg NIP-BSA +10 µg DNP-BSA	34	36
Anti NIP-OA + anti DNP-OA 10^6	Anti BSA 2×10^7	10 µg NIP-BSA +10 µg DNP-BSA	457	932

a) The indicated immune cell suspensions were injected into irradiated recipients together with the immunogen. One week later the spleens were tested for antibody forming cells. The figures are geometric means from 6 mice.

Fig. 3. Variation in the antibody response of mice immunized with 2.500 /ug of BSA: The effect of administered antibody.

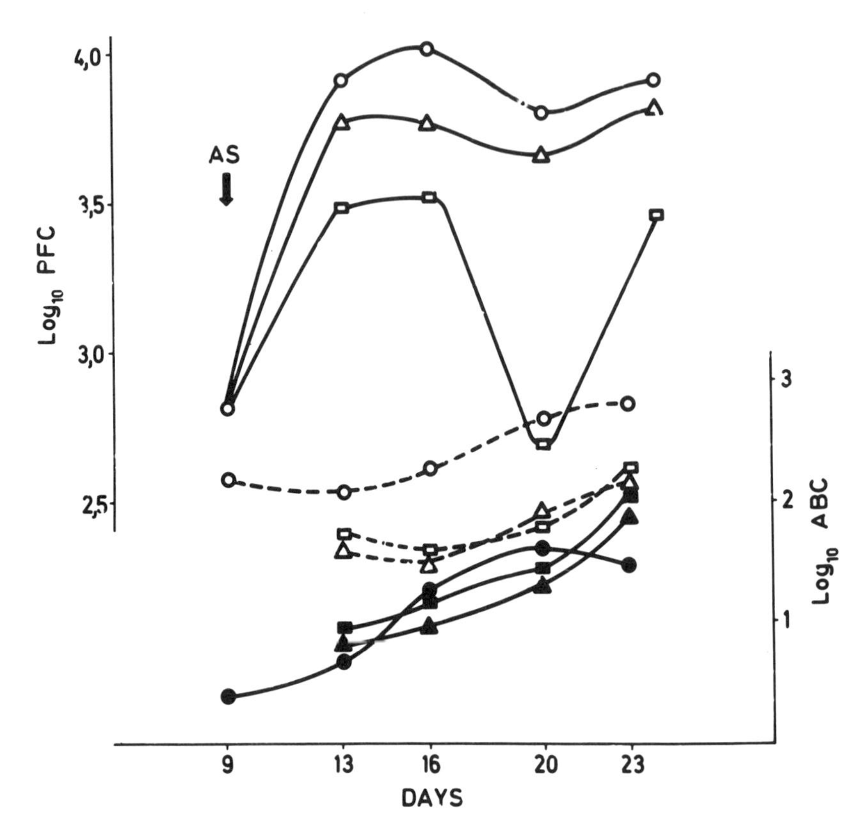

Fig. 3a. Quantitative changes in number of PFC:s in the regional lymph nodes and in the serum antibody levels.

○ ——— ○ PFC:s in control animals
△ ——— △ PFC:s in animals given antibody, ABC 20 /ug
□ ——— □ PFC:s in animals given antibody, ABC 200 /ug
○ – – – ○ ABC at 100 /ug in control animals
△ – – – △ ABC at 100 /ug in animals given antibody, ABC 20 /ug
□ – – – □ ABC at 100 /ug in animals given antibody, ABC 200 /ug
● ——— ● ABC at 1 /ug in control animals
▲ ——— ▲ ABC at 1 /ug in animals given antibody, ABC 20 /ug
■ ——— ■ ABC at 1 /ug in animals given antibody, ABC at 200 /ug

ABC is the antigen binding capacity (from Andersson 1971)

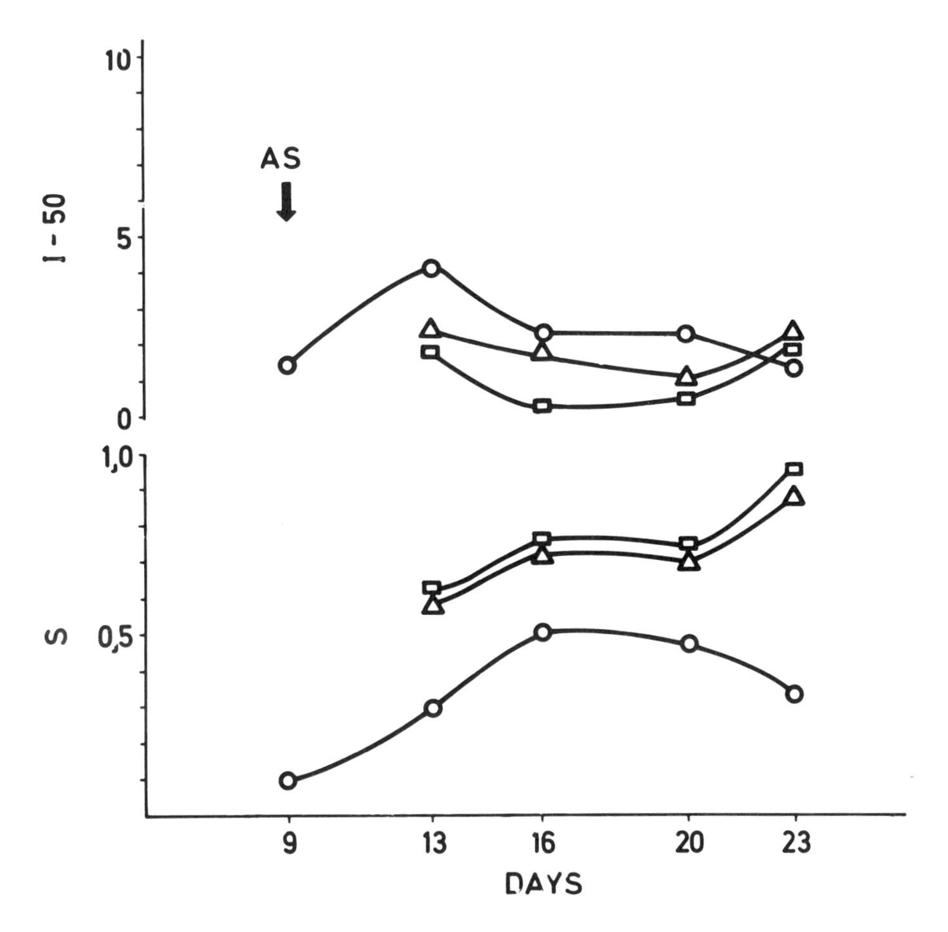

Fig. 3b. Qualitative changes in antibodies released by PFC:s and in serum antibody. I-50 is the concentration of free BSA giving 50% inhibition of PFC:s. S is the factor of avidity of serum antibodies. ABC is the antigen binding capacity (from Andersson 1971)

○ ——— ○ Control animals
△ ——— △ Animals given antibody, ABC 20 μg
□ ——— □ Animals given antibody, ABC 200 μg

Table 3
Modification of the anti-hapten response by induction of paralysis to the carrier protein.

Day 0	Day 7	Day 14 PFC Log_{10}	Day 14 I-50 M/liter
100 μg DNP-BSA	———	4.0	3.1×10^{-6}
100 μg DNP-BSA	10 mg BSA	2.8	9.5×10^{-6}

Table 4
Modification of the anti-BSA response of mice: comparison between specific and unspecific suppression.

Day 0	Day 9	Day 14 PFC Log_{10}	Day 14 suppression %	Day 14 I-50 μg/ml
100 μg BSA	———	3.35	-	1.04
100 μg BSA	200R	3.08	48	1.19
100 μg BSA	10 mg BSA	2.27	92	29.60

4. ANALYSIS OF THE AFFINITY OF CELL MEMBRANE RECEPTORS

Affinity chromatography of immunological memory cells.- In the previous sections is presented a series of indirect evidence for the same affinity of the cellular receptors for antigen and the antibody released from the cell after immunization. This section will provide direct evidence on this point. Cells immune to the DNP and NIP haptens were mixed and allowed to pass through a DNP coated immunoadsorbent column. Antibody affinity was tested after transfer to new irradiated mice. As can be seen in table 7 a selective retention of the high affinity DNP memory cells had taken place in the DNP column. The anti-NIP memory showed no affinity changes i.e. no selective retention of such cells occurred. Thus, the immunological memory cells have antigen specific surface receptors which express similar affinity for the antigen as the humoral antibody released after stimulation. It should be stressed here that the affinity specialization in the present system occurs in the B-cell line; an excess of anti-carrierprotein immune cells were added to the cell suspensions before transfer to the new hosts.

5. ISOLATION OF B-CELL CLONES

Several investigators have reported that clones of antibody forming cells can be obtained by limiting dilution techniques. Markers such as antibody allotype and affinity both showed homogeneity. It was therefore of interest to investigate whether the inhibition assay used in the present work could be used to characterize antibody forming cell populations with regard to homogeneity of binding affinity. In fig. 4 such an experiment is shown. Mice that had been previously irradiated were injected with gradually decreasing

Table 5
Modification of the anti-BSA response by pretreatment with various antigen doses.

Pretreatment[1] BSA x3/week	PFC[2] Log_{10}	Suppression %	I-50 µg/ml
100 mg	2.40	97	101.4
10 mg	2.95	90	10.4
1 mg	3.64	52	7.3
100 µg	3.63	53	6.3
10 µg	3.42	71	8.6
1 µg	3.82	28	10.0
0.1 µg	3.83	26	11.5
—	3.96	—	14.5

1) Soluble BSA was given repeatedly for 8 weeks.
2) The mice were immunized with 2.500 µg of BSA in Freund's complete adjuvant and tested 20 days later.

Table 6
Effect of the immunizing dose on the sensitivity to paralysis and anti - serum suppression.

	Response day 14			
	Immunizing dose 2.500 µg BSA		Immunizing dose 25 µg BSA	
	PFC Log_{10}	Suppression %	PFC Log_{10}	Suppression %
Control	3.40	-	3.15	-
BSA 10 mg day 7	2.26	93	1.18	99
Anti-BSA serum ABC[1] 10 µg day 7	2.53	87	2.79	57

1) ABC is the antigen binding capacity.

Table 7
Affinity chromatography of immunological memory cells.

Cells[a]	DNP-response PFC Log_{10}	DNP-response I-50 M/liter	NIP-response PFC Log_{10}	NIP-response I-50 M/liter	DNP/NIP
Control	4.10±0.10	1.68×10^{-8}	4.55±0.10	1.14×10^{-5}	0.35
Passed DNP-BSA column	3.46±0.12	3.37×10^{-5}	4.45±0.04	4.20×10^{-5}	0.10

a) Mixture of cells immune to DNP-OA and NIP-OA. 10^7 cells were injected into 500R X-irradiated recipients together with 10^7 cells immune to OA and 10 µg of DNP-OA and 10 µg of NIP-OA.

doses of anti-DNP-OA memory cells. An excess of anti-OA carrier cells was added to all cell suspensions before injection. It can be seen that PFC:s derived from the low cell doses, containing statistically seen only a single antibody forming unit, showed a hapten inhibition curve which was very sharp as compared to the inhibition curves from animals given higher cell doses. It can be concluded from this that the memory cells were specialized with regard to affinity. A single antibody forming unit being stimulated into antibody formation, proliferates into a clone of antibody forming cells, all of which produce antibody of the same affinity. This is also in perfect agreement with the fractionation experiments described in the previous section, where it was possible to eliminate clones of high affinity from the memory cell population. In this case only antibody of low affinity was obtained after stimulation with antigen.

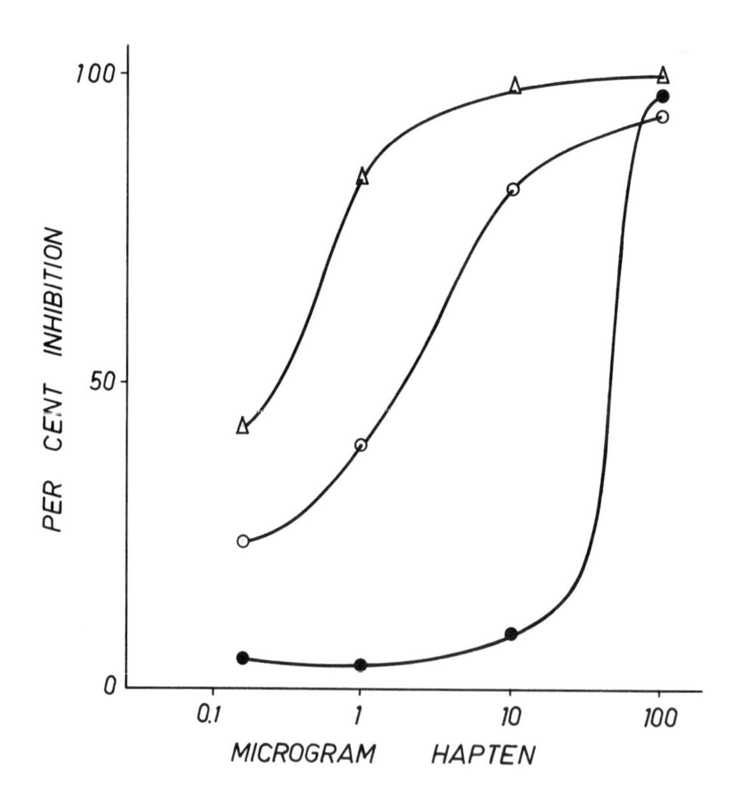

Fig. 4. Specific inhibition of DNP-PFC:s from animals repopulated with different doses of immune cells.

△ ——— △ 5×10^6 cells

○ ——— ○ 5×10^6 cells

● ——— ● 5×10^5 cells

REFERENCES

Andersson, B. (1970), Studies on the regulation of avidity at the level of the single antibody-forming cell, J.Exp.Med. 132, 77.
Andersson, B. and H. Wigzell (1971), Studies on antibody avidity at the cellular level, Eur.J.Immunol. in press.
Davies, A.J.S. (1969), The thymus and the cellular basis for immunity, Transpl.Rev. 1, 43.
Jerne, N.K. and A.A. Nordin (1963), Plaque formation in agar by single antibody producing cells, Science (Washington), 140, 405.
Mitchison, N.A. (1964), Induction of immunological paralysis in two zones of dosage, Proc.Roy.Soc. "B", 161, 275.
Mitchison, N.A. (1971), Cell interactions and receptor antibodies in immune responses, Acad.Press Inc. (London), p.249.
Siskind, G.W. and B. Benacerraf (1969), Cell selection by antigen in the immune response, Advanc.Immunol. 10, 1.
Wigzell, H. and B.Andersson (1969), Cell separation on antigen coated columns, J.Exp.Med. 129, 23.
Wigzell, H. (1970), Specific fractionation of immunocompetent cells, Transpl. Rev. 5, 76.

ACKNOWLEDGEMENT

These investigations were conducted under Contract No. NIH-69-2005 within the Special Virus Cancer Program of the National Cancer Institute, National Institutes of Health, USPHS, and the Swedish Cancer Society.
The author wishes to thank drs. F. Celada and H. Wigzell for valuable cooperation.

APPENDIX

Definitions and Abbreviations

Affinity is the binding constant in an antibody-hapten system. Avidity is the tendency of an antiserum to form stable complexes with a macromolecular antigen. ABC is the antigen binding capacity of antisera. BSA is bovine serum albumin. OA is ovalbumin. NIP and DNP are haptenic groups. T-cells are thymus-derived cells responsible for cell-bound immune reactions. B-cells are bone marrow derived humoral antibody forming cells and their precursors.

CELL INTERACTIONS IN THE IMMUNE RESPONSE

N.A. MITCHISON

Department of Zoology, University College London
(Imperial Cancer Research Fund, Tumour Immunology Unit)

This paper is concerned with one particular kind of cell interaction, that which takes place between T-cells and B-cells in the induction of the humoral immune response. We may begin by restating the conceptual framework within which ideas about these two kinds of cell are currently formulated. It is generally agreed that the cells which are triggered by antigen are lymphocytes and that they belong to two main kinds: T-cells, derived from the thymus, and B-cells derived directly from bursa or marrow. These cells can be distinguished by a variety of markers, such as theta and PHA responsiveness for T-cells, and surface Ig, endotoxin responsiveness, receptors for C' and Fc, and MBLA for B-cells. Each cell has its own characteristic physiology, e.g. T-cells recirculate more rapidly and by a different route from B-cells. It is generally agreed also that immune responses can be divided into two kinds, one humoral and the other cell-mediated. In both only immunoglobulins recognise antigen. In the humoral response T-cells do not function as antibody forming cell precursors (AFCP). In cell-mediated immunity, T-cells do, on the other hand, function both as initiators and mediators of cell-mediated immunity in at least the narrow sense in which this form of immunity is usually defined. Many, but not all, humoral responses are thymus dependent. The thymus-dependent humoral responses need both antibody forming cell precursors (B-cells) and helper cells (T-cells). In this sense then these responses are cooperative in origin and depend on synergy between two cell types. This is not, of course, the only type of synergy which occurs in the induction of immunity where, e.g., cooperation can also be mediated by humoral antibody.

Our present understanding of cooperation between T- and B-cells rests on two main studies. One is of the interaction between thymus cells and marrow cells. The usual experimental procedures here are neonatal thymectomy or adult thymectomy followed by irradiation and repopulation with bone marrow. An alternative approach is to transplant the two cell populations into animals whose own capacity to mount an immune response has been depleted by various procedures. The other study makes use of different antigenic determinants rather than different cell populations. The usual procedure here is to immunise with the two determinants carried on different molecules in order to study the response when they are carried on the same molecule.

We have been engaged in an investigation of the latter type in which hapten conjugates are used to elicit a secondary response (Boak, Mitchison and Pattisson, 1971; Britton, Mitchison and Rajewsky, 1971; Mitchison, 1971 a,b,c). This investigation has led to the following conclusions, which can be summarised as follows: (using DNP to denote a hapten and C to denote a protein) -

(1) Immunisation with C induces proliferation of C-helpers and C-precursors. These can cooperate in secondary responses.

(2) Immunisation with DNP-C induces proliferation of DNP- and C-helpers and precursors.

(3) DNP-helpers do <u>not</u> cooperate with DNP-precursors.

(4) Secondary stimulation of DNP-C_1-immune cells with DNP-C_2 therefore is not effective. If C_2-immune cells are added, stimulation becomes effective. (Anti-DNP response measures helper activity; anti-C_2 measures helper plus precursor activity of added cells).

(5) Secondary stimulation of C_1-immune cells by DNP-C_1 stimulates anti-C_1 response. This can be prevented by depleting helpers with ALS *in vivo*, or anti-theta *in vitro*.

(6) Secondary stimulation of depleted C_1-immune cells by DNP-C_1 becomes effective if DNP-C_2-immune cells are added. (Anti-C_1 response measures

helper plus precursor activity of added cells).

These conclusions have been interpreted in terms of a matrix of antigen bridging T- and B-cells (Mitchison, 1971b,d,e).

Let us now proceed to enumerate some of the current problems raised by cooperation between T- and B-cells.

(1) Cell numbers: helpers are in excess after conjugate priming (Mitchison, 1971b), but what are the actual numbers? We need to know whether the totals are compatible with the hypothetical collision between specifically committed T- and B-cells. We need also to know whether one T-cell can activate more than one B-cell.

(2) The helper population is radio-sensitive in most (Mitchison, 1971b; Cunningham and Sercarz, 1971) but not all systems (Katz, Paul, Goidl and Benacerraf (1970). Does the helper have to do something active while co-operating? Recently, experiments analogous to those of Katz _et al_. have been performed in mice (Yachnin and Mitchison - unpublished). That is, helper cells have been transferred into hapten-primed, non-irradiated hosts. Contrary to the finding previously made with guinea pigs, these experiments have shown that helper function is highly susceptible to irradiation of the transferred population. This raises the question whether the guinea pig experiments of Katz _et al_. operate through a T-cell mechanism strictly analogous to that which obtains in the mouse system. The fact that CGG tolerant lymphocytes are not effective in mediating helper function for attached CGG in the primary response does not necessarily imply that these cells have to play an active role: it may well be that this population is functionally inactive not so much because it cannot mediate cooperation but more because it does not contain T-cells capable of amplifying late helper function through multiplication (Miller, 1971).

(3) Helpers appear to become redundant at high concentrations of antigen (Mitchison, 1971 a,c): is help then truly optional, or do these high doses of antigen elicit a primary T-cell response which renders their help redundant? The evidence from mice depleted of T-cells by _in vivo_ treatment with anti-lymphocyte serum argues in favour of an optional role, but the finding has not been confirmed. Alternative possibilities such as activation of residual T-cells or a switch to an IgM response cannot be entirely discounted.

(4) Most thymus independent responses are IgM: is this merely because the IgG switch occurs late (implying that helpers play a role in antigen retention) and maybe uses less efficient receptors, or is cooperation fundamental to the switch?

(5) Is a physical link, presumably via an antigen bridge, needed for co-operation between T- and B-cells? What then about factors? What function is left for macrophages? The state of affairs at present is that several studies have shown that hapten has to be directly linked to the carrier recognised by T-cells for cooperation to work (Rajewsky, Schirrmacher, Nase and Jerne, 1969; Mitchison, 1971b; Hamaoka, Takatsu and Kitagawa, 1971). The bridge may be needed only to ensure that short-range soluble factors can operate, e.g. as suggested by the allogeneic effect (McCullagh, 1970; Katz, Paul, Goidl and Benacerraf, 1971), particularly in so far as activity has been detected in supernatants (Schimpl and Wecker, 1971). Alternatively, the bridge may be needed solely to promote low affinity receptors to make multipoint (high net avidity)binding (Mitchison, 1971e). The third possibility suggested by recent work on the mobility of B-cell surface immunoglobulin (Taylor, Duffus, Raff and dePetris, 1971) is that the bridge may encourage 'patching' of B-cell receptors (Mitchison, 1971d). Evidence in support of the latter two mechanisms comes from experiments demonstrating the high potency of lattice-bound mitogens (Yachnin, Allen, Baron and Svenson, 1971; Greaves and Bauminger, 1971) as well as from the potentiation of flagellin stimulation by lattice-forming antibody (Feldmann and Diener, 1971). As regards macrophages, an intriguing possibility is that antigen bound to macrophages promotes 'patching' on helper cells and

this in turn enhances their capacity to 'patch' B-cells.

(6) Are helpers a restricted subpopulation of T-cells? An argument in favour of this view is that solid phase antigen can deplete a cell population of uridine incorporators (Davie and Paul, 1970) but not of helpers (Wigzell, Andersson, Mäkelä and Walters, 1971).

(7) What receptors are used by T cells in general and helpers in particular? Evidence obtained mainly from rosette inhibition indicates that T cells bear IgM receptors (Greaves and Hogg, 1971), yet attempts to inhibit uptake of antigen by helpers with anti-IgM sera have so far proved disappointing (Mitchison, 1971d).

(8) T- and B-cells have difference ranges of preferred reactivity with antigens: for example, carbohydrates have long been known as poor inducers of delayed-type hypersensitivity. Does this reflect something fundamental about triggering? Does this explain the special problems encountered with hapten specific help (Paul, Katz, Goidl and Benacerraf, 1970). See also (3) above.

(9) Do T- and B-cells have different thresholds of response? If so, does this reflect differences at the receptor stage or later on? (Mitchison, 1971f; Möller, 1970; Chiller, Habicht and Weigle, 1971)

Inhibition of DNP-specific uptake onto helpers and precursors

Two recent investigations bear on these problems. In one, hapten-specific help was further examined, in order (i) to resolve the question whether what has hitherto been described by this term may not rather be help directed towards hapten-dependent determinants on the carrier protein ('neo-antigens'), and (ii) to estimate, albeit crudely, the relative avidity of T- and B-cells for the same determinant. Point (i) bears particularly on problem (8) in the preceding list, and point (ii) on problems (7) and (9).

In this experiment cells from the spleen of mice immunised with DNP_4OA (alum + pertussis) were incubated with DNP_6CGG, washed, and then adoptively transferred together with cells from the spleen of mice immunised with CGG (by the same method) and treated *in vivo* with ALS to deplete helpers. As described above, this system is designed to detect uptake of antigen onto DNP-specific helper cells by the subsequent production of antibody to CGG. Procedures were used as previously described (Mitchison, 1971a,c; Britton, Mitchison and Rajewsky, 1971), and the system was judged to have achieved its purpose, since it met the following criteria: (i) antigen taken up by normal spleen cells, under the same conditions of incubation, did not induce a full response, even when as much total I^{125}-labelled antigen was transferred; (ii) antigen taken up by irradiated DNP-immune cells (these cells took up approximately the same amount of antigen as non-irradiated immune cells) was unable to induce a full response, even when the antigen-exposed, incubated cells were mixed with unexposed, non-incubated immune cells; (iii) comparable incubation experiments could be performed with educated thymus cells (so far these have been performed only in a slightly different system, in which uptake onto carrier-immune helpers was tested).

Granted that the system is able to detect specific uptake onto DNP-specific helper cells, the question was then asked whether this uptake could be inhibited by DNP-ϵ-aminocaproate, and if so by how much? Incubation with antigen in the presence of 3×10^{-4}M DNP-EACA suppressed uptake entirely, as judged by the subsequent response, while the concentration needed for 50% inhibition was calculated as 1.3×10^{-5}M. Helper cells were tested 19-58 days post-immunisation, and no effect was detected of the interval upon the susceptibility to inhibition. The antigen, DNP_6CGG, was used at concentrations of 1 and 0.1 μg/ml (with little difference in outcome), so that the molar excess of DNP needed for 50% inhibition was 300-3000 fold.

In comparable experiments, non-ALS-treated CGG-primed cells were used in the mixture, and the anti-DNP response was measured, so that uptake onto precursors could be tested and inhibited. The same slope of inhibition was then obtained, but displaced so that x7 more DNP-EACA was required.

The conclusion may be drawn from this investigation that helpers and precursors cannot be very different in their avidity for antigen. If anything, these findings indicate a higher affinity on the part of the helper cell receptor. Alternatively, and this is the interpretation which we favour, the affinities are similar but the density of receptors on helpers is lower, so that in the competition with monovalent hapten the antigen is less able to take advantage of its potential polyvalency. These findings also confirm the view that helpers can bind the DNP group.

High and low dose tolerance

T- and B-cells can both be rendered tolerant of protein (Rajewsky, 1970) and hapten (Mitchison, 1971 g). For this T-cells need 100-1000 times less antigen (Mitchison, 1970, 1971f; Chiller, Habicht and Weigle, 1971). My estimate was based on the transfer of educated thymus cells to non-irradiated hosts, an insensitive and variable procedure. The point is further examined in the experiments shown in Table 1, in which cells were tested after transfer into irradiated (900 r Co^{60}) hosts. The hosts received irradiation, cells i.v., and

Table 1

	Exp.1	Exp.2
Marrow	0.2	0.3
Marrow + normal spleen cells	2.5	17.0
Marrow + high-dose tolerant spleen cells	0.4	0.2
Marrow + high-dose tolerant spleen cells + thymocytes	0.5	0.4
Marrow + low-dose tolerant spleen cells	0.5	1.0
Marrow + low-dose tolerant spleen cells + thymocytes	1.5	2.5

400 μg alum-precipitated BSA i.p. on day 0, 100 μg fluid BSA on day 20, and were assayed for anti-BSA antibody on day 30. Marrow cells (5×10^6) came from high-dose tolerant donors (10 mg BSA x3/week for 10 weeks); spleen cells came (40×10^6) from normal, high-dose tolerant, or low-dose tolerant donors (10 μg BSA x3/week for 14 weeks); thymocytes (50×10^6) came from normal 6-week old donors. Group size, titrations, etc., followed procedures already described (Mitchison, 1971f).

The levels of anti-BSA antibody attained in Table 1 show that the spleen cell population from low-dose but not high-dose tolerant donors could be reconstituted with thymocytes. This finding confirms the conclusion that low-dose tolerance resides primarily in T-cells.

Other examples of cooperation between determinants

Instances of cooperation between determinants which can provisionally be ascribed to T-B cooperation were catalogued two years ago (Mitchison, 1970). Recent developments have strengthened the evidence and increased the number of examples. Thus, Boone, Blackman and Brandchaft (1971) have further documented cooperation with viral determinants in the response to tumour-specific antigens. An area of interest for the understanding of cell interactions is cooperation between surface alloantigens and autoantigens in the mouse. Boyse, Bressler, Iritani and Lardis (1970) report that cytotoxic IgM autoantibodies commonly appear in response to immunisation with allogeneic thymocytes. These may well be identical with the autoantibodies, again resulting apparently from a cooperative response, which are important in determining the activity of anti-θ antisera

(Greaves and Raff, 1971). They are probably related to the autoantibodies encountered by Schlesinger (1971) in H-2 antisera, which are cytotoxic for neuraminidase-treated lymphocytes. An analogous instance occurs in the liver, where antibodies to the autoantigen F (probably not located on the cell surface) develop in response to cross-strain immunisation (Fravi and Lindenmann, 1968). M.Iverson and J. Lindenmann (personal communication) find that the helper determinant in this situation appears to be a monofactorially-determined alloantigen other than H-2.

In each of these instances a structure which would not otherwise induce a response becomes immunogenic when presented in company with another antigen. Quite apart from the potential importance of this phenomenon in cancer immunotherapy and the genesis of auto-immune disease, it may tell us something new about the cell surface. It implies, in the light of the well-established need for a physical link between hapten and carrier which has already been mentioned, that the two cell antigens are physically linked. If both are located on the cell surface (e.g. the thymocyte allo- and autoantigens), they must surely then inhabit the same "mobile unit" as defined by the capping experiments of Raff and DePetris (1971).

Negative T-B control

So far we have been concerned exclusively with cooperation whereby T-cells enhance the response of B-cells. The product of this form of cooperation is antibody, and antibody can of course exercise an inhibitory effect on the response via a feedback mechanism. Thus there is ample opportunity, via well-established mechanisms, for T-cells to exercise an inhibitory influence on the immune response. Nevertheless, Allison, Denman and Barnes (1971) and Gershon (1971) propose that T-cells may exercise a more direct role in inhibiting the B-cell response, not mediated by antibody. In support of this view they cite work of McCullagh (1970b) on tolerance of foreign erythrocytes in the rat, as well as original work on tolerance of native and foreign erythrocytes in the mouse. The evidence is now strong that neither the conventional clone-elimination mechanism of tolerance nor conventional antibody feedback can entirely account for unresponsiveness, particularly unresponsiveness to cellular antigens.

My own impression is that the conventional explanations are not yet exhausted. They have been given a new lease of life by the finding that antigen-antibody complexes (Sjögren, Hellström, Bansal and Hellström, 1971) can exercise a more powerful inhibition than had been expected. For this reason I am reluctant to accept the hypothesis of direct negative control by T-cells.

REFERENCES

Allison, A.C., Denman, A.M. and Barnes, R.D.,(1971),Cooperating and controlling functions of thymus-derived lymphocytes in relation to autoimmunity, Lancet ii, 135.

Boak, J.L., Mitchison, N.A. and Pattisson, P.H., (1971), The carrier effect in the secondary response to hapten-protein conjugates. III. The anatomical distribution of helper cells and antibody forming cell precursors, Eur.J. Immunol. 1, 63.

Boone, C., Blackman, K. and Brandchaft, P., (1971), Tumour immunity induced in mice with cell-free homogenates of influenza virus-infected tumour cells, Nature,Lond. 231, 265.

Boyse, E.A., Bressler, E., Iritani, C. and Lardis, M., (1970), Cytotoxic γM autoantibody in mouse alloantisera, Transplantation 9, 339.

Britton, S., Mitchison, N.A. and Rajewsky, K., (1971), The carrier effect in the secondary response to hapten-protein conjugates. IV. Uptake of antigen in vitro, Eur.J.Immunol. 1, 65.

Chiller, J.M., Habicht, G.S. and Weigle, W.O., (1971), Kinetic differences in unresponsiveness of thymus and bone marrow cells, Science 171, 813.

Cunningham, A. and Sercarz, E., (1971) - in press.
Davie, J.M. and Paul, W.E., (1970), Receptors on immunocompetent cells. I. Receptor specificity of cells participating in a cellular immune response, Cell.Immunol. 1, 404.
Feldmann, M. and Diener, E., (1971), Antibody-mediated suppression of the immune response in vitro. III. Low zone tolerance in vitro, Immunology 21, 387.
Fravi, G. and Lindenmann, J., (1968), Induction by allogeneic extracts of liver-specific precipitating autoantibodies in the mouse, Nature,Lond. 218, 141.
Gershon, R., (1971), Immunology - in press.
Greaves, M.F. and Bauminger, S., (1971), Activation of T and B lymphocytes by insoluble phytomitogens - implications for theories of cell triggering, Nature,Lond. - in press.
Greaves, M.F. and Hogg, N.M., (1971), Antigen binding sites on mouse lymphoid cells. In: Cell Interactions in Immune Responses (Academic Press) p.145.
Greaves, M.F. and Raff, M.C., (1971), The specificity of anti-θ sera in cytotoxicity and functional tests on T lymphocytes, Nature,Lond. - in press.
Hamaoka, T., Takatsu, K. and Kitagawa, M., (1971), Antibody production in mice. IV. The suppressive effect of anti-hapten and anti-carrier antibodies on the recognition of hapten-carrier conjugate in the secondary response, Immunology 21, 259.
Katz, D.H., Paul, W.E., Goidl, E.A. and Benacerraf, B., (1970), Radioresistance of cooperative function of carrier-specific lymphocytes in anti-hapten antibody responses, Science 170, 462.
Katz, D.H., Paul, W.E., Goidl, E.A. and Benacerraf, B., (1971), Carrier function in anti-hapten antibody responses. III. Stimulation of antibody synthesis and facilitation of hapten-specific secondary antibody responses by graft-versus-host reactions, J.exp.Med. 133, 169.
McCullagh, P.J., (1970a), The abrogation of sheep erythrocyte tolerance in rats by means of the transfer of allogeneic lymphocytes, J.exp.Med. 132, 916.
McCullagh, P.J., (1970b), The transfer of immunological competence to rats tolerant of sheep erythrocytes with lymphocytes from normal rats, Austr. J.exp.Biol.Med.Sci. 48, 351.
Miller, J.F.A.P., (1971), Interaction between thymus-dependent (T) cells and bone marrow-derived (B) cells in antibody responses, In: Cell Interactions and Receptor Antibodies in Immune Responses (Academic Press) p.293.
Möller, G., (1970), Editorial: Immunocyte triggering, Cell.Immunol. 1, 573.
Mitchison, N.A., (1970), An immunological approach to cancer, Transpl.Proc. 2,92.
Mitchison, N.A., (1971a), The carrier effect in the secondary response to hapten-protein conjugates. I. Measurement of the effect and objections to the local environment hypothesis, Eur. J. Immunol. 1, 10.
Mitchison, N.A., (1971b), The carrier effect in the secondary response to hapten-protein conjugates. II. Cellular cooperation, Eur. J. Immunol. 1, 18.
Mitchison, N.A., (1971c), The carrier effect in the secondary response to hapten-protein conjugates. V. Use of antilymphocyte serum to deplete animals of helper cells, Eur. J. Immunol. 1, 68.
Mitchison, N.A., (1971d), Control of the immune response by events at the lymphocyte surface, In: In Vitro (Waverley Press) - in press.
Mitchison, N.A., (1971e), Cell cooperation in the immune response: the hypothesis of an antigen presentation mechanism, Immunopathology 6, 52.
Mitchison, N.A., (1971f), The relative ability of T and B lymphocytes to see protein antigen, In: Cell Interactions and Receptor Antibodies in Immune Responses (Academic Press) p. 249.

Mitchison, N.A., (1971g), Tolerance in T and B lymphocytes: evidence from hapten-specific tolerance, In: Immunological Tolerance of Tissue Antigens (Proc. 4th Symposium of Charles Salt Research Centre, Oswestry) p.67.

Paul, W.E., Katz, D.H., Goidl, E.A. and Benacerraf, B., (1970), Carrier function in anti-hapten immune responses. II. Specific properties of carrier cells capable of enhancing antibody responses. J.exp.Med. 132, 283.

Raff, M.C. and dePetris, S., (1971), Antibody-antigen reactions at the lymphocyte surface: implications for membrane structure, lymphocyte activation and tolerance induction, - this Symposium.

Rajewsky, K., Schirrmacher, V., Nase, S. and Jerne, N.K., (1969), The requirement of more than one antigenic determinant for immunogenicity, J. exp. Med. 129, 1131.

Schimpl, A. and Wecker, E., (1971), - Communication to the Third Meeting of the Gesellschaft für Immunologie, Marburg.

Schlesinger, M., (1971), Transpl.Proc.- in press.

Sjögren, H.O., Hellström, I., Bansal, S.C. and Hellström, K.E., (1971), Suggestive evidence that the 'blocking antibodies' of tumour-bearing individuals may be antigen-antibody complexes, Proc.Nat.Acad.Sci.,U.S. 68, 1372.

Taylor, R.B., Duffus, W.P.H., Raff, M.C. and dePetris, S., (1971), Redistribution and pinocytosis of lymphocyte surface immunoglobulin molecules induced by anti-immunoglobulin antibody, Nature,Lond. - in press.

Wigzell, H., Andersson, B., Mäkelä, O. and Walters, C.S., (1971), Characteristics of surface-attached antibodies as analysed by fractionation through antigen-coated columns, In: Cell Interactions and receptor antibodies in immune responses (Academic Press) p.231.

Yachnin, S., Allen, L.W., Baron, J.M. and Svenson, R., (1971), Proc. Fourth ann. Leukocyte Culture Conference (Appleton Century Crofts, N.Y.) p.37.

IgM MOIETIES ON MALIGNANT LYMPHOID CELLS.

Dick Killander, Eva Klein, Bo Johansson and Arthur Levin

Radiumhemmet, Karolinska Sjukhuset, 104 01 Stockholm 60
Dept. of Medical Cell Research and Dept. of Tumor Biology,
Karolinska Institutet, 104 01 Stockholm 60, Sweden

Considerable interest has been focused recently on the properties of lymphocytes bearing immunoglobulins (Ig) as evidence is accumulating that the Ig moieties function as antigen receptors. The Ig is also regarded as marker for the bone marrow derived lymphocyte population (Raff et al., 1970). Malignant lymphoid cells are useful tools in studies of the surface localised Ig as they may represent clonal amplification of a certain cell type. Proof that the Burkitt lymphomas are of single cell origin is provided by the X-linked glucose-6-phosphate dehydrogenase cell marker (Fialkow et al., 1970). The majority, but not all of Burkitt lymphoma biopsies were found to have variable amounts of cell membrane bound Ig (Klein et al., 1968; Klein et al., 1971). The Ig was in all cases mu chain with or without kappa light chain establishing a link between the malignant cells and their normal counterparts as IgM was found to be the predominant Ig class present on lymphocytes (Warner et al., 1970; Pernis et al., 1970). Similarly peripheral lymphocytes from chronic lymphocytic leukemia patients were also found to react with anti mu and anti kappa reagents (Johansson and Klein, 1970; Klein and Eskeland, 1971; Wilson and Nossal, 1971). Among the patients considerable quantitative differences were found and repeated samples from the same patient did not differ. Similarly to the polarized distribution of the immunoglobulin on normal lymphoid cells due probably of the so called capping phenomenon brought about by the anti-Ig reagent (Taylor et al., 1971), the Ig on the malignant cells are also localised usually to a certain area when visualized with fluorescein conjugated reagents.

The IgM on cells of a CLL patient (T.P.) and on a Burkitt derived cell line (Daudi), could be released by homogenization and was found to consist of 7S subunits (Eskeland et al., 1971). Free kappa chains were also found. About 80.000 7S IgM molecules were calculated to be present on the cells. On the basis of quantitative absorption of anti IgM serum and staining with fluorescein conjugated reagents using both viable and frozen or fixed cells it was concluded that most of the Ig structure is exposed on the cell surface. The liberation of free Ig molecules was achieved under conditions which are not favorable to disrupt covalent bonds. Part of the mu and kappa structures remained on the membranes. As a large proportion of the molecules was found to be liberated by freezing and thawing it seems that the binding requires an intact membrane. Similarly to the human leukemic cells the Ig on mouse splenic lymphocyte surfaces was also found to be represented predominantly by IgM monomers (Vitetta et al., 1971).

On the basis of the quantities of cell membrane Ig Wilson and Nossal (1971) proposed that the CLL represent a B lymphocyte disease. However, Catovsky and Holt (1971) pointed out that some of the properties of B cells are not valid for the CLL cells, e.g. lack of response to PHA, poor survival *in vitro* and ultrastructure. Since our first described CLL patient (T.P.) carried exceptionally high amounts of membrane bound IgM we were uncertain whether the reactivities with anti IgM serum of lymphocytes derived from other CLL patients could be considered significant. However, since then we have tested additional 40 cases and we are now of the opinion that a certain amount of IgM can be detected on lymphocytes of all CLL patients, although only one of these patients' cells had as high amount of

IgM, as the T.P. cells. The heterogeneity in the amount of surface bound IgM observed among the CLL patients may represent different subclasses of lymphocytes in the different patients even if the cells are morphologically indistinguishable. In order to investigate this further we have studied the degree of heterogeneity of membrane bound Ig in populations of lymphocytes derived from different CLL patients. This was done by measuring the fluorescence intensities of individual cells in a microspectrofluorimeter after exposure to FITC-conjugated anti IgM and anti kappa serum. In previous model experiments we have shown that membrane bound IgM can be quantitated in this way (Killander et al., 1970). In all of 8 analyzed CLL cases the IgM fluorescence intensity was higher than the values taken as controls i.e. cells exposed to anti IgA serum. The mean cellular intensities, as well as the distribution of the values for different patients varied. The profile obtained after anti kappa "staining" did not always follow that of anti mu "staining" indicating that the relative quantities of heavy and light chains on the cells also vary among the different CLL patients. Typical profiles of fluorescence intensities of individual lymphocytes of 3 CLL patients are presented in Fig. 1. All three patients had high white counts with over 90 per cent lymphocytes. Therefore it is unlikely that the broad distribution of the values for patients J. and O. is due to admixture of non-leukemic cells. The population of J. cells was classified as consisting of 94 per cent poorly differentiated lymphocytes.

In this context it may be mentioned that while the majority of CLL cases have mature lymphocytes the histological picture of BL consists of immature lymphoid cells. Thus if the affected cell type is common for CLL and BL the product of the immunological differentiation is expressed on both mature and immature cells.

The amounts of mu and kappa judged by visual evaluation of tho reactivities with the conjugated reagents on tumor cells from 15 patients with solid lymphocytic lymphoma and reticulum cell sarcoma varied also. Generally those with poorly differentiated morphology lacked completely Ig while the differentiated ones showed often a brilliant staining usually more intense than commonly seen in CLL.

A number of cell cultures derived from Burkitt lymphoma maintain membrane bound Ig which, in fact, proved that the Ig was synthesized by the cell. Characteristics of the Daudi line were described previously (Hammond, 1970; Klein et al., 1970). A second Burkitt line, Namalwa, contains only mu chains and in lower amounts as found in Daudi cells. Comparing the ability of cells and known amounts of purified IgM to absorb a given amount of anti IgM serum the amount of IgM is 24 x 10^{-6} ng/Daudi cell and 6 x 10^{-6} ng/ Namalwa cell (Table 1).

Table 1
Quantitative estimation of IgM on Daudi and Namalwa cells

	Cell number x 10^4 used for absorption					
Cells	250	125	63	31	15	-
Daudi	0.03	0.07	0.11	0.26	0.45	
Namalwa	0.10	0.35	0.52	0.72	0.73	
	IgM ng					
Inhibition by soluble IgM	63	31	15	8	4	-
	0.04	0.07	0.10	0.34	0.56	0.75

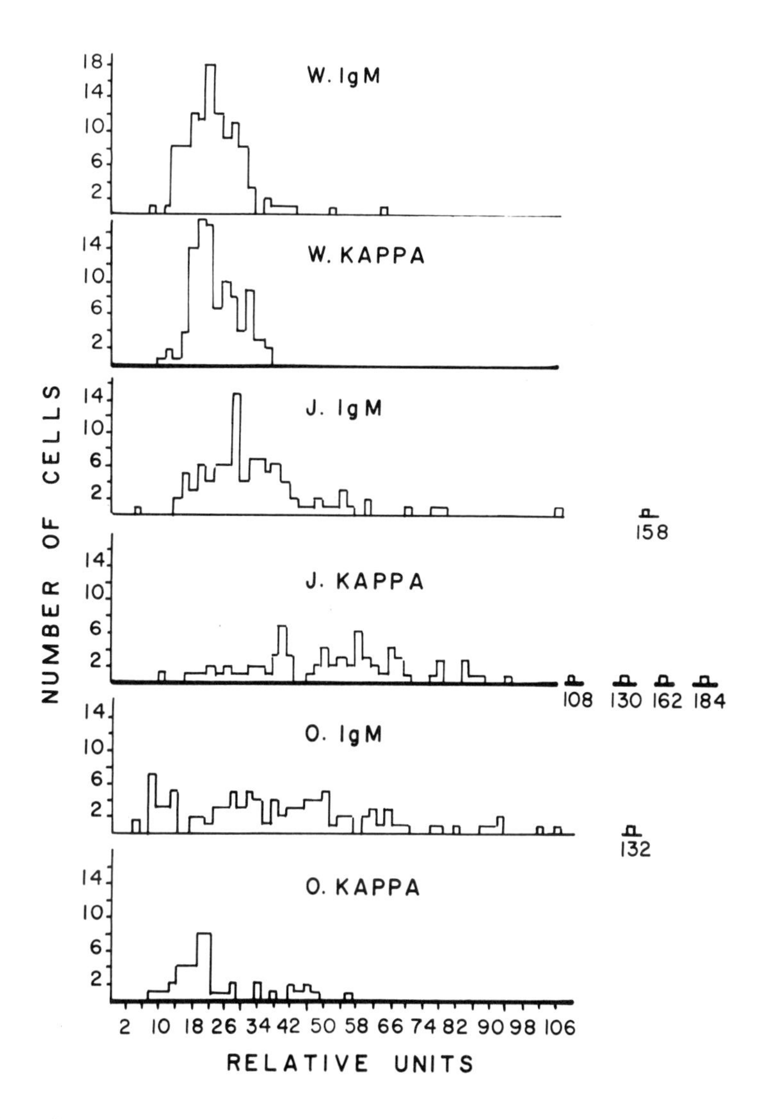

Fig. 1. Frequency distributions of fluorescence intensities of individual peripheral blood lymphocytes sampled from 3 CLL patients after staining with FITC conjugated anti IgM and anti kappa serum (goat sera, Hyland Laboratories, anti IgM: 30.3 mg protein/ml F/P ratio 4.3; anti kappa: 100.9 mg protein/ml F/P ratio 5.8; anti IgA 62.8 mg/ml F/P 3.49). The sera were used in 1:15 dilutions. The mean fluorescence intensities of cells "stained" with anti IgA (n=10-25) were 9.5 (W.), 19.6 (J.) and 15.3 (O.) respectively. The corresponding values of cells immersed in BSS (n=10) were 3.2 (W.) 6.1 (J.) and 9.6 (O.). Conditions for microfluorimetry are given in the legend of Fig. 3.

In this experiment rabbit anti IgM (Brostex) serum was used. As described previously (Klein et al., 1970) 50 ul aliquots of 1:250 diluted serum were absorbed with the indicated cell numbers or the indicated amounts of IgM. Residual activity was tested on Daudi cells in the cytotoxic test. The results are expressed in cytotoxic index, C.I. = the difference between the percentage of unstained cells in the control and antibody-treated sample divided by the percentage of unstained cells in the control sample.

We observed that the reactivity of membrane bound IgM of the Daudi cells declined in aging cultures. This was reflected both by a decrease in the sensitivity of the cells towards the cytotoxic effect of anti IgM and anti kappa serum and a decrease in fluorescence intensity in cells "stained" with conjugated anti IgM (Fig. 2). After refeeding and seeding the culture at low initial cell concentration the reactivity due to IgM recovered (Fig. 2). Thus, the reactivity of membrane bound IgM seems to be dependent on the state of culture growth activity i.e. reaching minimal values in late log-phase and stationary phase and recovering during lag/early log-phase attaining maximal values during the first half of log-phase.

According to studies on Ig secreting cells the quantity of Ig present in the cytoplasm and its secretion fluctuate in relation to the cell cycle (Takahashi et al., 1969). It was also reported that the expression of alloantigens and virally determined antigens on the cell surface show similar cell cycle dependent changes (Cikes and Friberg, 1971), being low during DNA synthesis. If the behaviour of the Ig on this lymphoma cell reflects the behaviour of the normal lymphocytes it seems essential to study the relationship between DNA synthesis and the Ig expression. We quantitated therefore the IgM present on log-phase Daudi cells and related the values to the position of cells in interphase. We were able to carry out the determinations on asynchronously proliferating cells by means of cytophotometry and autoradiography. Possible artefacts due to synchronisation procedures were thus avoided. IgM reactivities were determined by measuring the fluorescence intensities on individual cells after exposure to FITC conjugated anti IgM. The cell cycle position of the same cells was then determined by microinterpherometry, Feulgen microspectrophotometry and H^3-thymidine autoradiography. Since the cellular dry mass (measured by microinterferometry) increases throughout interphase the dry mass is a measure of the cell cycle position (Killander and Zetterberg, 1965). Furthermore, labelled cells after pulse incubations with H^3-thymidine are, by definition, in the S-phase of the cell cycle. The non-labelled cells are either in G1 or in G2 which is further determined by Feulgen DNA measurements. The results are depicted in Fig. 3. The top graph shows that the DNA values of H^3-thymidine incorporating cells fall between two groups of non-labelled cells (G1 and G2 cells) having DNA values close to the ratio 1:2. The expression of IgM (=fluorescence intensity) is highest in G2, lowest in G1 and intermediate in S. The middle graph illustrates the increase in dry mass as cells pass from G1 to S and to G2. There is a strong correlation between fluorescence intensity and dry mass (bottom graph, coefficient of correlation (r) = 0.79). Similar results were obtained with regard to the expression of HLA antigens in the same experiment.

Taken together these results indicate that the expression of surface bound IgM increases continously throughout interphase in close relationship to the accumulation of total cellular proteins. The results are thus in contrast with those regarding Ig production in synchronized cells which secrete Ig (Takahashi et al., 1969). Accordingly the regulatory mechanisms for synthesis of Ig for export and for membrane Ig may differ. It has to be excluded, however, whether methods, cell materials and experimental conditions are responsible for the difference found.

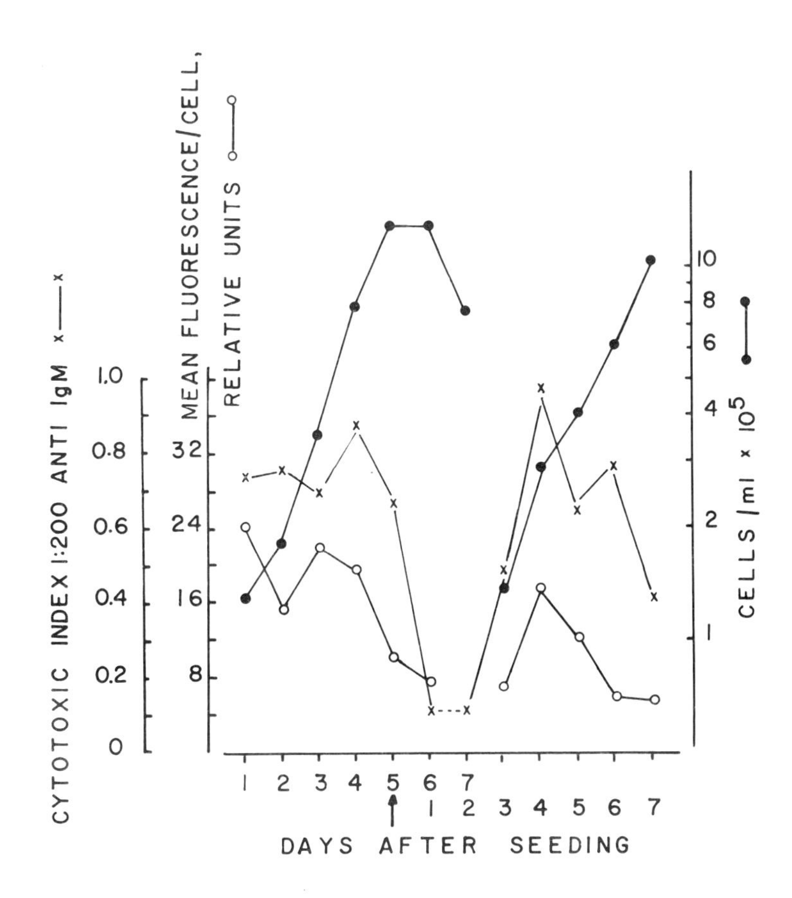

Fig. 2. Fluorescence intensities of Daudi cells "stained" with FITC-conjugated anti IgM (means of 20-50 cells for each day) and their sensitivity to the cytotoxic effect of anti IgM serum in relation to the culture ages. Cells were seeded in MEM supplemented with 20 % FCS with an initial concentration 10^5 cells/ml. On day 5 a new culture was started with 10^5 cells/ml using fresh medium. Otherwise no medium changes were made. The cytoxic test was performed as described previously (Klein et al., 1970). Conditions for microfluorimetry are given in the legend of Fig. 3.

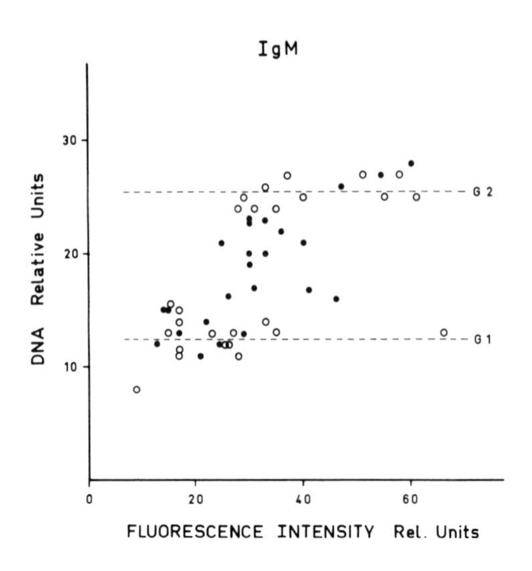
IgM
DNA Relative Units
G 2
G 1
FLUORESCENCE INTENSITY Rel. Units

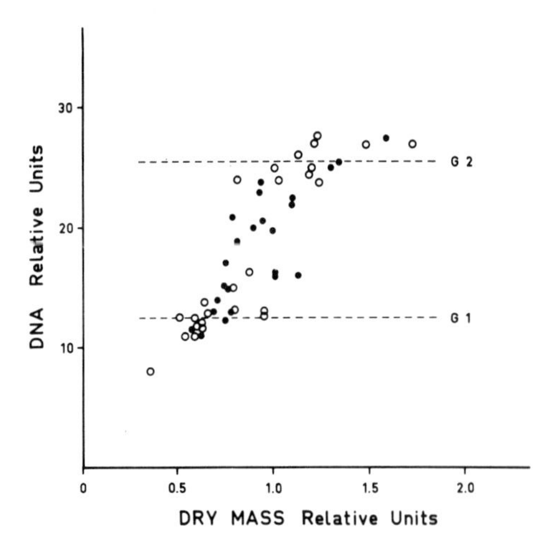
DNA Relative Units
G 2
G 1
DRY MASS Relative Units

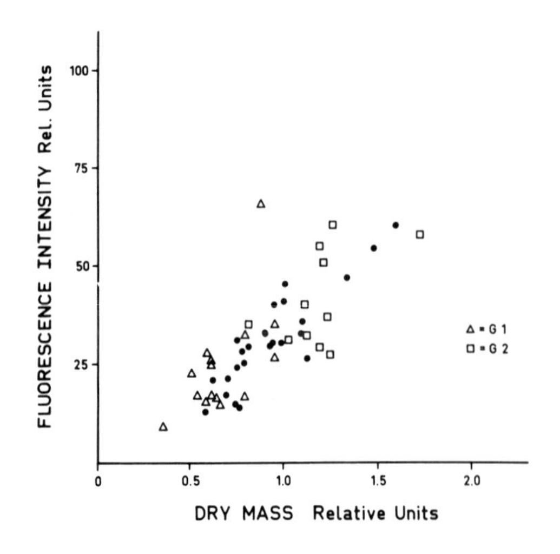
FLUORESCENCE INTENSITY Rel. Units
△ = G 1
□ = G 2
DRY MASS Relative Units

Fig. 3. Fluorescence intesities (after staining with FITC-conjugated anti IgM), dry mass and Feulgen DNA determined on the same Daudi cells. Solid symbols represent cells labelled with H^3-thymidine (S-cell and open symbols represent non-labelled cells. Mean DNA values of G1 and G2 cells are indicated (------).

Daudi cells in the logarithmic phase of growth were exposed to a pulse of H^3-thymidine (1 μC/ml for 20 min), washed twice in PBS and "stained" with FITC-conjugated anti IgM serum. All cytophotometric measurements and autoradiography were made on the same cells. For identification cell maps were used prepared by photographing in phase contrast using a Polaroid camera. Fluorescence measurements were made in incident light using a Zeiss photomicroscope equipped with a photomultiplier. For excitation a BG 38 heat filter and a BG 12 + KP 500 filter were used. The emitted light was measured using a 530 nm interference filter (Levin et al., 1971). The dry mass was then measured in a microinterferometer (Caspersson and Lomakka, 1962) and, after Feulgen staining the amount of DNA was determined in a microspectrophotometer (Caspersson and Lomakka, 1962). The preparations were finally processed to autoradiography using Kodak AR 10 stripping film.

ACKNOWLEDGEMENTS

These investigations were conducted under Contract No. NIH-69-2005 within the Special Virus Cancer Program of the National Cancer Institute, National Cancer Institute, National Institutes of Health, USPHS. Grants were also received from the Swedish Cancer Society and from funds of Karolinska Institutet.

We thank for the excellent technical assistance of Miss Lena Lundin, Mrs. Karin Kvarnung and Miss Hélène Theodoridis.

REFERENCES

Caspersson, T. and G. Lomakka, (1962), Scanning microscopy techniques for high resolution quantitative cytochemistry, Ann. N.Y. Acad. Sci. 97, 449.

Catovsky, D. and P.J.L. Holt, (1971), T or B lymphocytes in chronic lymphocytic leukaemia, Lancet 7731, 976.

Cikes, M. and S. Friberg, (1971), Expression of H-2 and Moloney leukemia virus-determined cell-surface antigens in synchronized cultures of a mouse line, Proc. Nat. Acad. Sci. 68, 566.

Eskeland, T., E. Klein, M. Inoue and B. Johansson, (1971), Characterization of immunoglobulin structures from the surface of chronic lymphocytic leukemia cells, J. Exp. Med. 134, 265.

Fialkow, P.J., G. Klein, S.M. Gartler and P. Clifford, (1970), Clonal origin for individual Burkitt tumours, Lancet 7643, 384.

Hammond, E., (1970), Ultrastructural characteristics of surface IgM reactive malignant lymphoid cells, Exp. Cell Res. 59, 359.

Johansson, B. and E. Klein, (1970), Cell surface localized IgM-kappa immunoglobulin reactivity in a case of chronic lymphocytic leukemia, Clin. Exp. Immunol. 6, 421.

Killander, D., A. Levin, M. Inoue and E. Klein, (1970), Quantification of immunofluorescence on individual erythrocytes coated with varying amounts of antigen, Immunology 19, 151.

Killander, D. and A. Zetterberg, (1965), Quantitative cytochemical studies on interphase growth. I. Determination of DNA, RNA and mass content of age determined mouse fibroblasts in vitro and of intercellular variation in generation time, Exp. Cell Res. 38, 272.

Klein, E., G. Klein, J.S. Nadkarni, J.J. Nadkarni, H. Wigzell and P. Clifford, (1968), Surface IgM-kappa specificity on Burkitt lymphoma cells in vivo and derived culture lines, Cancer Res. 28, 1300.

Klein, E., T. Eskeland, M. Inoue, R. Strom and B. Johansson, (1970), Surface immunoglobulin-moieties on lymphoid cells, Exp. Cell Res. 62, 133.

Klein, E., R. van Furth, B. Johansson, I. Ernberg and P. Clifford, (1971), Immunoglobulin synthesis as cellular marker of malignant lymphoid cells. In: Proc. Cambridge Symp. on Oncogenesis and Herpes Type Viruses, to be published.

Levin, A., D. Killander, E. Klein, B. Nordenskjöld and M. Inoue, (1971), Applications of microspectrofluorometry in quantitation of immunofluorescence on single cells, Ann. N.Y. Acad. Sci. 177, 481.

Pernis, B., L. Forni and L. Amante, (1970), Immunoglobulin spots on the surface of lymphocytes, J. Exp. Med. 132, 1001.

Raff, M.C., M. Sternberg and R.B. Taylor, (1970), Immunoglobulin determinants on the surface of mouse lymphoid cells, Nature 225, 553.

Takahashi, M., Y. Yagi, G.E. Moore and D. Pressman, (1969), Immunoglobulin production in synchronized cultures of human hematopoetic cell lines, J. Immunol. 103, 834.

Taylor, R.B., W.P.H. Duffus, M.C. Raff and S. de Petris, (1971), Redistribution and pinocytosis of lymphocyte surface immunoglobulin molecules induced by anti-immunoglobulin antibody, Nature 233, 225.
Vitetta, E.S., S. Baur and J. Uhr, (1971), Cell surface immunoglobulin. II. Isolation and characterization of immunoglobulin from mouse splenic lymphocytes, J. Exp. Med. 134, 242.
Warner, N.L., P. Byrt and G.L. Ada, (1970), Blocking of the lymphocyte antigen receptor site with anti-Ig sera in vitro, Nature 226, 942.
Wilson, J.D. and G.J.V. Nossal, (1971), Identification of human T and B lymphocytes in normal and in chronic lymphocytic leukemia, Lancet 7728, 788.

THE SPECIFICITY OF T AND B LYMPHOCYTES

STUART F. SCHLOSSMAN
Department of Medicine, Harvard Medical School
and Beth Israel Hospital, Boston, Mass., U.S.A.

INTRODUCTION

An overwhelming body of evidence exists to support the view that two types of lymphocytes are involved in the recognition of antigen and the production of conventional antibody (Claman et. al. 1966, Davies 1969, Miller and Mitchell 1969). The bone marrow derived (bursa-equivalent), thymus independent B cell is the precursor of the antigen reactive antibody forming cell whereas the thymus dependent T cell cooperates with B cells in the production of antibody. In studies of the secondary responses to hapten-protein conjugates T cells are relatively carrier specific whereas B cells are more hapten specific (Mitchison et. al. 1970, Rajewsky et. al. 1969, Katz et. al. 1970). Moreover, T cells are effector cells in cellular immune reactions. Several questions relating to cellular cooperation in antigen recognition remain unanswered. What is the precise specificity of each class of cells and how do they interact? Is collaboration important only in the formation of antibody (in relation to which it had chiefly been studied) or does it extend to primarily T cell activities such as delayed hypersensitivity and its in vitro correlates.

It is the purpose of this article to review the specificity of antigen-reactive cells involved in cellular immunity and antibody production using defined DNP oligolysines, to suggest that the antigen receptors on both T and B are identical and to provide an explanation as to why the receptors appear different in functional tests.

T AND B LYMPHOCYTE SPECIFICITY

Firstly, what is the nature of the antigen receptors on the thymus-derived cells which participate in cellular immune responses among which are; delayed skin reactivity, allograft immunity, acquired microbial resistance and their invitro correlates. Most studies indicate that the recognition of antigen by these immunologically committed cells involves the participation of an exquisitely specific receptor system which can discriminate among closely related antigens. The very specificity of the cellular immune responses studied suggests that either antibody or an antibody-like molecule will account for antigen recognition at the surface of the cell. For example, precisely the same chemical characteristics of antigen are necessary to induce the immune responses (i.e. immunogenicity) as are required to elicit, desensitize to the delayed skin reaction and to provoke cellular immune reactions in vitro as measured by the incorporation of thymidine or the production of macrophage inhibiting factor (Schlossman et. al. 1965, 1966, 1967, 1967a, David and Schlossman 1968, Stulbarg and Schlossman 1968). Immunogenic DNP-oligolysines containing 7 or more L-lysines in sequence could, for example, trigger these cellular immune responses. Nonimmunogenic peptides, on the other hand, containing fewer than 7-L-lysines or α DNP-Lys$_9$ (L_4DL_4) neither trigger nor prevent immunogenic peptides from triggering the cell. If T cell activation were simply a consequence of ligand-antibody interaction at a cell surface, nonimmunogenic α ,DNP-lys$_{3-6}$ or α ,DNP-Lys$_9$ (L_4DL_4) should have triggered the α ,DNP-Lys$_9$ cell since these peptides react strongly with anti α ,DNP-Lys$_9$ antibody (Levin et. al. 1970). Similar studies of the cellular and humoral immune response in guinea pigs with another series of mono substituted DNP-oligolysine peptides have confirmed these observations (Stupp et. al 1971, Paul et. al 1971).

This degree of carrier specificity suggests that; 1, Effector T cells require oligolysine helper cells and that such cells are not available for nonimmunogens (Schlossman and Levine 1970, Stupp et. al. 1971a). This T-T help would be analogous to T-B help in antibody forming systems; or 2, That the T cell receptor is antibody, that a precise fit between ligand and receptor is necessary to initiate a response. Under these circumstances, nonimmunogens or cross-reactive antigens are either incapable or very inefficient in producing the appropriate allosteric or conformational change in the receptor to trigger the cell.

To determine whether the specificity of the effector T cell resulted from the interaction of two cells, one oligolysine specific and the other DNP-oligolysine specific, a series of DNP-nonalysines was prepared. The peptides differed in hapten position and by substitution of D for L-lysine residues. It was shown that lymph node cells obtained from strain 2 guinea pigs immunized with α,DNP-Lys$_9$, α,DNP-Lys$_9$ (L_7DL), α,DNP-Lys$_9$ (LDL_7), α,DNP-Lys$_9$ (L_4DL_4), 5, ε ,DNP-Lys$_9$) 9, ε ,DNP-Lys$_9$ and Lys$_9$ could discriminate among these peptides and were maximally stimulated by the homologous immunizing antigen to incorporate, thymidine, produce M.I.F. or evoke a delayed skin reaction (Schlossman et. al. 1969). Cross reactions were often minimal and could be accounted for by the common oligolysine group but not the haptenic group. The carrier specificity of these T cell responses was not overcome by increasing the amount of cross-reactive antigens to test the response. These results and others indicated that antigen-binding receptors on T cells had an extremely restricted specificity for the determinant used to induce the responses (Davies and Paul 1970, Paul et. al 1971). In addition, there was little to suggest that separate cells bearing hapten and carrier oriented receptors were involved in a cooperative act of antigen recognition prior to cell triggering. Some experiments support the concept T cell synergy in immune responses (Cantor H. and Asofsky 1970) but it has been more difficult to demonstrate T-T cell cooperation in the recognition of an antigenic determinant. In antibody formation, on the other hand, the evidence for T cells cooperating with and triggering B cell precursors of antibody forming cells is very impressive. In a further attempt to demonstrate T-T cell cooperation a series of DNP-oligolysines was prepared which differed from one another only by the insertion of alanine spacers to separate the haptenic group from the remainder of the lysine carrier (Dunham et. al. 1971). Thus, ε,DNP-Lys Ala$_3$Lys$_n$ and ε ,DNP-Lys-Lys$_n$ peptides were used to sensitize animals. If cooperation of T lymphocytes with specificity for different portions of an antigenic determinant occured, i.e. hapten and carrier T cells, we anticipated that lymph node cells from animals sensitized to ε,DNP-Lys-Lys$_8$ would be triggered readily by ε ,DNP-Lys-Ala$_3$-Lys$_8$ and vice versa. Again cross reactions were minimal and could be attributed to the common oligolysine determinant, i.e., heterologous DNP-oligolysines were no better than oligolysines alone. It should be noted that when α,DNP-oligolysines or ε ,DNP-oligolysines are used to prime guinea pigs for a secondary antibody responses, both homologous and heterologous DNP-oligolysines can trigger the primed antibody forming cells in vivo (Schlossman and Levine 1970, Levin et. al. 1971). Immunogenic oligolysines are, however, required to either prime or elicit these responses wheras nonimmunogenic DNP-oligolysines do neither. These observations suggest that oligolysine helper cells are required for secondary antibody formation. More important, the specificity of antibody forming cell precursors appear less discriminating than do cells involved in cellular immune responses. It is not that antibody is not highly specific for antigen but that both the antibody produced as well as the precursors of antibody forming cells are more reactive with heterologous antigens than are cells mediating cellular immune reactions. Perhaps these differences do not reflect differences in T and B cell receptors but can be attributed to the failure of T cells to provide help or amplify the activity of effector T cells. The T cell may be visualized as broadening the specificity of B cells by supplying mediators which decrease its threshhold for activation by heterologous antigens. In most cases the T cell is attracted to the microenvironment of the B cell by specificity for carrier determinants of the antigens bound to B cells. More important, recent studies have shown that even allogeneic cells without specificity for carrier determinants can help trigger B cells (Katz et. al. 1971, Kreth and Williamson 1971).

If one accepts the premise that an analysis of antibody molecules in solution is a reliable measure of of the specificity of the receptor on the precursor of the antibody forming cell, then antibody has sufficient specificity to account for the discriminatory ability of T cells. Detailed analysis of antibodies raised to α,DNP-Lys$_{7-10}$ α,DNP-Lys$_{12}$ α,DNP-Lys$_{16-30}$ α,DNP-Lys$_{10}$ 8, ε,DNP-Lys$_{8}$, 10,ε,DNP-Lys$_{10}$, 14, ε,DNP-Lys$_{14}$, 15, ε,DNP-Lys$_{15}$ indicated that all responder (polylysine gene positive) animals could distinguish the precise DNP-oligolysine chain length used to induce the immune response, plus or minus one lysyl residue as reflected by a greater $-\Delta F$ for the interaction of antibody with the homologous immunizing antigen than for its interaction with others (Levin et. al. 1971). The capacity of antibody to recognize the conformation of each peptide as reflected by alteractions of chain length or hapten position was more than we expected at the outset of these experiments. Although we wanted to demonstrate carrier specificity of antihapten antibody we did not expect it to be this precise.

Antibody made by nonresponder animals by use of high doses of adjuvant to the identical antigens, in contrast, shows no greater $-\Delta F$ with the homologous immunizing antigen and cannot distinguish the precise DNP-oligolysine chain length used to induce the response. Not only does the nonresponder lack specific T cells capable of reacting with DNP-oligolysines to generate a delayed response but they seem to lack specific B cells capable of elaborating antibody to the same determinant. Circumventing a deficient T cell step in a nonresponder did not unmask the same precursors af antibody forming cells as exists in responder guinea pigs. On the other hand, if nonresponder and responder antibodies were identical one would have strong support for the view that T and B cell receptors were different. The similar specificity or lack of it in T and B cell populations of responders and nonresponders respectively to the same antigen suggests that the receptors for antigen in both T and B cell populations may be identical.

CONCLUSION

In the final analysis are the specificities of cells mediating cellular immune reactions very different from the specificity of precursors of antibody forming cells? Functionally yes, but such differences can occur subsequent to antigen interaction with the cellular receptor and not to differences in the receptor itself. The role of helper cells in amplifying a B cell response and the failure to readily demonstrate T-T help may then account for observed differences in T and B cell specificity.

This work was supported in part by National Science Foundation grant GB-25474, U.S. Public Health Service grants AI09003 and a Career Development Award (K03 HE11666).

REFERENCES

Cantor, H. and Asofsky, R., (1970), Synergy among lymphoid cells mediating the graft-versus-host response. II Synergy in graft-versus-host reactions produced by Balb/c lymphoid cells of differing anatomic origin, J. Exp. Med. 131, 235.

Claman, H.N., Chaperon, E.A. and Triplett, R.F., (1966), Thymus-marrow cell combinations. Synergy in antibody production, Proc. Soc. Exp. Biol. 122, 1167.

David, J.R. and Schlossman, S.F., (1968), Immunochemical studies on the specificity of cellular hypersensitivity: The in vitro inhibition of peritoneal exudate cell migration by chemically defined antigens, J. Exp. Med. 128, 1451.

Davie, J.M. and Paul, W.E., (1970), Receptors on immunocompetent cells. 1. Receptor specificity of cells participating in a cellular immune response, Cell. Immunol. 1, 404.

Davies, A.J.S., (1969), The thymus and the cellular basis of immunity, Transplant. Rev. 1, 43.

Dunham, E.K., Yaron, A. and Schlossman, S.F., (1971), Unpublished observations.

Katz, H., Paul, W.E., Goidl, E.H. and Benacerraf, B., (1970), Carrier function in anti-hapten immune responses. I. Enhancement of primary and secondary anti-hapten antibody responses by carrier preimmunization, J. Exp. Med. 132, 261.

Katz, D.H., Paul, W.E., Goidl, E.A. and Benacerraf, B., (1971), J. Exp. Med. 133, 169.

Kreth, H.W. and Williamson, A.R., (1971), A cell surveillance model for lymphocyte cooperation. The role of allogeneic lymphocytes in stimulating a single antibody forming cell clone, Nature, in press.

Levin, H.A., Levine, H., and Schlossman, S.F., (1970), Studies on the specificity and affinity of α ,DNP-oligolysine antibody: A basis for questioning the role of cell-bound antibody in cellular recognition of antigen, J. Immunol 104, 1377.

Levin, H.A., Levine, H. and Schlossman, S.F., (1971), Antigen recognition and antibody specificity: Carrier specificity and genetic control of anti-DNP-oligolysine antibody, J. Exp. Med. 133, 1199.

Levin, H.A., Herman, J., Levine, H. and Schlossman, S.F. (1971), The secondary immune response: The specificity of lymph node cells and antibody, J. Immunol., in press.

Miller, J.F.A.P. and Mitchell, G.F., (1969), Thymus and antigen-reactive cells, Transplant. Rev. 1, 3.

Mitchison, N.A., Rajewsky, K. and Taylor, R.B., (1970), Cooperation of antigenic determinants and of cells in the induction of antibodies. In: Sterzl and Rhia, Developmental aspects of antibody formation and structure, Vol. 2, 547, Academic Press, New York.

Paul, W.E., Stupp, Y., Siskind, G.W. and Benacerraf, B., (1971), Structural control of immunogenicity, IV. Relative specificity of elicitation of cellular immune responses and of ligand binding to anti-hapten antibody after immunization with mono, ε, DNP-nona-L-Lysine, Immunology 21, 605.

Rajewsky, K., Shirrmacher, V., Nase, S. and Jerne, N.K., (1969), The requirement of more than one antigenic determinant for immunogenicity, J. Exp. Med. 129, 1131.

Schlossman, S.F., Yaron, A., Ben-Efraim, S. and Sober, H.A., (1965), Immunogenicity of a series of α ,N-DNP-L-lysines, Biochemistry 4, 1638.

Schlossman, S.F., Ben-Efraim, S., Yaron, A. and Sober, H.A., (1966), Immunochemical studies on the antigenic deteminants required to elicit delayed and immediate hypersensitivity reactions, J. Exp. Med. 123, 1083.

Schlossman, S.F. and Levine H., (1967), Immunochemical studies on delayed and Arthus-type hypersensitivity reactions. I. The relationship between antigenic determinant size and antibody combining site size, J. Immunol. 98, 211.

Schlossman, S.F. and Levine, H., (1967a), Desensitization to delayed hypersensitivity reactions: With special reference to the requirement for an immunogenic molecule, J. Immunol. 99, 111.

Schlossman, S.F., Herman, J., and Yaron, A., (1969), Antigen recognition: In vitro studies on the specificity of the cellular immune response, J. Exp. Med. 130, 1031.

Schlossman, S.F. and Levine, H., (1970), The specificity of the secondary immune response: Characterization of the antibody produced to cross reactive antigens, Cell. Immunol. 1, 419.

Stulbarg, M. and Schlossman, S.F., (1968), The specificity of antigen-induced thymidine-2-^{14}C incorporation into lymph node cells from sensitized animals, J. Immunol 101, 764.

Stupp, Y., Paul, W.E. and Benacerraf, B., (1971), Structural control of immunogenicity II. Antibody synthesis and cellular immunity in response to immunization mono- -DNP-oligo-L-lysines, Immunology 21, 583.

Stupp, Y., Paul, W.E. and Benacerraf, B., (1971a), Structural control of immunogenicity, III Preparation for and elicitation of anamnextic antibody responses by oligo-and poly-L-lysines and their DNP derivatives, Immunology 21, 595.

HETEROGENEITY OF THYMUS-DEPENDENT LYMPHOID CELL FUNCTIONS IN THE MOUSE

HENRY N. CLAMAN and JOHN W. MOORHEAD
Division of Clinical Immunology
Department of Medicine and Microbiology
University of Colorado School of Medicine
Denver, Colorado 80220 U.S.A.

Abstract: Mouse spleen cells responsive to phytohemagglutinin (PHA) have a large component which is corticosteroid sensitive (T_{CSS}) and a smaller component which is corticosteroid-resistant (T_{CSR}). Since the spleen cells responsible for initiating graft-vs-host (GvH) reactions are corticosteroid-resistant, this indicates that the population of PHA-responsive cells is not co-existent with the population of GvH cells.

Normal bone marrow cells respond weakly to PHA, but bone marrow cells from hydrocortisone (OHC)-treated mice have a greater response than normal to PHA. Irradiation experiments with selective shielding strongly suggest that the increased PHA responsiveness of marrow cells of OHC-treated mice is due to the redistribution of peripheral T_{CSR} cells to the marrow under the influence of hydrocortisone.

The PHA response of normal spleen cells is greatly decreased by exposure to dilute anti-θ serum plus complement. The PHA-response of normal marrow cells is resistant even to more concentrated anti-θ. The enhanced PHA response of OHC-treated marrow cells, however, is sensitive to dilute anti-θ. These data indicate a third population of PHA-responsive cells in the normal marrow which are resistant to anti-θ (T_{CSRa}). Treatment with OHC results in additional PHA-responsive cells in the bone marrow (T_{CSR}) which are sensitive to anti-θ and probably come, at least in part, from the spleen.

INTRODUCTION

Although mouse lymphoid cells have been thought to be uniformly killed by corticosteroids, it is now known that considerable heterogeneity exists in this respect. Mouse thymus medullary cells are resistant to steroids [Blomgren (1969), Cohen (1970)], and following treatment, the resistant population contains cells able to do the following; initiate graft-vs-host (GvH) reactions [Blomgren (1969), Cohen (1970)], cooperate with bone marrow cells in antibody formation (helper cell activity) [Cohen (1971)] and respond to phytohemagglutinin (PHA) stimulation *in vitro* [Blomgren (1971a)]. Thymus-derived cells (T-lymphocytes) in the mouse spleen which have GvH and helper cell activity also appear to be resistant to hydrocortisone [Cohen (1970)]. Heterogeneity with regard to corticosteroid sensitivity also exists in B-lymphocytes, since the precursors of antibody-forming cells in the mouse bone marrow are resistant to steroids [Levine (1970)], while precursors in the spleen are sensitive [Cohen (1971)].

These experiments investigated heterogeneity within populations of cells which respond to PHA.

The stimulation of mouse cells by phytohemagglutinin (PHA) is regarded as a function of thymus-derived cells (T cells) [Doenhoff (1970)]. PHA-responsive cells are numerous in the spleen. Bone marrow cells of mice, although mainly noted for their content of B-lymphocytes, also have a small component of T-lymphocytes and can react weakly to PHA. There is a paradoxical relation between the PHA response of spleen cells and the PHA response of marrow cells. Spleen cells taken from mice heavily pretreated with corticosteroids in vitro respond poorly if at all to PHA, while marrow from the same mice give a better-than-normal response to PHA [Levine (1970)].

The purpose of these experiments was to explore the following questions.

1. What are the quantitative relations between doses of corticosteroids and PHA responsiveness of spleen and marrow cells from steroid-treated mice?

2. What is the explanation for the enhanced PHA response of marrow from steroid-treated mice?

3. Are there other qualitative differences between PHA-responsive cells of spleen and PHA-responsive cells of marrow?

MATERIALS AND METHODS

Adult male LAF_1 mice were used. Treated donors were given various doses of hydrocortisone acetate suspension (OHC) or vehicle two days before sacrifice and culture of cells (4×10^6 per ml) for 3 days, with or without PHA. Stimulation was measured by uptake of ^3H-thymidine during a terminal 5 hour pulse. Anti-θ serum (AKR anti-C_3H thymus serum [Raff (1969(]) or normal mouse serum was added to selected cultures for 15 minutes, followed by guinea-pig complement (absorbed with agarose) for 30 min. at 37°. After washing, the cells were recultured. Irradiation (1000r) was carried out with 250 KV and shielding was done with 1/4 inch lead shields.

RESULTS

PHA response of normal and hydrocortisone-treated spleen and bone marrow. Figure 1 shows the PHA response of spleen cells from mice given various doses of OHC two days before sacrifice and culture of 4×10^6 cells. It is obvious that most of the cells which respond to PHA are quite sensitive to OHC since at 1 mg OHC/mouse, a considerable decrease in ^3H-thymidine uptake occurs. At higher doses of OHC, the fall in PHA response tapers off more slowly, indicating some PHA-responsive cells which are more resistant to OHC.

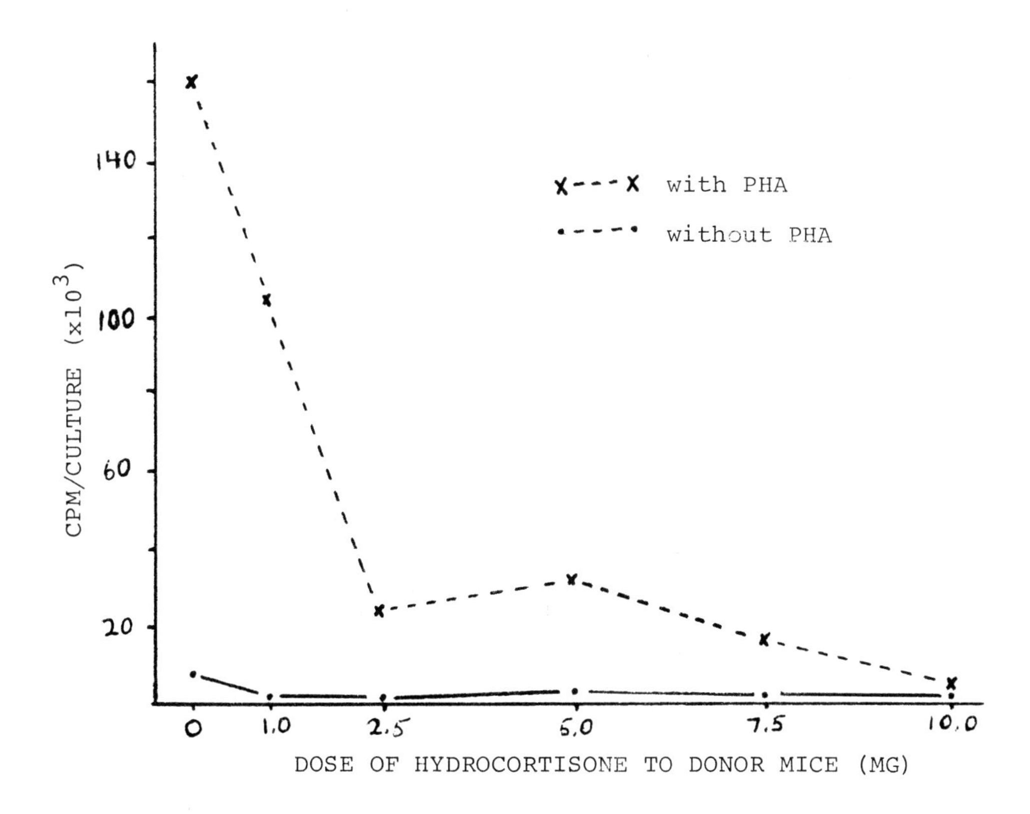

Fig. 1. PHA response of 4 x 10^6 spleen cells of mice given various doses of hydrocortisone acetate 2 days before sacrifice and culture with or without PHA

Figure 2 shows the PHA response of bone marrow cells from mice given various doses of OHC two days before sacrifice and culture of 4 x 10^6 cells. This response is nearly the reverse of that seen in Figure 1. Normal marrow cells have a small (and variable) response to PHA but cells taken from OHC-treated donors give <u>increased</u> PHA responses.

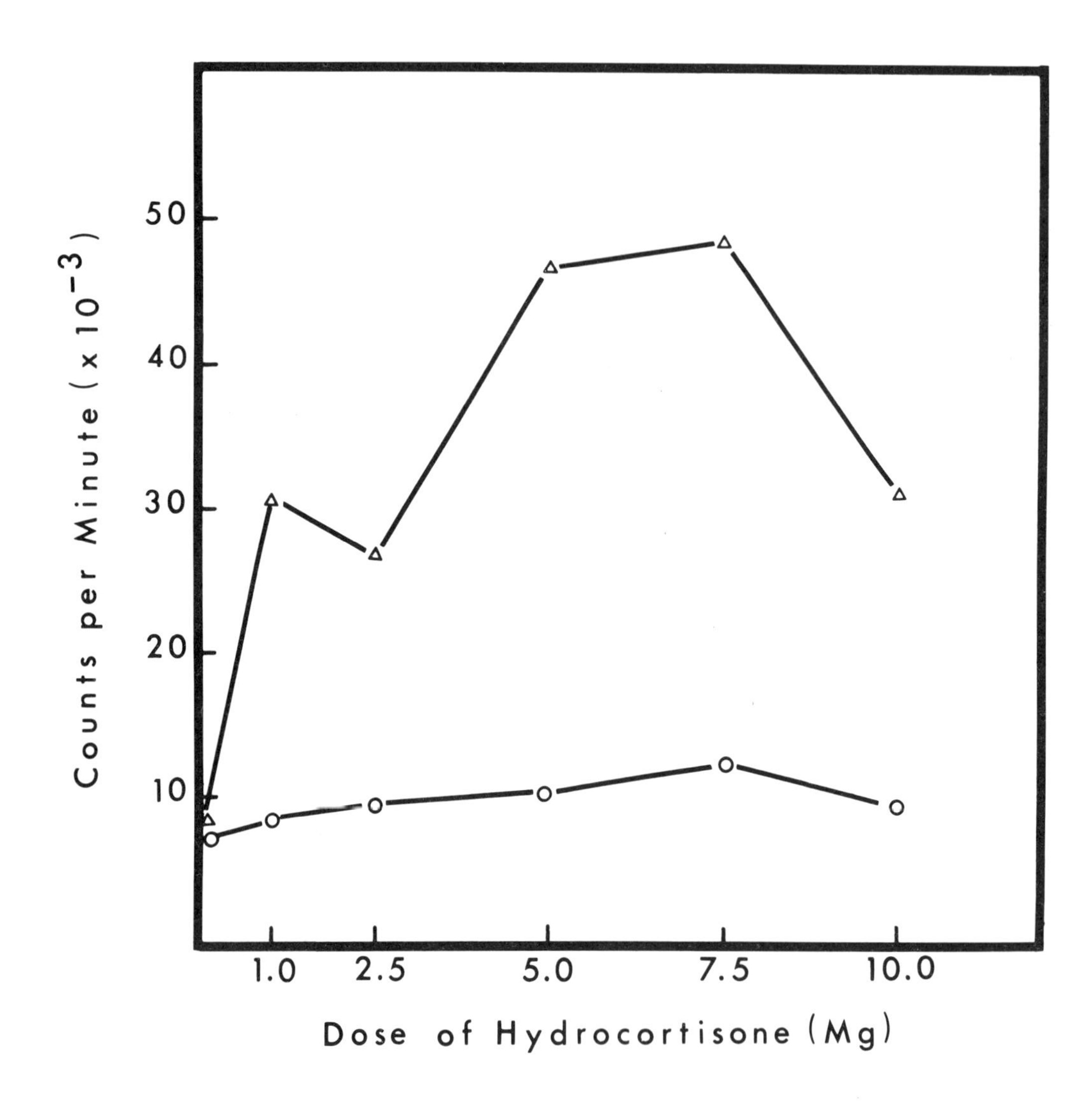

Fig. 2. Phytohemagglutinin response of bone marrow cells from LAF_1 mice treated with various doses of hydrocortisone acetate 2 days before culture. 4×10^6 bone marrow cells per ml were cultured for 3 days with (Δ) or without (o) PHA. Each point represents the mean of triplicate cultures.

PHA response of bone marrow cells from irradiated mice. These experiments were done to test whether the increased PHA response of marrow cells from OHC-treated mice might be due to redistribution of OHC-resistant PHA-responsive cells from the peripheral lymphoid tissues to the marrow. Therefore, we tested the PHA response of tibial marrow from mice given OHC and whole body irradiation (tibias shielded) 2 days before sacrifice and culture. In this case, therefore, OHC was active during the time that the peripheral lymphoid tissues were depleted by x-ray. The results show that the increased PHA response induced by prior OHC-treatment (Group B is greater than Group A) was not present if x-ray was used (Group D is the same as Group C).

Table 1. Phytohemagglutinin Response of Bone Marrow Cells from Hydrocortisone-Treated Irradiated Mice

	Donor Treatment[a]	Counts per Minute[b] No PHA	PHA	Delta[c]
A	-	10,008(±731)	21,347(±952)	11,339
B	5 mg OHC	9,828(±97)	30,613(±888)	20,785
C	x-ray (tib.sh.)	11,863(±142)	20,588(±396)	8,775
D	x-ray (tib.sh.) + 5 mg OHC	8,955(±739)	16,145(±264)	7,190

a Two days before sacrifice and cell culture, mice received no treatment, 5 mg hydrocortisone acetate suspension (OHC), 1000r x-ray (tibia shielded) or both OHC and x-ray. On day 0, 4 x 10^6 marrow cells were cultured per tube.

b Mean CPM of triplicate cultures ± S.E.

c CPM of stimulated cultures minus CPM of unstimulated cultures.

A second irradiation experiment was done to see if some of the increased PHA response of OHC-treated bone marrow cells could be from spleen cells moving to the marrow. In this experiment, on day -2 mice were given OHC and whole body irradiation with both spleen and tibias shielded. On day 0, cells from the tibia were cultured as above. Table 2 shows that enhanced PHA responsiveness of marrow cells is now seen after x-ray if both spleen and tibia of OHC-treated mice are shielded. (Group D is similar to Group B and both are significantly greater than Group A).

Table 2. Phytohemagglutinin Response of Shielded Bone Marrow Cells from Hydrocortisone-Treated Irradiated Mice

	Donor Treatment[a]	Counts per Minute No PHA	PHA	Delta
A	-	11,293(±599)	25,426(±304)	14,133
B	5 mg OHC	10,993(±717	41,115(±521)	30,122
C	x-ray (tib.+sp. sh.)	9,755(±1036)	23,158(±1220)	13,402
D	x-ray (tib.+sp. sh.) + 5 mg OHC	10,846(±350)	36,603(±1381)	25,757

a Conditions as in Table 1, except that both tibia and spleen were shielded during irradiation.

Heterogeneity of PHA-responsive cells with regard to anti-Θ sensitivity. Since PHA-responsiveness is a function of T-lymphocytes, it was reasonable to expect that cultures of cells treated with anti-Θ plus complement would no longer respond to PHA. This indeed was the case, since treatment of spleen cells with fairly dilute anti-Θ abolished almost the entire response to PHA (Table 3, Group B vs Group A). It was surprising that neither this concentration nor a 3-fold higher concentration of anti-Θ had any effect on the PHA response of normal bone marrow cells (Groups C, D and E are not different). When marrow cells from OHC-treated mice were exposed to anti-Θ, a different result was observed. The enhanced PHA response of OHC-treated marrow cells was eliminated by prior treatment with anti-Θ (Group G vs Group F) and the remaining PHA response is identical to that seen in normal marrow cells (Groups G and H are the same as Group C).

TABLE 3. EFFECT OF ANTI-Θ SERUM ON THE PHYTOHEMAGGLUTININ RESPONSE OF SPLEEN AND BONE MARROW CELLS FROM NORMAL AND HYDROCORTISONE TREATED MICE

Group	Donor Treatment	Cell Source	Treatment In Vitro	Counts per Minute[a] No PHA	Counts per Minute[a] PHA	Delta[b]
A	None	Spleen	NMS[c] (1:30)[d]	639(±12)	21,765(±750)	21,126
B			Anti-Θ[e] (1:30)	191(±132)	3,558(±660)	3,367
C	None	Bone Marrow	NMS (1:10)	8,762(±562)	15,290(±340)	6,528
D			Anti-Θ (1:10)	8,602(±417)	14,581(±976)	5,979
E			Anti-Θ (1:30)	11,149(±641)	19,782(±1088)	8,633
F	5.0mg OHC	Bone Marrow	NMS (1:10)	6,164(±779)	25,616(±495)	19,452
G			Anti-Θ (1:10)	6,893(±6)	13,131(±66)	6,238
H			Anti-Θ (1:30)	7,715(±420)	17,307(±2231)	9,592

a Mean of triplicate cultures ± S.E.

b Counts per minute stimulated cultures minus counts per minute unstimulated cultures

c Normal AKR serum

d Final serum dilution

e AKR anti-C_3H Θ serum

DISCUSSION

The following inferences can be made from the above results, together with previous experiments.

1. The spleen contains at least 2 populations of PHA-responsive cells. This is based on an inspection of Figure 1, which appears to be a composite of two lines, one indicating a sharp drop in PHA response of spleen cell populations exposed to low doses of OHC (2.5 mg or less) and the other showing a slower fall in PHA responses with higher doses. Thus, one can identify two populations of spleen cells which respond to PHA (both presumably T lymphocytes). One is corticosteroid sensitive (T_{CSS}) and another is (relatively) corticosteroid resistant (T_{CSR}).

2. The splenic PHA-responsive cell is not the same as the splenic GvH initiator cell, because most of the splenic PHA-responsive cells are sensitive to OHC (T_{CSS}) and splenic GvH cells are resistant to OHC (Cohen, 1970). There are some spleen PHA-responsive cells which are OHC-resistant, so the spleen GvH cell may be included in this subpopulation (T_{CSR}).

3. The bone marrow contains a third population of PHA-responsive cells. Spleen PHA-responsive cells, whether OHC-sensitive or OHC-resistant (T_{CSS} or T_{CSR}) are sensitive to dilute anti-θ. The small normal bone marrow response to PHA is resistant both to OHC and to even higher concentrations of anti-θ and is therefore considered a subset of T_{CSR}, called T_{CSRa}.

4. Treatment of mice with hydrocortisone is accompanied by the appearance of a new population of PHA-responsive cells in the marrow, namely cells which are resistant to OHC (by their survival following in vivo OHC) and which are, unlike normal resident marrow cells, sensitive to anti-θ. Therefore, these results are best interpreted as indicating that in vivo hydrocortisone treatment causes the appearance of T_{CSR} in the marrow. The shielding experiments indicate that the spleen may be at least one source of these immigrant cells.

These results are diagrammatically shown in Figure 3.

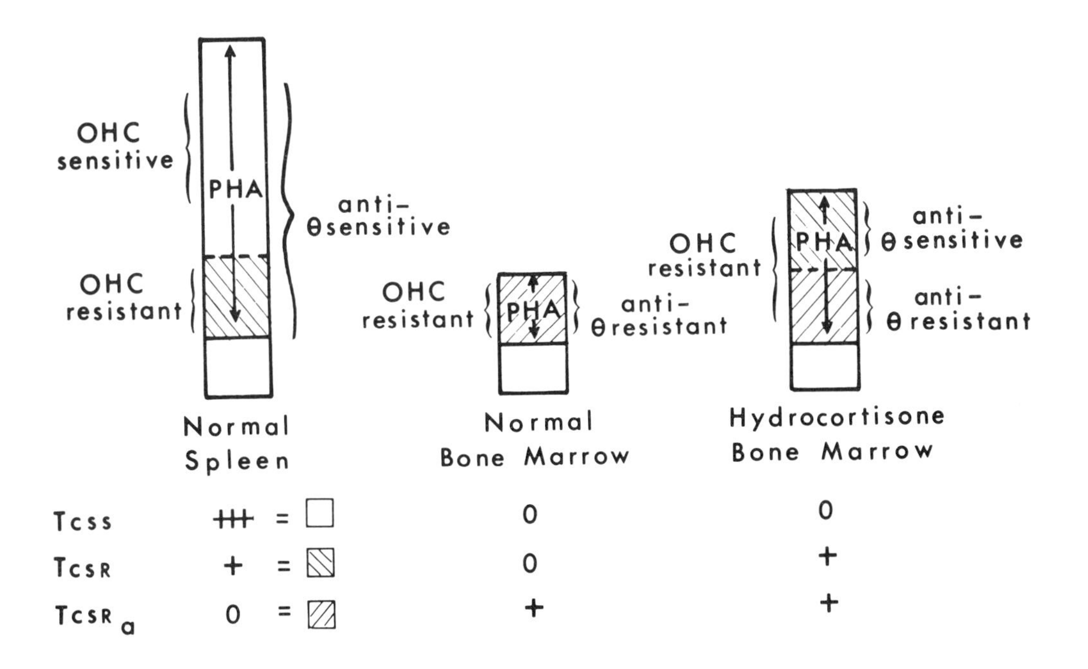

Fig. 3. Schematic representation showing heterogeneity of PHA responsive cells in the spleen and bone marrow of normal and hydrocortisone-treated mice.

It is evident that testing of lymphoid cells with different sorts of reagents (e.g. PHA, anti-θ, OHC) reveals increasing degrees of heterogeneity. It is not yet clear how much overlap there is in various subsets of lymphocytes. For instance, it is not known whether the failure of normal resident bone marrow cells to be affected by anti-θ indicates that they have no θ antigen and, therefore, may be thymus-independent (Blomgren, 1971b) or whether, as we think, that they are probably T cells with too little θ to be detected with anti-θ and complement.

Another implication of these experiments is that treatment of animals with various reagents (e.g. hydrocortisone) may not only cause intrinsic change within cells and cell populations (e.g. destruction of certain subsets) but may also result in redistribution of cells within the body.

Supported in part by USPHS grants AI-TI00013 and AM10145

REFERENCES

Blomgren, H. and B. Andersson, (1969), Evidence for a small pool of immunocompetent cells in the mouse thymus, Exp. Cell Res. 57, 185.
Blomgren, H. and E. Svedmyr, (1971a), In vitro stimlulation of mouse thymus cells by PHA and allogeneic cells, Cell Immunol. 2, 285.

Blomgren, H. and E. Svedmyr,(1971b), Evidence for thymic dependence of PHA-reactive cells in spleen and lymph nodes and independence in bone marrow, J. Immunol. 106, 835.

Cohen, J. J., M. Fischbach and H. N. Claman, (1970), Hydrocortisone resistance of graft vs. host activity in mouse thymus, spleen and bone marrow, J. Immunol. 105, 1146.

Cohen, J. J. and H. N. Claman, (1971), Thymus-marrow immunocompetence V. Hydrocortisone-resistant cells and processes in the hemolytic antibody response of mice, J. Exptl. Med. 133, 1026.

Doenhoff, M. J., A.J.S. Davies, E. Leuchars and V. Wallis, (1970), The thymus and circulating lymphocytes of mice, Proc. Roy. Soc. Lond. B. 176, 69.

Levine, M. A. and H. N. Claman,(1970), Bone marrow and spleen: dissociation of immunologic properties by cortisone, Science 167, 1515.

Raff, M.C., (1969), Theta isoantigen as a marker of thymus-derived lymphocytes in mice, Nature 224, 378.

ACTIVATION OF LYMPHOCYTES BY PHYTOMITOGENS AND ANTIBODIES TO CELL SURFACE COMPONENTS—A MODEL FOR ANTIGEN INDUCED DIFFERENTIATION

M.F. GREAVES
and
G. JANOSSY*

National Institute for Medical Research,
Mill Hill, London.

*Present Address: Clinical Research Centre,
Northwick Park, London.

T AND B LYMPHOCYTES

During the past decade a mounting body of evidence has accumulated testifying to the heterogeneity of lymphocytes. What now seems clear is that the older empirical subdivision of immunity into cell-mediated and humoral categories has a rational basis in the different origin and properties of the two major classes of lymphocytes mediating or initiating these responses (Moller 1968, Roitt et al 1969, Meuwissen et al 1969). Lymphocytes derived from the thymus (in all vertebrates) are termed T cells and are responsible for initiating and, to some extent, directly effecting 'cell-mediated' immunity, e.g. delayed hypersensitivity, graft rejection. B cells are derived in birds from the bursa of Fabricius and we assume from a functionally analogous but as yet undetermined site in mammals. Their primary function is antibody secretion. Much current interest in immunology is dictated by the observation originally made by Claman et al (1966) and subsequently extensively confirmed, that T cells somehow assist (= 'helper' activity or cell cooperation), B cells to produce antibody.

Both T and B cells have the capacity to recognise antigens. This stereospecific phenomenon is mediated by immunoglobulin-like receptor molecules on the surface of lymphocytes and its exquisite specificity and essentially clonal nature (different cells recognise different determinants*) forms the central feature of adaptive immunity.

The initiation, enhancement or depression of cellular activity following ligand-receptor interactions is a fundamental control feature of biological systems encompassing ovum fertilisation, embryonic induction, hormone and drug effects etc. and the opinion has been frequently voiced that antigen induced proliferation and differentiation of lymphocytes provides a useful model system to analyse the nature of this phenomenon. The same philosophy applies also to the studies of cell interactions.

A severe limitation on the capacity of the lymphoid system to fulfill this promise is paradoxically its extreme sophistication. In contrast to almost all other ligand/target cell systems, responsiveness of lymphocytes is 'clonal' and involves what appears, in the present confusion of phenomenology, to be a miriad of cellular interactions and soluble 'factors' modulating or influencing overall responses.

An alternative approach which we and others have been pursuing is to investigate the response of lymphocytes *in vitro* to so called 'non-specific' stimulants.

STIMULATION OF T AND B LYMPHOCYTES BY VARIOUS 'NON-SPECIFIC' MITOGENS

The capacity of phytomitogens-glycoproteins extracted from plants of the Leguminosae family to stimulate lymphocyte proliferation has been extensively documented (Nowell 1960, Ling 1968). Many other substances induce a similar response. These include antibodies to lymphocyte surface antigens including immunoglobulins (Grasbeck et al 1964, Sell 1970, Greaves 1970) and bacterial products (e.g. lipopolysaccharide endotoxins, Peavy et al 1970). The mitogenicity

* Estimates range from 1 in 10^3 to 10^6 cells recognising one antigenic determinant. Makela and Cross (1970).

of anti-immunoglobulin sera is of particular interest since it is possible that it may well operate via the same molecular species of receptors involved in specific antigen recognition. These reactions are generally considered 'non-specific' or'non-immunological' since the majority of lymphocytes appear to respond even when they are derived from foetuses. This contrasts with the clonal response to antigens which are poor stimulators of unsensitised lymphocytes in suspension cultures and stimulate responses in 'primed' populations which are considerably smaller than those induced by phytomitogens. In considering the possible significance of non-specific lymphocyte activation an important early finding with PHA implying an immunological relevance was that responsiveness paralleled immunocompetence of the lymphocyte donor for cell-mediated immune reactions. For example cells from patients with immune deficiency diseases responded poorly (Oppenheim 1969). There were many circumstantial pointers to T cells as being the primary targets of phytomitogens. However, it was not until the T/B dichotomy was fully appreciated and marker systems became available that the response capacity of these cells could be clearly determined.

We have screened a range of non-specific stimulants for their capacity to activate mouse T and B cells. One standard procedure has been to culture purified T and B lymphocytes (Table 1). Wide ranges of mitogen concentrations and varied culture conditions have been used. Our results are summarised in Table 2 where it can be seen that some stimulants are T specific, others B specific and at least one (Pokeweed) activates both T and B cells.

MODE OF ACTION OF MITOGENS

There is little doubt that the initial reaction of lymphocytes with the mitogens listed above takes place via cell surface receptors. Binding of radio-labelled (Powell and Leon 1970, Greaves et al 1971) or fluoresceinated (Smith and Hollers 1970, Greaves et al 1971) phytomitogens, ALS (Woodruff et al 1967) and anti-immunoglobulin sera (Raff et al 1969) to the lymphocyte surface has been well established. In addition binding sites for phytomitogens (Kornfeld and Kornfeld 1969, Allen et al 1971) and immunoglobulins (Vitetta et al 1971) have been solubilised from plasma membranes of lymphocytes. Several recent findings seem to be of prime importance in determining the possible mode of action of lymphocyte mitogens.

(1) The density and distribution of binding sites on T and B cells in relation to the observed selectivity of response: T and B cells appear to have approximately the same average density of receptors for phytomitogens (Greaves et al 1971, Stobo et al 1971). However they differ markedly in binding sites for anti-immunoglobulins; B cells having considerably more exposed cell surface immunoglobulins than T cells (Raff 1970).

(2) Binding of phytomitogens and anti-immunoglobulin sera to the lymphocyte surface involves an active process of receptor rearrangement resulting in a polarisation of binding sites to one side of the cell (usually corresponding with the uropod) and eventual pinocytosis (Taylor et al 1971). Similar effects have also been observed with antigens (Taylor, personal communication). At high concentrations both anti-immunoglobulins and phytomitogens bind but fail to induce a rearrangement of receptor pattern.

(3) Both the initial receptor topography changes and subsequent lymphocyte activation require that the stimulants are at least divalent. Thus monomeric (Fab) anti-immunoglobulin does not alter receptor distribution (Taylor et al 1971) neither does it (or Fab' ALS) trigger a proliferative response (Fanger et al 1970, Woodruff et al 1967, Riethmuller et al 1968). These monomers do however bind to the cell surface and if they are 'bridged' with anti-globulin a response is induced (Fanger et al 1970). Similarly acid-dissociated monomeric PHA appears to bind but fails to activate (Lindahl-Kiessling, personal communication). These results suggest that a critical degree of cross linkage of receptors is obligatory for initiation of lymphocyte responses. It is highly probable that the same requirement is made of antigens since free haptens uniformly fail to activate and monosubstituted hapten-carrier protein conjugates bind but fail to alter receptor distri-

Table 1

Origin of 'selected' T and B cells

1. T Cells

Thymocytes from mice injected 2/3 days previously with 2.5 mg cortisone acetate. These cortisone-resistant cells constitute a small (1-5%) intrinsic thymic population with effectively all the properties of 'peripheral' T cells including immunocompetence (Blomgren and Anderson 1970)

2. B Cells

i. Spleen cells from mice thymectomised, lethally X-irradiated and reconstituted with syngeneic bone marrow pretreated with anti-θ serum, to kill residual T cells (Raff 1971) (Janossy and Greaves 1971).
ii. Spleen cells from congenitally athymic mice (='nudes') (Pantelouris 1968)
iii. Spleen cells from 'normal' mice treated with anti-θ serum *in vitro* (to kill T cells)

3. Control, unselected (T + B) cells. Spleen cells from normal mice.

Note: The 'purity' of these cell populations is verified by cell surface marker analysis (θ on T cells, immunoglobulin and mouse B-lymphocyte specific antigen-MBLA on B cells—Raff 1971). All cell suspensions are filtered through cotton wool to eliminate adhesive macrophages and granulocytes. Eluted suspensions used for cultures contain over 98% viable lymphocytes.
Details given in Janossy and Greaves 1971, 1972.

Table 2
Selective activation of mouse T and B lymphocytes by soluble 'non-specific' mitogens

Stimulant	Origin		Responsive Cell[1]	
			T	B
1. Phytomitogens				
Phytohaemagglutinin(PHA)	*Phaseolus vulgaris*		+	-
Concanavalin A (ConA)	*Canavalia ensiformis*		+	-
Lentil mitogen (LM)	*Lens culinaris*		+	-
Pokeweed mitogen (PWM)	*Phytolacca americana*		+	+
2. Bacterial mitogens				
Endotoxin	*E. coli*		-	+
3. Anti-lymphocyte sera[2] (ALS)	Rabbits, calf	i.	+	-
	Rabbit	ii.	-	+
4. Anti-immunoglobulin sera	Rabbit		(-	+ ?)

1. Response assayed by measuring uptake of (^{14}C)-uridine, (^{3}H)-thymidine, or (^{3}H)-leucine into RNA, DNA or protein respectively. Details given in Janossy and Greaves 1971, 1972. (+) signifies a positive incremental response 10-50 times background (-) signifies less than 2-fold increment. We do not assume that in (+) situations that all cells within a category (T or B) are in fact responding although it is possible that the majority are.

2. Devoid of anti-immunoglobulin activity. Of 15 ALS investigated 5 were T-specific and 1 B-specific. The remaining 9 had little or no mitogenicity. All bound to both T and B cells and were immunosuppressive *in vivo*. Anti-θ (T-specific) serum did not stimulate.

3. We have not yet investigated the specificity of this reaction, however, data from experiments with chicken lymphocytes suggests that anti-immunoglobulin sera are likely to activate only B cells (Alm and Peterson 1969, Ivanyi et al 1969) particularly if they are specific for the Fc piece (Greaves 1970).

bution on anti-hapten reactive cells (Taylor, personal communication). This requirement is also reflected in the observation that the size and density of antigenic determinants has a profound influence on immunogenicity in vivo and in vitro (Feldman and Basten 1971), and may provide the rationale of cell interactions in immunity.

(4) Insoluble covalently linked phytomitogens activate lymphocytes. Moreover an important change in the selectivity of these stimulants occurs when they are presented to lymphocytes as a cross-linked lattice on a bead (Sepharose) surface. PHA (Greaves and Bauminger 1971) and ConA (Anderson and Moller 1971) which are absolutely T specific when in a soluble form activate both T and B cells when insoluble. The response of T and B cells to phytomitogens is therefore determined by the form of the stimulant and possibly by the degree of receptor cross-linkage and subsequent membrane conformational changes. This result provides an explanation of mitogen selectivity, suggests a model to cell-cell cooperation and in addition tells us something very important about the mode of action of phytomitogens. Since the stimulants were effectively insoluble they presumably were unable to be internalised by the cells. It would therefore seem that cell activation in this system is initiated entirely at the level of the plasma membrane and that the stimulants have no obligatory intracellular function.

(5) Neither PWM, ALS or endotoxin seemed to trigger B lymphocytes by binding directly to or even in close proximity to immunoglobulin molecules on the cell surface (Greaves et al 1971 and unpublished observations). Thus, when the binding of these substances and anti-immunoglobulins is compared there is effectively no cross-inhibition. Preliminary analysis of active receptor rearrangement with these stimulants suggests that the B-specific ALS may in fact influence the distribution of cell surface immunoglobulin molecules, however, we have not so far observed this effect with PWM or endotoxin.

IMPLICATIONS FOR CELL TRIGGERING

A consideration of the above information leads us to conclude that lymphocyte activation by 'non-specific' stimulants and by implication, antigens also, is essentially a cell surface phenomenon involving membrane conformational changes induced by cross-linkage of receptors. We assume that such alterations might activate plasma membrane associated enzyme systems (e.g. adenyl cyclase Pastan and Perlman 1971) which would then be responsible for the intracellular generation of response. This interpretation focuses attention on the early cell surface events in lymphocyte triggering and carries with it the implication that the decision between effective response and unresponsiveness (-tolerance) may relate to the pattern of ligand-receptor interaction (e.g. extent of cross-linkage).

There is evidence to suggest that receptor cross-linkage could have a general biological significance. Thus antigen and antiglobulin induced release of histamine and vaso-active amines from mast cells and platelets and also immune complex phagocytosis by macrophages probably involves lattice formation on the cell surface (Greaves 1971). Similarly the capacity of anti-blood group-L-Antiserum to stimulate active potassium transport in LK (low potassium type) sheep erythrocytes depends on the divalency of the antibody (Lauf et al 1970). The capacity of Long Acting Thyroid Stimulator (LATS)—on antibody directed towards thyroid cell constituents to mimic TSH in triggering cellular activity (McKenzie 1968) is particularly intriguing in the context of cell surface recognition phenomena* and raises the question of whether some hormone effects might be initiated by cross-linkage of receptor sites on the cell surface. A suggestion that this may in fact occur is the recent demonstration by Oka and Topper (1971) of the considerably superior efficacy of insoluble, sepharose-linked insulin molecules over soluble insulin in activating mammary epithelial cells from mature virgin mice. The authors, however, propose an alternative interpretation which supposes that receptor-ligand dissociation might constitute a trigger signal. Finally it may not be unreasonable to suppose that many cell-cell and cell-substrate interactions requiring direct surface contact involve integrated effects of multipoint attachments.

*Monomeric Fab_{γ} (papain digested IgG) appears to have some biological activity in this system.

PATTERN OF RESPONSE

It has often been considered that the response of lymphocytes in vitro to 'non-specific' mitogens is essentially 'sterile' or 'abortive' in that no differentiated immunological functions of lymphoid cells are expressed. This view which clearly challenges the validity of the model would now seem to be incorrect since the major effector activities of the immune apparatus, i.e. 'mediator' release by T (and possibly B) cells and immunoglobulin secretion by B cells can be induced by phytomitogens. PHA and ConA activated cells become indiscriminantly cytotoxic for various target cells and release macrophage inhibition factor—MIF (Pick et al 1970), reactions generally considered to be analogues in vitro of cell-mediated hypersensitivity (primarily T cell) reactions. The question of immunoglobulin synthesis and secretion by non-specifically activated cells has been controversial. Early findings were interpreted to suggest that considerable immunoglobulin synthesis occurred in PHA stimulated human blood lymphocytes (Ripps and Hirschhorn, 1967). This was not however confirmed under more controlled experimental conditions (Greaves and Roitt, 1968). The key to this problem lies in the selectivity of mitogens. Soluble PHA is specific for T cells and one would therefore a priori predict that little or no active immunoglobulin synthesis and secretion would occur in cells directly activated by this stimulant, although some enhancement of B cell immunoglobulin production might occur indirectly. Pokeweed endotoxin and some ALS directly stimulate B cells and might in contrast to PHA be expected to induce immunoglobulin synthesis. An early clue that this might in fact occur (before the B cell specificity of these stimulants was appreciated) was the well documented appearance in PWM cultures of human lymphocytes of plasma cells and plasmablasts with well-developed rough endoplastic reticulum (Douglas et al 1967). Curiously enough, the authors of much of this ultrastructural descriptive work considered the response to be an abortive one not involving immunoglobulin synthesis. However, immunofluorescent analysis showed that a small but significant proportion of cells activated by PWM (in contrast to less than 0.1% of cells stimulated by PHA) contain demonstrable intracellular immunoglobulin (Greaves and Roitt, 1968). We have resurrected this controversy and used the mouse lymphocyte model where the activities of T and B cells can be clearly defined.

ULTRASTRUCTURAL CHANGES IN T AND B CELLS ACTIVATED BY MITOGENS IN VITRO

We have analysed the patterns of ultrastructural development of lymphocytes in relation to the nature of the stimulant used and the origin of lymphoid cell (T or B). The results which will be published in detail elsewhere were in accord with our prediction that the response pattern of the cell is a reflection of the origin and immunological function of the activated cells rather than the stimulant used (Fig. 1). When T cells were activated by PWM or ConA relatively little ultrastructural development occurred, the enlarged cells being relatively devoid of rough endoplasmic reticulum but containing abundant free ribosomes or polyribosome aggregates and closely resembling those cells observed in cell-mediated immune reactions in vivo (Turk 1967). In marked contrast stimulation of B cells with PWM or endotoxin induced the appearance of cells with features characteristic of some antibody secreting cells (Harris et al 1966) with well-developed rough endoplasmic reticulum often with dilated cisternae. Some activated B cells had a very similar appearance to T lymphoblast cells which was not altogether surprising since such cells have been found to secrete IgM antibody (Harris et al 1966). Ultrastructural analysis offers only an empirical guide to functional activity and we conclude that a stimulant such as PWM added to T and B cells induces morphological changes closely paralleling those occurring in vivo when these populations respond to specific antigens.

Fig 1a Electron Microscopy of Pokeweed Activated B Cells

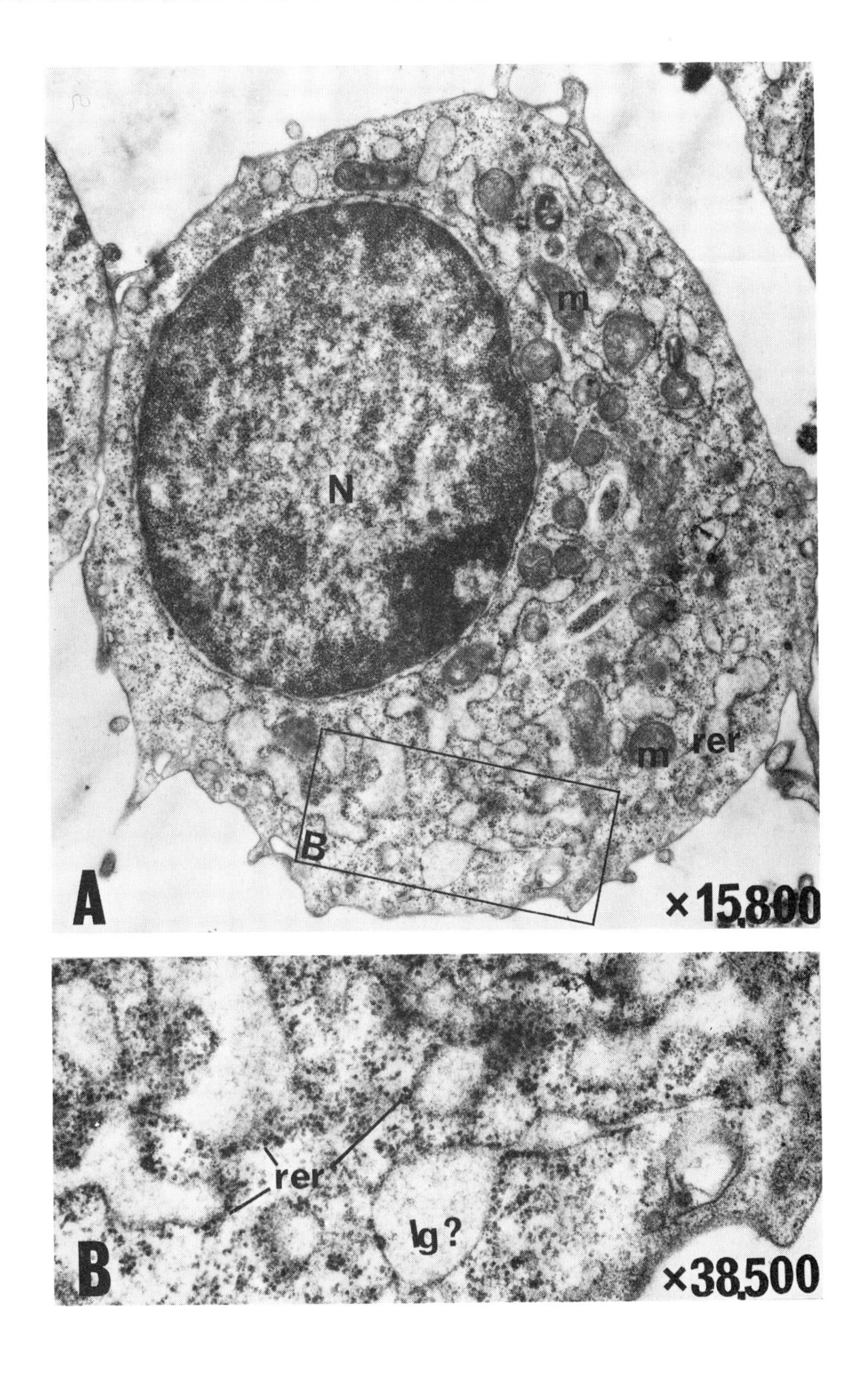

Legend. See Fig. 1b

Fig. 1b Electron Microscopy of Pokeweed Activated T cells

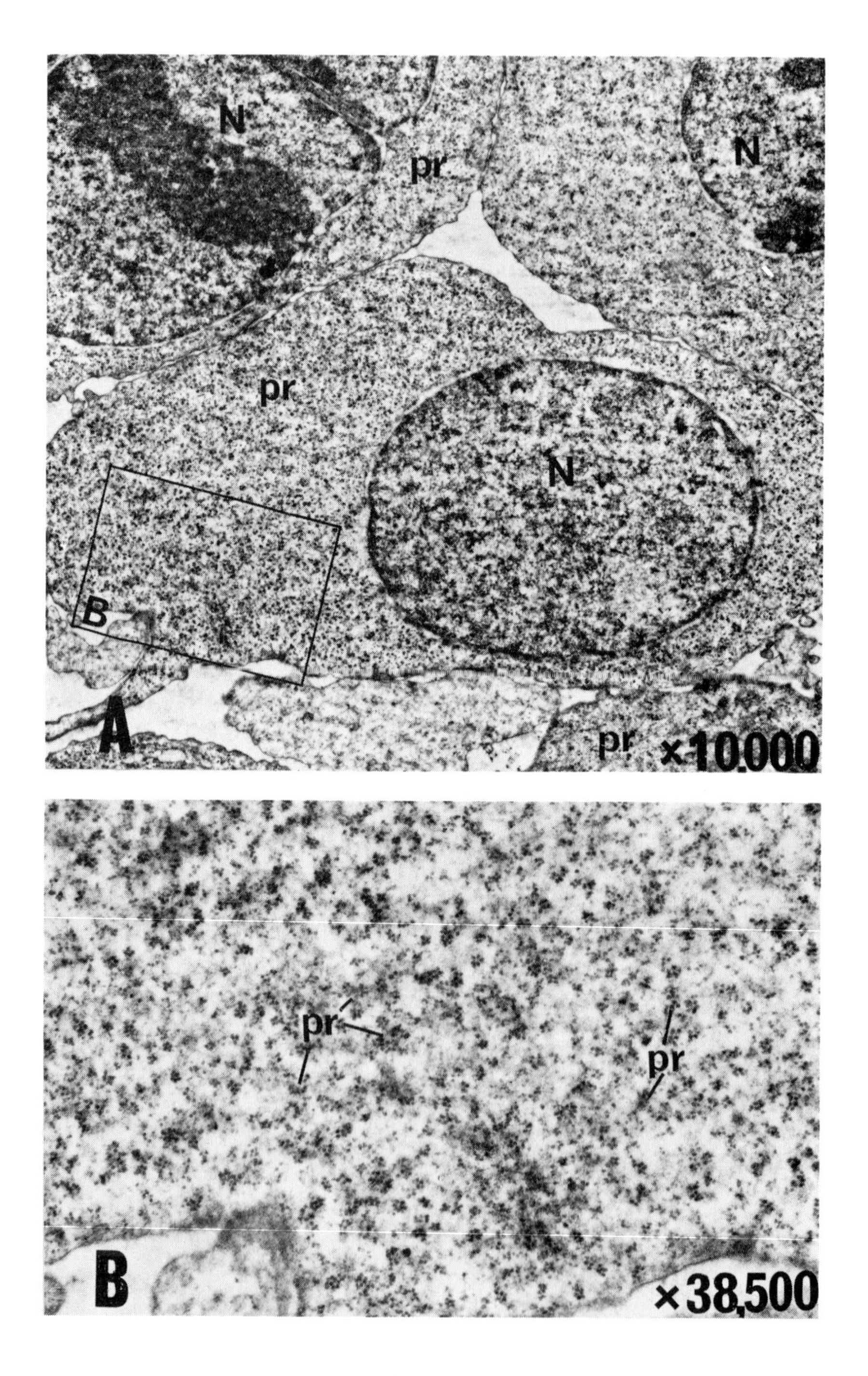

Ig = Immunoglobulin; M = Mitochondrium; N = Nucleus; pr = polyribosomes; rer = rough endoplastic reticulum

IMMUNOGLOBULIN SYNTHESIS

The synthesis and secretion of immunoglobulin was assayed by pulsing activated mouse spleen cell cultures with (^3H)-leucine for 2 hours (Parkhouse et al 1971). Cell culture fluids and cell extracts (nonidet detergent) were analysed for radioactive immunoglobulin by precipitation with anti-immunoglobulin sera and the amount of immunoglobulin detected expressed as the percentage of total protein synthesized. Fig. 3 shows the result of one such experiment and demonstrates that when B cells are activated a considerable selective increase in immunoglobulin synthesis and secretion occurs. Considerable protein synthesis also occurs when T cells are stimulated but there is no selective increase in immunoglobulin synthesis. The nature of the immunoglobulin synthesized by mitogen activated B cells has been analysed by preprepative acrylamide gel electrophoresis and appears to be exclusively 19S IgM with intracellular 7S IgMs subunits. In both these qualitative and quantitative aspects PWM stimulated B cells produce immunoglobulin in the same fashion as IgM synthesizing myeloma cells (e.g. MOPC 104E, Parkhouse and Askonas 1969). These results have been confirmed by immunofluorescent analyses of activated cultures (Fig. 3) PWM, endotoxin or ALS stimulated B cells contain cytoplasmic IgM almost entirely with Kappa light chains. In marked contrast, T cells stimulated by PHA, ConA or PWM contained no demonstrable intracellular immunoglobulin. We have not so far detected the synthesis of any immunoglobulin class other than IgM which therefore imposes some restrictions on our attempts to generalize from non-specific B cell activation to overall immunological activity of B cells.

CONCLUSIONS

Our interpretation of both the ultrastructural changes and immunoglobulin synthesis data in conjunction with considerations on the mode of action of non-specific stimulants provide the basis for the contention we wish to forward that the response induced by phytomitogens and antibodies to cell surface components can incorporate specific differentiation changes similar, if not identical to those resulting normally from antigenic activation. Since the stimulants we have described initiate responses at various distinct receptor sites on the plasma membrane and appear to bypass immunoglobulin receptors for antigens we conclude that some or possibly all lymphocytes may be relatively ignorant of precisely which binding sites on their surface have been occupied and that when an appropriate degree of cell surface cross-linkage and consequent membrane activity has been effected, a genetically predetermined response pattern is evoked.

The idea of a cell surface active inducer or releasor of the expression of a cells intrinsic response potential has both antiquity and respectibility in biology and there is a strong temptation to suppose that the same principle holds for embryonic induction phenomena (Holtzer 1970), polypeptide hormone action (Hechter and Halkerston 1964) and possibly drug action (Porter and O'Conner 1970).

The concept has the major appeal of evolutionary efficiency in that a common pool of intracellular 'second messengers' (Sutherland and Robison 1966) can be used to inter-communicate cell surface and genome. It is particularly interesting to note in this context that adrenergic receptor mechanisms and probably the adenyl cyclase-cyclic AMP system are almost certainly operative in lymphocyte responses (Hadden et al 1971, Smith et al 1971).

The work was supported by the Medical Research Council of Great Britain and The Wellcome Trust. We are grateful to Miss Prafulla Haria and Miss Frances Rose for excellent technical assistance. We thank Miss M. Shohar (Clinical Research Centre, Northwick Park, London) for preparing the electron micrographs.

Fig. 2 Effect of Phytohaemagglutinin and Pokeweed mitogen on Immunoglobulin Synthesis

	PHA	PWM	C
N Spleen			x5
B Spleen			

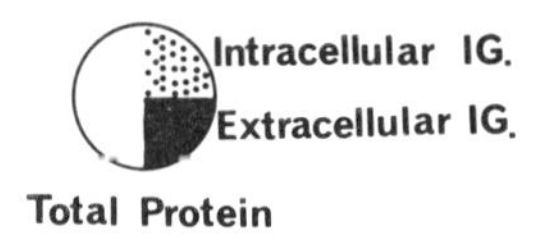

Pulse labelled 3-day cultures (see text). C = Unstimulated control.

Fig. 3 Immunofluorescent Staining of Pokeweed Activated B cells for Intracellular Immunoglobulin

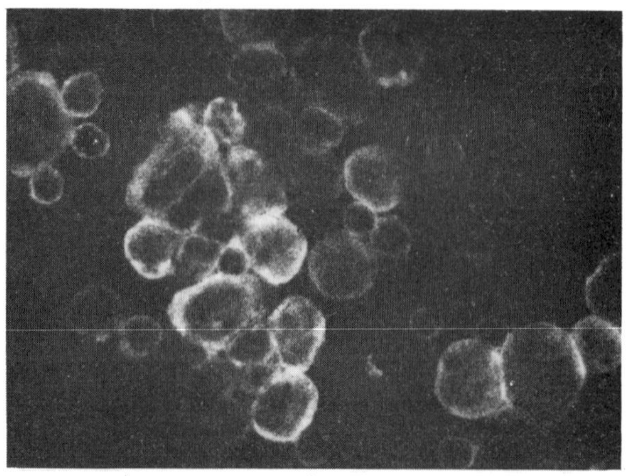

3-day cultures B cells (with Pokeweed) fixed and treated with rabbit anti-mouse IgM serum, followed by goat anti-rabbit IgG conjugated with fluorescein isothiocyanate.

REFERENCES

Allan, D., Auger, J. and Crumpton, M.J., (1971), Interaction of the plasma membranes of pig lymphocytes and thymocytes with phytohaemagglutinin, Exp. Cell Res. 66, 362.
Alm, G.V. and Peterson, R.D.A., (1969), Antibody and immunoglobulin production at the level in bursectomised-irradiated chickens, J. exp. Med. 129, 1247.
Andersson, J. and Moller, G., (1971), personal communication.
Blomgren, H. and Anderson, B., (1970), Characteristics of the immunocompetent cells in the mouse thymus: Cell population changes during cortisone induced atrophy and subsequent regeneration, Cell. Immunol. 1, 545.
Claman, H.N., Chaperon, E.A. and Tripetti, R.F., (1966), Thymus-marrow cell combinations synergism in antibody production, Proc. Soc. exp. Biol. Med. 122, 1167.
Douglas, S.D., Hoffman, P.F., Borjeson, J. and Chessin, L.N., (1967), Studies on human peripheral blood lymphocytes in vitro. III. Five structural features of lymphocyte transformation by pokeweed mitogen, J. Immunol. 98, 17.
Fanger, M.W., Hart, D.A., Wells, J.V. and Nisonoff, A., (1970), Requirement for cross-linkage in the stimulation of transformation of rabbit peripheral lymphocytes by anti-globulin reagents, J. Immunol. 105, 1484.
Feldman, M. and Basten, A., (1971), The relationship between antigenic structure and the requirement for thymus-derived cells in the immune response, J. exp. Med. 134, 103.
Gräsbeck, R., Nordman, C.T. and De La Chapelle, A. (1964), The leucocyte-mitogenic effect of serum from rabbits immunised with human leucocytes, Acta Med. Scand. Suppl. 412, 39.
Greaves, M.F.,(1970), Biological effects of anti-immunoglobulins, Transpl. Rev. 5, 45.
Greaves, M.F., (1971), Antigen recognition mechanisms, Haematologia (in press).
Greaves, M.F. and Bauminger, S., (1971), Activation of T and B lymphocytes by insoluble phytomitogens—implications for theories of cell triggering, Nature (in press).
Greaves, M.F., Bauminger, S. and Janossy, G., (1972), Lymphocyte activation. III. Binding sites for phytomitogens on lymphocyte subpopulations, Clin. exp. Immunol. (in press).
Greaves, M.F. and Roitt, I.M., (1968), The effect of phytohaemagglutinin and other lymphocyte mitogens on immunoglobulin synthesis by human peripheral blood lymphocytes in vitro, Clin. exp. Immunol. 3, 393.
Hadden, J.N., Hadden, E.M., Middleton, E. and Good, R.A., (1971), Lymphocyte blast transformation. II. The mechanism of action of alpha adrenergic receptor effects, Int. Arch. Allergy 40, 526.
Harris, T.N., Hummeler, K. and Harris, S., (1966), Electron microscopic observations on antibody producing lymph node cells, J. exp. Med. 123, 161.
Hechter, O. and Halkerston, I.D.K., (1964), On the action of mammalian hormones in 'The Hormones' 5, 697.
Holtzer, H., (1970), Proliferative and quantal cell cycles in the differentiation of muscle, cartilage and red blood cells.In 'Control mechanism in the expression of cellular phenotypes', Symp. Int. Soc. Cell Biol. 9, 69.
Ivanyi, J., Marvanova, H. and Skamene, E., (1969), Immunoglobulin synthesis and lymphocyte transformation by anti-immunoglobulin sera in bursectomised chickens, Immunology 17, 325.
Janossy, G. and Greaves, M.F., (1971), Lymphocyte activation. I. Response of T and B lymphocytes to phytomitogens, Clin. exp. Immunol. 9, 483.
Janossy, G. and Greaves, M.F., (1972), Lymphocyte activation. II. Discriminating stimulation of lymphocyte subpopulations by phytomitogens and heterologous antilymphocyte sera, Clin. exp. Immunol. (in press).
Kornfeld, R. and Kornfeld, S., (1970), The structure of a phytohaemagglutinin receptor site from human erythrocytes, J. biol. Chem. 245, 2536.
Ling, N.R., (1968), Lymphocyte Stimulation, North Holland Publ. Co., Amsterdam.
Lwoff, p.K., Rasmusen, B.A., Hoffman, P.G., Dunham, P.F., Cook, P., Parmelee, M.L. and Tosteson, D.C., (1970), Stimulation of active potassium transport in LK

sheep red cells by blood group-L-antiserum, J. Memb. Biol. 3, 1.
Mäkelä, O. and Cross, A.M., (1970), The diversity and specialisation of immunocytes, Progr. Allergy 14, 145.
McKenzie, J.M., (1968), Humoral factors in Graves' disease, Physiol. Rev. 48, 252.
Meuwissen, H.J., Stutman, O. and Good, R.A., (1969), Functions of the lymphocytes, Seminars in Haematology 6, 28.
Möller, G., (1970), Editor Antigen binding lymphocyte receptors, Transpl. Rev. 5.
Nowell, P.C., (1960), Phytohaemagglutinin: an initiator of mitosis in cultures of normal human leucocytes, Cancer Res. 20, 462.
Oka, T. and Topper, Y.J., (1971), Insulin-Sepharose and the dynamics of insulin action, Proc. natl. Acad. Sci. 68, 2066.
Oppenheim, J.J., (1968), Relationship of in vitro lymphocyte transformation to delayed hypersensitivity in guinea-pigs and man, Fed. Proc. 27, 21.
Pantelouris, E.M., (1968), Absence of thymus in a mouse mutant, Nature 217, 370.
Parkhouse, R.M.E. and Askonas, B.A., (1969), Immunoglobulin M biosynthesis intracellular accumulation of 7S subunits, Biochem. J. 115, 163.
Parkhouse, R.M.E., Janossy, G. and Greaves, M.F., (1971), Selective stimulation of IgM synthesis in mouse B lymphocytes by pokeweed mitogen, Nature (in press).
Pastan, I. and Perlman, R.L., (1971), Cyclic AMP in metabolism, Nature New Biology 229, 5.
Peavy, D.L., Adler, W.H. and Smith, R.T., (1970), The mitogenic effects of endotoxin and stephylococcal enterotoxin B on mouse spleen cells and human peripheral lymphocytes, J. Immunol. 105, 1453.
Pick, E., Brostoff, J., Kresjci, J. and Turk, J.L., (1970), Interaction between 'sensitised lymphocytes' and antigen in vitro. II. Mitogen-induced release of skin reactive and macrophage migration inhibitory factors, Cell. Immunol. 1, 92.
Porter, R. and O'Conner, M. (Editor), 1970, Molecular properties of drug receptors Ciba Foundn. Churchill Ltd.
Powell, A.E. and Leon, M.A., (1970), Reversible interactions of human lymphocytes with the mitogen Concanavalin A, Exp. Cell. Res. 62, 315.
Raff, M.C., (1970), Two distinct populations of peripheral lymphocytes in mice distinguishable by immunofluorescence, Immunol. 19, 637.
Raff, M.C., (1971), Surface antigenic markers for distinguishing T and B lymphocytes in mice, Transpl. Rev. 6, 52.
Raff, M.C., Sternberg, Taylor, R.B., (1970), Immunoglobulin determinants on the surface of mouse lymphoid cells, Nature 225, 553.
Riethmüller, G., Riethmüller, D., Stein, M. and Mausen, P., (1968), The in vivo and in vitro properties of intact and pepsin-digested heterologous antimouse thymus antibodies, J. Immunol. 100, 969.
Ripps, C.S. and Hirschhorn, K., (1967), The production of immunoglobulin by human peripheral lymphocytes in vitro, Clin. exp. Immunol. 2, 377.
Roitt, I.M., Greaves, M.F., Torrigiani, G., Brostoff, J. and Playfair, J.H.L., (1969), The cellular basis of immunological responses, Lancet ii, 367.
Sell, S., (1970), Development of restrictions in the expression of immunoglobulin specificities by lymphoid cells, Transpl. Rev. 5.
Smith, C.W. and Hollers, S.C., (1970), The pattern of binding of fluorescein-labelled Concavavalin A to the motile lymphocyte, J. Reticulo. End. Soc. 8, 458.
Smith, J.W., Steiner, A.L., Newberry, W.M., and Parker, C.W., (1971), Cyclic adenosine 3', 5'-monophosphate in human lymphocytes. Alterations after phytohaemagglutinin stimulation, J. clin. Invest. 50, 432.
Stobo, J.D., Rosenthal, A.S., and Paul, W.E., (1971), Functional heterogeneity of murine lymphoid cells. I. Responsiveness to and surface binding of Concanavalin A and phytohaemagglutinin, J. Immunol.(in press).
Sutherland, E.W. and Robison, G.A., (1966), The role of cyclic-3', 5'-AMP in responses to catecholamines and other hormones, Pharmacol. Rev. 18, 145.
Taylor, R.B., Duffus, P., Raff, M.C. and DePetris, S., (1971), Redistribution and pinocytosis of lymphocyte surface immunoglobulin molecules induced by anti-

immunoglobulin antibody, Nature New Biology 233, 225.
Turk, J., (1967), Delayed Hypersensitivity. North Holland Research Monograph 4.
Vitetta, E.S., Baur, S. and Uhr, J.W., (1971), Cell surface immunoglobulins. II. Isolation and characterisation of immunoglobulin from mouse splenic lymphocytes, J. exp. Med. 134, 242.
Woodruff, M.F.A., Anderson, N.F. and Abaza, H.M., (1967), Experiments with anti-lymphocytic serum. In 'The Lymphocyte in Immunology and Haemopoiesis", Ed. J.M. Jaffey. E. Arnold Publ., Lond, p. 286.
Woodruff, M.F.A., Reid, B. and James, K., (1967), Effect of antilymphocytic antibody and antibody fragments on human lymphocytes in vitro, Nature, 215, 591.

ANALOGIES BETWEEN TRIGGERING MECHANISMS IN IMMUNE AND OTHER CELLULAR REACTIONS

A.C. ALLISON
Clinical Research Centre, Watford Road, Harrow, Middlesex

One of the reasons for holding a meeting on "Cell Interactions" is to find out whether cross-fertilization is possible. Can anything be learned about the functioning of the nervous system, for example, from what is known about cellular interactions in immune responses? In view of the obvious differences, this seems at first sight an unprofitable exercise; but since all cells have a similar basic construction and a restricted repertoire of reactions, the search for analogies may be illuminating. My colleagues and I have been analysing trigger mechanisms in three cell types: lymphocytes, macrophages and mast cells. Some generalizations are emerging, and we believe that they may apply also to other cells, including those of the nervous and endocrine systems.

Cross-linking of protein units in membranes

The first phenomenon that I want to discuss is the cross-linking of protein units in membranes, of which the patch formation in lymphocytes described by Raff and his associates at this meeting is an example. Recent evidence suggests that cell membranes consist of a lipid bilayer, which is a moderately viscous fluid at body temperature, with intercalated protein units. These float like icebergs, retained within the membrane by laterally directed hydrophobic groups interacting with hydrocarbon chains of lipids. The hydrophilic surface groups of membrane proteins, including attached carbohydrates such as blood-group antigens, extend outwards into the extracellular fluid. The protein units have translational mobility: they are free to diffuse laterally in the planar fluid domain of the membrane. Three examples of such translational mobility illustrate the principle. One is the mixing of surface antigens following fusion of two different cell types described by Frye and Edidin (1970). The second is the observations of da Silva et al. (1971) on particles 85 Å in diameter visualized by freeze-etching of human erythrocyte ghosts. The particles appear to consist of glycoproteins containing the A antigenic sites. In untreated cells the particles are distributed at random over the erythrocyte membrane, but after various treatments they are found to be aggregated. This presumably results from translational movement of membrane protein units which come together to form patches, as described also for immunoglobulin molecules on lymphocyte membranes by Taylor et al. (1971). This process, illustrated in fig. 1, could be brought about by bivalent antibody (but not univalent Fab) or by any other agent that cross-links membrane proteins or glycoproteins.

Patch formation occurs in a few minutes at room temperature or in the presence of metabolic inhibitors, so there is no reason to believe that it requires anything more than diffusion of protein units within membranes and trapping when the units collide. However, the coming together of patches to form a cap at one side of the cell occurs more slowly, and is facilitated by cell mobility. This involves gliding movement of the cell in relation to its membrane and the substratum, so that membrane constituents come together towards the uropod of a lymphocyte or the equivalent polar region in other cells. Since cell movements appear to depend on contraction of actomyosin-like microfilaments which lie beneath the membrane, and may be inserted into membrane proteins, it is conceivable that contraction of the microfilaments may help to bring together patches, as illustrated in fig. 1d. However they are brought together, patches are further cross-linked to form the cap, and endocytosis of the cap follows.

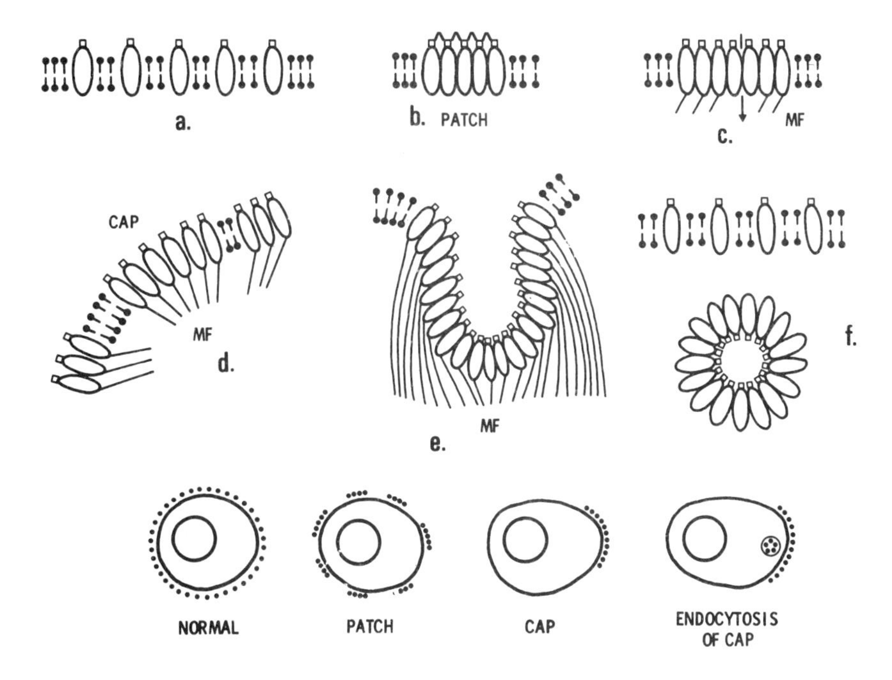

Fig. 1. Diagram of plasma membrane proteins with hydrophilic groups (squares) facing outwards, arranged at random in a lipid bilayer (a). When the proteins are crosslinked they come together to form a patch (b). This may increase permeability of Ca^{2+} and other ions (c). Patches come together to form a cap at one pole (d). The cap is pulled inwards by microfilaments (MF) (e) to form an endocytic vacuole containing membrane proteins (f).

Treatment of cells with cytochalasin B, which inhibits microfilament contraction, retards cap formation (but does not abolish it) and prevents endocytosis of the cap (Taylor et al., 1971).

Agents able to cross-link membrane protein units

Bivalent antibodies with their Fab moieties directed towards membrane protein or glycoprotein units could cross-link them, as illustrated in fig. 2a.

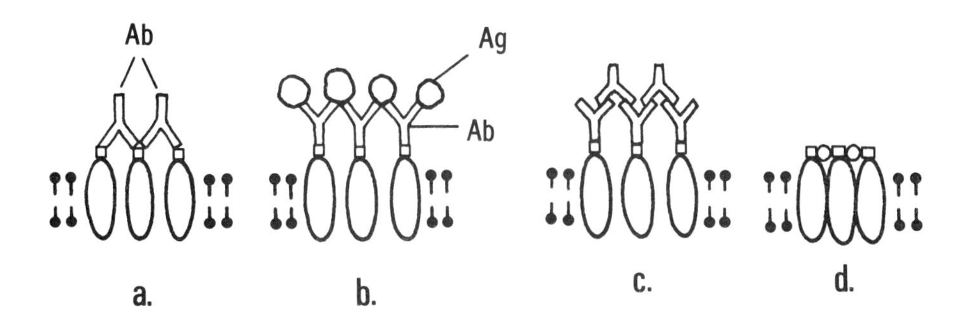

Fig. 2. Mechanisms of cross linkage of membrane proteins: (a) antibody combining through Fab with membrane proteins; (b) cell-bound antibody (Ab) cross-linked by antigen (Ag); (c) cell-bound antibody cross-linked by anti-immunoglobulin; (d) proteins cross-linked by small molecules.

The same is true of bivalent fragments, $F(ab')_2$, prepared from these antibodies, but not of univalent, Fab, fragments. Immunoglobulin molecules in the membrane, or cell-bound immunoglobulins attached through their Fc ends to membrane receptors, would be cross-linked only in the presence of antigen or of a second antibody reacting with the surface immunoglobulin (fig. 2b and 2c).

Evidence is accumulating that cross-linking membrane proteins or glycoproteins can trigger transformation of lymphocytes, release of histamine and other mediators from mast cells and endocytosis in macrophages. Human and mouse peripheral blood and thymus lymphocytes are transformed in the presence of heterologous anti-lymphocytic sera, or $F(ab')_2$ fragments but not by Fab (Woodruff et al., 1967; Riethmuller et al., 1968). Goat anti-rabbit immunoglobulin or the corresponding $F(ab')_2$ brings about transformation of rabbit lymphocytes but Fab does so only when subsequently exposed to divalent antibody against goat immunoglobulin (Fanger et al., 1970). Cell-bound antibody does not trigger histamine release from mast cells unless antigen is added; however chemically cross-linked IgE molecules are able to induce acute hypersensitivity reactions in the absence of antigen (Ishizaka and Ishizaka, 1970). Divalent antibodies against IgE induce histamine release (reversed-type erythema-wheal reactions) whereas univalent antibody fragments do not (Ishizaka and Ishizaka, 1969). The formation of patches and caps in human basophil membranes, demonstrated by immunoferritin labelling in the presence of antibody against IgE, has been described by Sullivan et al. (1971). Attachment of immunoglobulin molecules to macrophages is not followed by endocytosis, but attachment of antigen-antibody complexes, or of immunoglobulin molecules polymerized by heat, stimulates endocytosis (Allison and Davies, 1971).

We have observed the formation of patches and caps in human peripheral blood lymphocytes exposed to concanavalin A, which stimulates transformation. The basic compound 48/80, which in the presence of Ca^{2+} stimulates histamine release from mast cells, appears to induce the same changes in the surface of these cells, suggesting that compounds of low molecular weight can cross-link membrane protein or glycoprotein units (fig. 2d). Since the release mediated by cell-bound antibody or compound 48/80 is inhibited by cytochalasin (Orr et al., 1971), we conclude that it depends on contraction of microfilaments beneath the plasma membrane. A simple model would be that aggregation of membrane protein units to form a patch on the surface of a mast cell allows Ca^{2+} ions to enter the peripheral cytoplasm and initiate contraction of microfilaments. The contraction draws apart infoldings of the membrane retaining granules or creates spaces in a "cage" of microfilaments beneath the plasma membrane, so allowing the granules to fuse with the membrane and be discharged.

Dual effector systems

Elsewhere (Allison, 1972) evidence has been presented in support of the view that there are two major effector systems for hormones and drugs. In one, attachment of an agonist to receptors on the plasma membrane of a target cell brings about an increase in the concentration of Ca^{2+} in the cytoplasm (the ions entering from the extracellular medium or being released from the sarcoplasmic reticulum). Contraction of an actomyosin system follows, leading to shortening of smooth muscle fibres, exocytosis of hormones or other pharmacologically active mediators or endocytosis of colloid by thyroid epithelial cells, of particles coated with antibody by macrophages, and so forth. This can conveniently be termed the contractile effector system, which results in movement of the target cell. This can also be termed the A effector system.

The second effector system involves activation of adenyl cyclase, with the production of e.g. 3',5'-cyclic adenosine monophosphate (cAMP). This nucleotide in turn activates protein kinases which phosphorylate a variety of substrates and produce a wide range of metabolic changes in target cells, including synthesis of hormones. Turning on cAMP synthesis can be conveniently termed the metabolic effector system, which sometimes works synergistically with the contractile effector system (for example, in the thyroid gland stimulated by thyrotrophic hormone). At other times the effector systems are antagonistic; for instance, in mast cells formation of cAMP inhibits granule release, which is brought about by the A effector system.

If aggregation of membrane protein or glycoprotein units is a widespread trigger mechanism, it should lead to activation of the A or B effector systems. Activation of the A system is analogous to the membrane depolarization in smooth muscle following attachment of acetylcholine or another agonist to membrane receptors. The question can therefore be asked whether formation of aggregates could play any role in facilitating cation permeability through the plasma membrane, which is responsible for membrane depolarization.

Cross-linking of neuromuscular receptors for transmitter substances

Dr Kuffler at this meeting has reminded us that in normal muscle and ganglion cells receptors for neurotransmitters are concentrated in the postjunctional and postsynaptic regions whereas after denervation they are also found spread over the plasma membrane. A possible explanation would be that newly synthesized receptor proteins are randomly distributed in the membrane, but they are gathered together in a patch by a cross-linking agent in the postjunctional and postsynaptic regions (fig. 3).

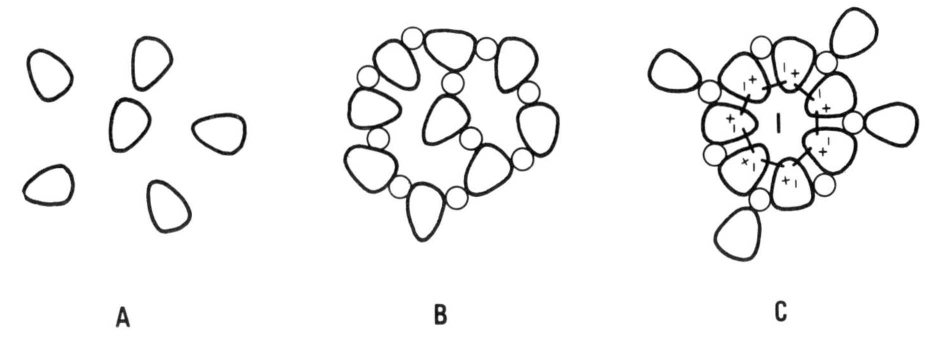

Fig. 3. (A) Cholinesterase receptors randomly distributed in the membrane of nerve or muscle. (B) Receptors cross-linked in the postsynaptic or postjunctional regions. (C) Further cross-linking of receptors by acetylcholine to form an ion-conducting channel (I).

Since the patch of concentrated receptors in these regions persists after denervation, it seems likely that the cross-linking agent is of a more durable material than neurotransmitter. I suggest that a specific product, probably a protein, with the capacity to react specifically with and cross-link receptor proteins, is secreted at the nerve ending. The receptors would then become concentrated beneath the ending, but oriented there more or less at random (fig. 3b). Alternatively, cholinesterase molecules, which are thought to be located in the basement membrane overlying the muscle surface rather than in the membrane itself, could interact with and immobilize receptor molecules in the membrane; this might account for the identity in the number of cholinergic receptor sites and cholinesterase active sites reported in mammalian muscle, the electric organs of electric fish and electric eels (see Nature 234, 173, 1961).

A further speculation would be that acetylcholine (or another neurotransmitter in other synapses) is able to make additional cross-links of receptor proteins and so establish a conducting channel through which ions can pass rapidly to generate the end-plate or postsynaptic potential (fig. 3c). This argument is based on increased ion transport through lipid membranes in the presence of peptide ionophores (reviewed by Cherry, 1972). In most cases, the electrical properties of the membrane in the presence of these compounds are more or less linear, e.g. with the potassium ionophore valinomycin. However, with some ionophores highly non-linear properties are produced and under suitable conditions many of the prominent characteristics of nerve membranes can be observed in the model system. From a kinetic analysis of one of these ionophores, alamethicin, Cherry concludes that six molecules of the ionophore have to come together, probably in the plane of the membrane, to form a conducting channel.

Counts of cholinergic receptors show that they form a fairly dense mosaic in the postjunctional membrane (Barnard et al., 1971). Acetylcholine has a molecular structure well suited to cross-linking membrane proteins at the interface between hydrophilic and hydrophobic regions. Clothia (1970) has presented evidence that the essential structure of nicotinic agonists is a quaternary group and a carbonyl group, while the essential structure for muscarinic agonists includes a quaternary group and an appropriately located methyl group, separated by a short hydrocarbon chain.

Changeux *et al.* (1969) have proposed that acetylcholine induces a configurational change in a receptor protein, converting it into an ionophore. Further observations are required to decide between the single receptor and multi-receptor models. Suitable experimental material is being developed. Del Castillo *et al.* (1967) and Parisi *et al.* (1970) have prepared membranes containing proteolipids from brain and the electric organ of *Electrophorus*, and found transient increases in ion conductances after addition of acetylcholine. Refinement of such systems should establish whether their electrical properties are linear or non-linear. Cross-linking receptor proteins by another mechanism, such as antibody, might result in increased cation permeability and postsynaptic and postjunctional potentials.

The presence of such sterically complementary products in nerve and muscle cell membranes could play a major role in the morphogenesis of the nervous system. Thus, growing axons of nerve cell A have in their membranes groups that can interact specifically with the constituents of the membrane of nerve cell B. Contact is established, the axon remains apposed and cross-linking substance from nerve cell A gathers together complementary receptors in the membrane of cell B, so that a functional synapse is established. If cell B synthesizes two or more receptors for different neurotransmitters, each one would become concentrated in the appropriate postsynaptic region. If cell B lacks any groups complementary to axon A, axon migration continues, as in the case of nerve fibres which fail to make contact with ganglion cells.

Endocytosis as a trigger

Cap formation in lymphocytes is normally followed by endocytosis. D. Webster and I have found that cytochalasin B, which inhibits endocytosis, blocks transformation of human peripheral blood lymphocytes by phytohaemagglutinin or concanavalin A. Hence endocytosis itself, rather than cap formation, may play a role in triggering transformation. Observations that stimulating agents attached to large particles trigger transformation are not necessarily in conflict with this view, since this need not block endocytosis. Perhaps endocytosis activates adenyl cyclase in lymphocytes, and this mediates the metabolic effects of transformation, as several authors have suggested. Further evidence in support of both of these propositions is, however, required.

REFERENCES

Allison, A.C. (1971) Role of membranes in effector systems for hormone and drug action. Phys. Chem. Lipids (in press).

Allison, A.C. and Davies, P. (1971) The control of lysosomal enzyme synthesis and effects of steroids. In: Effects of drugs on cellular control mechanisms (ed. R.B. Freedman) Churchill: London.

Allison, A.C., Davies, P. and de Petris, S. (1971) Role of contractile microfilaments in macrophage movement and endocytosis. Nature New Biology, 232, 153.

Barnard, E.A., Wisckowski, J. and Chu, T.H. (1971) Cholinergic receptor molecules and cholinesterase molecules at mouse skeletal muscle junctions. Nature, Lond., 234, 207.

Changeux, J.P., Podleski, T. and Meunier, J.C. (1969) On some structural analogies between acetylcholinesterase and the macromolecular receptor of acetylcholine J. gen. Physiol., 54, 225S.

Cherry, R. (1972) Model membranes and excitability. Phys. Chem. Lipids (in press)

Clothia, C. (1970) Interaction of acetylcholine with different cholinergic nerve receptors. Nature, Lond., 225, 36.

da Silva, P.P., Douglas, S.D. and Branton, D. (1971) The localization of A antigen sites on human erythrocyte ghosts. Nature, Lond. (in press).

del Castillo, J., Rodriguez, A. and Romero, C.A. (1967) Pharmacological studies on an artificial transmitter-receptor system. Ann. N.Y. Acad. Sci., 144, 803.

Fanger, M.W., Hart, D.A., Wells, V.J. and Nisonoff, A. (1970) Requirement for cross-linkage in the stimulation of transformation of rabbit peripheral lymphocytes by antiglobulin reagents. J. Immunol., 105, 1484.

Frye, C.D. and Edidin, M. (1970) The rapid intermixing of cell surface antigens after formation of mouse-human heterokaryons. J. cell Sci., 7, 319.

Ishizaka, K. and Ishizaka, T. (1969) Immune mechanisms of reversed type reaginic hypersensitivity. J. Immunol., 103, 588.

Ishizaka, K. and Ishizaka, T. (1970) Biological function of γE antibodies and mechanisms of reaginic hypersensitivity. Clin. exp. Immunol., 6, 25.

McFarland, W. and Schechter, G.P. (1969) E-M studies of lymphocytes in immunological reactions in vitro. Blood, 34, 832.

Miledi, R. and Potter, L.T. (1971) Acetylcholine receptors in muscle fibres. Nature, Lond., 233, 599.

Nayman, J., Datta, S.P. and de Boer, W.G.R.M. (1967) Renal transplantation in cattle twins. Nature, Lond., 215, 741.

Orr, T.S.C., Allison, A.C. and Hall, D.E. (1971) A role of contractile microfilaments in the release of histamine from mast cells. Nature, Lond. (in press).

Osunkoya, B.O., Williams, A.I.O., Adler, W.H. and Smith, R.T. (1970) Studies on the interaction of phytomitogens with lymphoid cells. Afr. J. med. Sci., 1, 3.

Parisi, M., Rivas, E. and de Robertis, E. (1971) Conductance changes produced by acetylcholine in lipidic membranes containing a proteolipid from Electrophorus Science, 172, 56.

Riethmüller, G., Riethmüller, D., Stein, H. and Hansen, P. (1968) In vivo and in vitro properties of intact and pepsin-digested heterologous anti-mouse thymus antibodies. J. Immunol., 100, 969.

Smith, C.W. and Hollers, J.C. (1970) The pattern of binding of fluorescein-labelled concanavalin A to the motile lymphocyte. J. Reticuloendothel. Soc., 8, 458.

Sullivan, A.L., Grimley, P.M. and Metzger, H. (1971) Electron microscopic localization of immunoglobulin E on the surface membrane of human basophils. J. exp. Med. (in press).

Taylor, R.B., Duffus, W.P.H., Raff, M.C. and de Petris, S. (1971) Redistribution and pinocytosis of lymphocyte surface immunoglobulin molecules induced by anti-immunoglobulin antibody. Nature New Biology, 233, 225.

Woodruff, M.F.A., Reid, B. and James, K. (1967) Effect of antilymphocytic antibody and antibody fragments on human lymphocytes in vitro. Nature, Lond., 215, 591.

Manuscript received after December 1, 1971

CHANGES IN LYMPHOCYTE CIRCULATION AFTER HYDROCORTISONE TREATMENT

J.JOHN COHEN

National Institute for Medical Research, Mill Hill,

London NW7 1AA, England

Hydrocortisone is becoming an important tool for the immunologist interested in cell interactions, since it appears to have differential effects upon the various populations of lymphocytes, at least in vivo, in the mouse (see the paper by Claman and Moorhead, this volume). It is currently believed that mature T-cells, whether in the thymus or the periphery, are hydrocortisone-resistant. Most of the lymphocytes in mouse peripheral blood are T-cells (Raff and Owen, 1971) and thus would be expected to be hydrocortisone-resistant; nevertheless, within 2 hours of an intraperitoneal injection of 5 mg of hydrocortisone acetate, blood lymphocyte counts are 10% or less of normal. Thus it was considered probable that these cells were sequestered, rather than killed, by hydrocortisone.

Examination of the bone marrow of hydrocortisone-treated, but not of normal mice, revealed considerable activity of the type associated with T-cells, that is the ability to respond to PHA (Claman and Moorhead, this volume) and to initiate graft-versus-host reactions (Cohen, Fischbach and Claman, 1970). Furthermore, bone marrow of treated mice transferred to irradiated syngeneic hosts produced large numbers of plaque-forming cells (PFC) in the recipient spleens (Cohen, in press). Thus, bone marrow behaves as if it had acquired T-cells, either endogenously (through induced differentiation) or from without. That the ability to transfer PFC is in fact due to T-cells was demonstrated by treating the bone marrow with anti-θ serum and complement, after which it was inactive; full activity was restored with normal thymus. To establish that T-cells enter the bone marrow from the circulation after hydrocortisone treatment, blood lymphocytes were labelled with ^{51}Cr and injected into normal recipients, some of which then received hydrocortisone. In 24 hours, no cells were found in the femurs of normals, while up to 5% had homed to the femurs of treated mice.

If sequestration of T-cells after corticosteroid treatment is a phenomenon which occurs in man (and the course of lymphopenia and recovery in man suggests strongly that it might) then it is essential to know what role sequestered cells may play in the immunological responses of the patient undergoing immunosuppressive therapy. For example, can ALS kill sequestered T-cells, or do they escape its effects, possibly to reappear after its concentration has fallen below effective levels?

It is concluded from these experiments that T-cells of the recirculating pool are induced by hydrocortisone to enter the bone marrow. Experiments are in progress to determine whether this is an effect on the lymphocyte itself, or perhaps on some bone marrow discretionary site equivalent to the post-capillary venule of the lymph node. The fate of the T-cell within the marrow is also being investigated. It appears that T-cell-B-cell interaction in the usual sense cannot take place within the bone marrow itself, so that even with repeated immunization, PFC are not found in the marrow of treated animals. This may be due to the presence of an inhibitor cell (R. Miller, personal communication). It appears that the marrow of hydrocortisone-treated mice may provide an excellent site in which to study the interactions of stem, precursor, helper and inhibitor cells.

REFERENCES

Cohen, J.J., Fischbach, M. and Claman, H.N. (1970) Hydrocortisone resistance of graft-vs-host activity in mouse thymus, spleen and bone marrow. J. Immunol. 105, 1146.

Raff, M.C. and Owen, J.J.T. (1971) Thymus-derived lymphocytes: their distribution and role in the development of peripheral lymphoid tissues of the mouse. Eur. J. Immunol. 1, 27.

Manuscript received after December 1, 1971

IMMUNOCYTE INTERACTIONS IN VITRO

G. DORIA, G. AGAROSSI, and S. DI PIETRO
CNEN-Euratom Immunogenetics Group, Laboratory
of Animal Radiobiology, C.S.N.-Casaccia (Rome), Italy

The existence of cell cooperation in the immune response has been well established since the finding (Claman, 1966) that in lethally irradiated mice a mixed inoculum of bone marrow and thymus cells allows greater production of hemolysins against sheep red blood cells (RBC) than can be accounted for by summating the activities of either cell population alone. In mice immunized with sheep RBC it has been demonstrated that: 1) Antibody synthesis occurs in marrow-derived (B) cells but not in thymus-derived (T) cells (Davies, 1967; Mitchell, 1968; Miller, 1968). 2) T cells have to react specifically with antigen before the cooperation with B cells can take place (Mitchell, 1968; Miller, 1969). The nature of the interaction between T and B lymphocytes is not known. Several mechanisms have been proposed (Doria, 1971a). Thus far, experimental evidence (Mitchison, 1971) favors the possibility that the T cell binds antigen by receptor sites specific for some antigenic determinants and focusses other determinants of the same antigen molecule onto specific receptors of the B cell. Hence, the former cell traps antigen so as to form a local concentration at some critical site on the latter cell which is then triggered to antibody synthesis.

It was felt that the interaction between T and B immunocytes could be highlited by studies in vitro. The Mishell and Dutton technique (Mishell, 1967), whereby unprimed mouse spleen cells can be stimulated in vitro with sheep RBC to give rise to antibody-forming cells, was used to investigate whether the interaction between T and B cells can occur in vitro. The immune response was evaluated by the Jerne and Nordin technique (Jerne, 1963) and expressed as number of hemolytic plaque-forming cells (PFC) per culture. Using C3HeB/Fe or DBA/2 mice, cooperation between thymocytes from normal mice and splenocytes from neonatally thymectomized mice or from thymectomized chimeras (mice thymectomized in adult life, lethally irradiated, and grafted with isogenic bone marrow cells) could be demonstrated in vitro (Doria, 1970). Similar results were obtained with (C57Bl/10xDBA/2)F1 mice in the present study. As shown in Fig. 1, the addition of sheep RBC to mixed cell cultures of $1x10^7$ nucleated thymocytes from normal mice and $1.5x10^7$ nucleated splenocytes from neonatally thymectomized mice elicited an immune response comparable to that of $1.5x10^7$ nucleated normal spleen cells. When the two cell populations were stimulated in separate cultures no response of thymocytes was ever observed, while the response of splenocytes from thymectomized mice was within the range of the PFC background in unstimulated control cultures. Appropriate controls ruled out that higher responses in

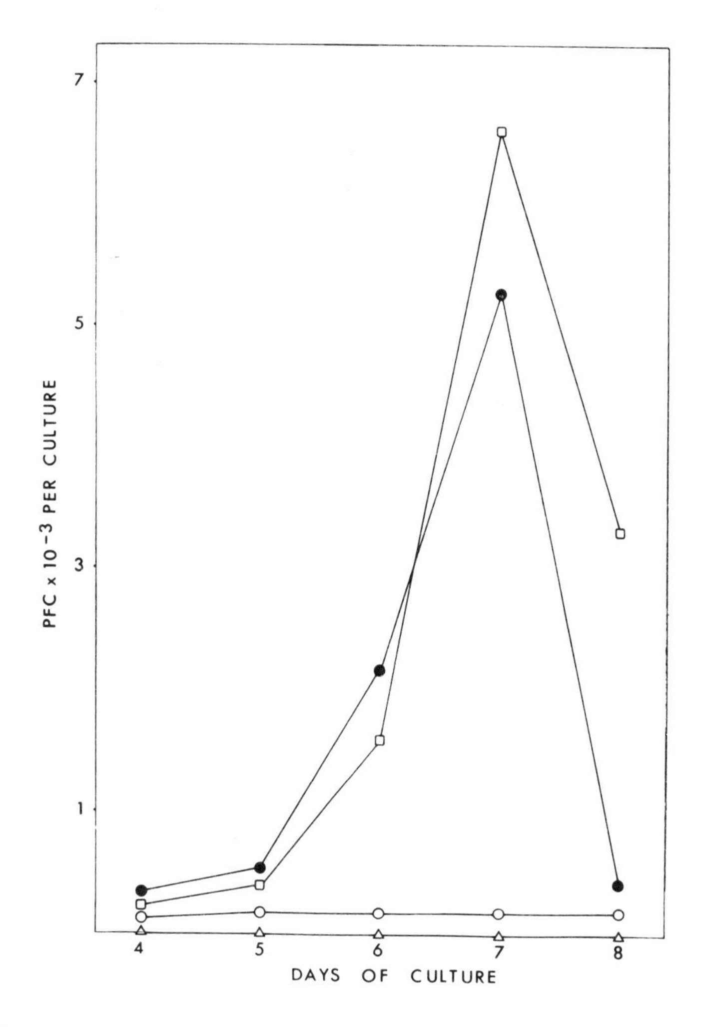

Fig. 1 Immune response resulting from in vitro cell cooperation. Spleen cells from normal mice, ● ; spleen cells from neonatally thymectomized mice, ○ ; thymus cells from normal mice, △; spleen cells from neonatally thymectomized mice and thymus from normal mice, □ . Sheep RBC added to all cultures.

mixed cultures were simply the result of greater cell density.

If receptor sites are involved in the cooperation between T and B cells, treatment of these cells with anti-receptor antibodies prior to culture may prevent cell interaction and the resulting immune response. Assuming that cell receptors are molecules shared by the serum, rabbits were immunized with normal mouse whole serum. Preparation of the rabbit antiserum and cell treatment have been described in detail elsewhere (Doria, 1971b). Cells were incubated with complement-inactivated rabbit antiserum and then extensively washed prior to culture with sheep RBC. The ability of the rabbit antiserum to impair the response of normal spleen cells is shown in Fig. 2. This

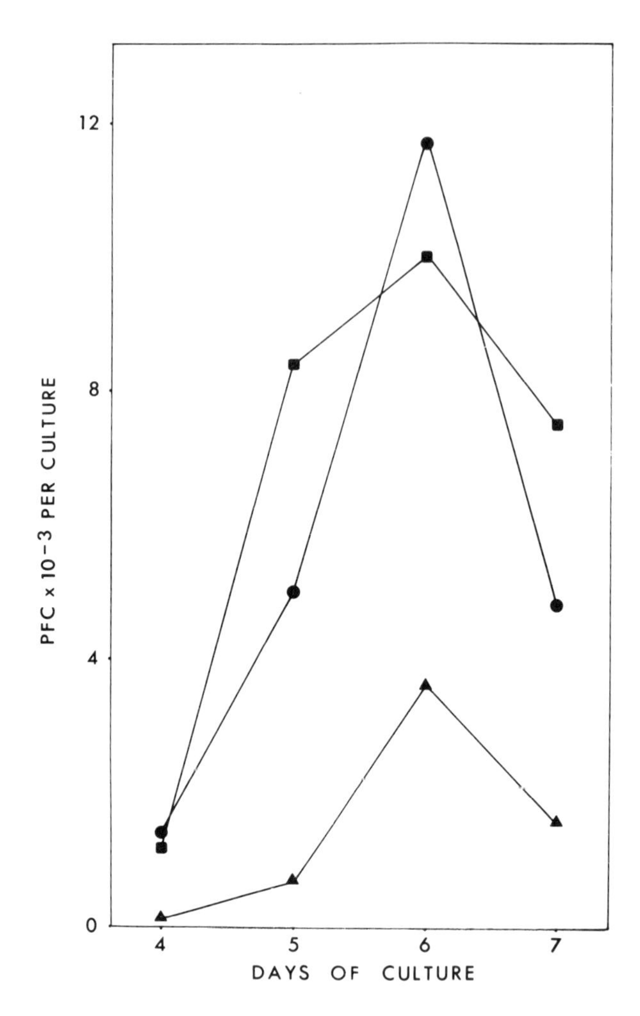

Fig. 2 Inhibition of the immune response in vitro. Spleen cells from normal mice were pretreated with balanced salt solution, ● ; rabbit normal serum, ■ ; or rabbit antiserum anti-mouse whole serum, ▲ . Sheep RBC added to all cultures.

indicates that spleen cells possess receptor sites antigenically identical to or crossreactive with molecules of the normal serum. Fig. 3 shows the results of experiments in which thymocytes, splenocytes from neonatally thymectomized mice, or both were similarly treated with the rabbit antiserum prior to culture. The immune response could be inhibited only when the B cells were pretreated with the antiserum. Incubation of T cells with the antiserum did not interfere with the cell cooperation events leading to antibody formation. The same pattern was observed when T and B cells were pretreated with rabbit anti-mouse Ig antisera obtained from other laboratories. Thus, T cells seem to lack demonstrable receptor molecules such as those detected by the same reagent on B cells and present in the normal serum.

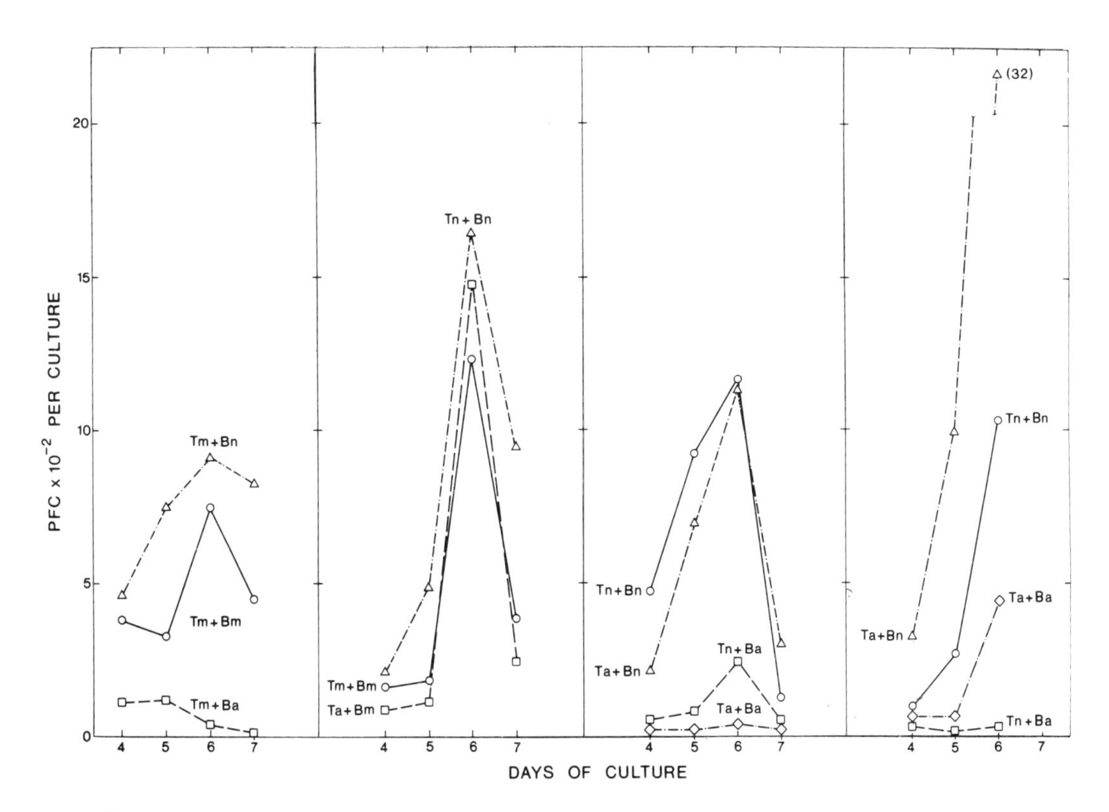

Fig. 3 Immune response resulting from in vitro cell cooperation. T: thymus cells from normal mice; B: spleen cells from neonatally thymectomized mice. Cells were pretreated with balanced salt solution, m; rabbit normal serum, n; or rabbit antiserum anti-mouse whole serum, a. Sheep RBC added to all cultures.

The present findings, although in agreement with other results showing absence of Ig on thymocytes (Raff, 1970; Pernis, 1970; Rabellino, 1971; Paraskevas, 1971), do not rule out the possibility that T cells possess receptor sites for antigen but interact with B cells by some entangled mechanism that does not involve receptors. This hypothesis is corroborated by the results of the following experiment. The rabbit antiserum anti-mouse whole serum was absorbed with thymus cells, spleen cells from normal mice, or spleen cells from neonatally thymectomized mice. Then, normal spleen cells were treated with the unabsorbed or absorbed rabbit antiserum prior to culture and antigenic stimulation. Preliminary results showed that the thymocytes are as efficient as the splenocytes in absorbing those rabbit antibodies that inhibit the immune response of normal spleen cells. This indicates that in fact T cells possess receptor sites, which may account for the thymocyte ability to bind antigen in a specific way (Dwyer, 1970; Modabber, 1970; De Luca, 1970).

The helper function of the T cell might not be tied to a mechanism of cell to cell contact. The T cell may release a soluble substance which promotes triggering of the B cell by antigen. According to this hypothesis it was investigated whether cell-free medium from thymocyte cultures can enhance the in vitro immune response of spleen cells from neonatally thymectomized mice. The following experiment will show that such a cell-free medium can replace thymocytes and display a similar helper effect. As described in detail elsewhere (Doria, 1971c), spleen cells from neonatally thymectomized mice were suspended in fresh medium at the concentration of $3x10^7$ nucleated cells per ml and then diluted with an equal volume of cell-free medium from cultures of $3x10^7$ nucleated thymocytes maintained in vitro for 24 hrs. One ml of the final spleen cell suspension, containing $1.5x10^7$ nucleated cells, was cultured with sheep RBC. The results reported in Fig. 4 demonstrate a great enhancing effect of

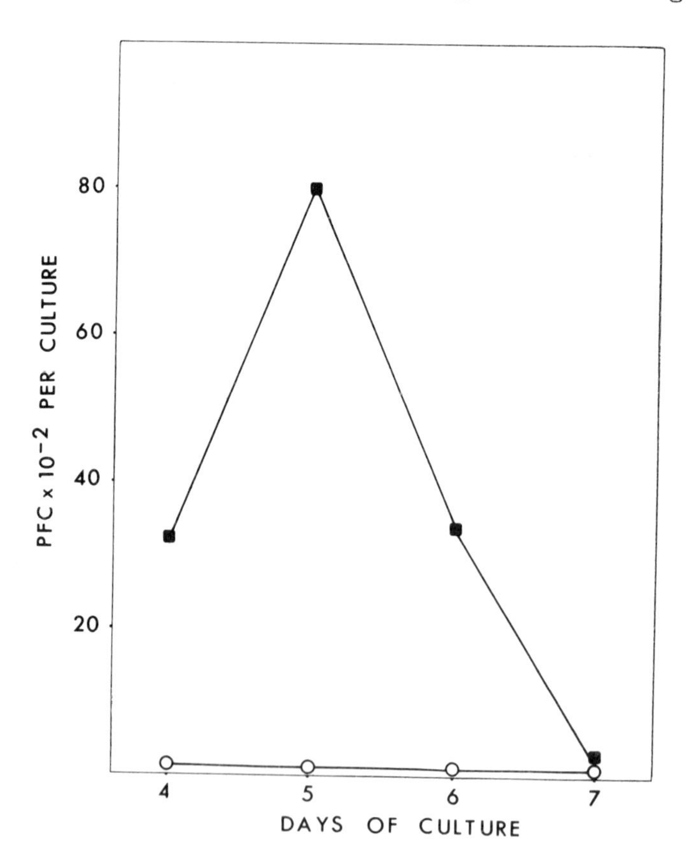

Fig. 4 Enhancing activity of thymocyte culture cell-free medium on the in vitro immune response of spleen cells from neonatally thymectomized mice. Spleen cells suspended in fresh medium, O ; or in thymocyte culture cell-free medium, ■ . Sheep RBC added to both cultures.

the thymocyte culture cell-free medium on the immunologic capacity of spleen cells from neonatally thymectomized mice. In other experiments this effect was found of lower magnitude but consistently reproducible. The activity of the cell-free medium was not reduced by centrifugation at 12000xg for 30 min (Table 1) nor was decreased

TABLE 1

ENHANCING ACTIVITY OF THYMOCYTE CULTURE CELL-FREE MEDIUM ON THE IN VITRO IMMUNE RESPONSE OF SPLEEN CELLS FROM NEONATALLY THYMECTOMIZED MICE TO SHEEP RBC (PFC PER CULTURE)

THYMOCYTE CULTURE CELL-FREE MEDIUM	DAYS OF SPLEEN CELL CULTURE			
	4	5	6	7
None	200	365	220	66
T1: Supernatant of thymocyte culture fluid centrifuged at 900xg for 20 min at 4°C and added to spleen cells immediately	783	1370	1280	691
T2: Same supernatant as T1 but added to spleen cells after storage at 4°C for 30 min	283	1020	1040	514
T3: Supernatant of T1 centrifuged at 12000xg for 30 min at 4°C and added to spleen cells immediately	498	858	1445	964

by heating at 56°C for 30 min. Presence of sheep RBC in the thymocyte culture for 24 hrs did not change the enhancing activity of the cell-free medium. Thus far, the T cell seems to secrete a promoting factor that lacks antigen specificity. Thymus specificity of the active factor is suggested by the finding that cell-free medium in which adult liver cells had been cultured for 24 hrs had no enhancing activity on the immune response of spleen cells from neonatally thymectomized mice.

The present demonstration that the helper function of T cells can be fully accounted for by a soluble substance released from unstimulated T cells in culture seems at variance with the results of in vivo experiments (Mitchell, 1968; Miller, 1969) implying that T and B cell cooperation is antigen specific. However, it remains to be determined whether the soluble factor acts directly on B cells or amplifies a small population of T cells that escaped neonatal thymectomy (Raff, 1971) and that could specifically interact with antigen and B cells.

Work supported by CNEN-Euratom Association Contract. Publication No. 762 of the Euratom Biology Division.

REFERENCES

Claman, H.N., E.A. Chaperon, and R.F. Triplett (1966), Immunocompetence of transferred thymus-marrow cell combinations. J. Immun. 97, 828.

Davies, A.J.S., E. Leuchars, V. Wallis, R. Marchant, and E.V. Elliott (1967), The failure of thymus-derived cells to produce antibody. Transplantation 5, 222.

De Luca, D., J. Decker, and E.E. Sercarz (1970), Binding of beta-galactosidase by normal thymus and bone marrow cells. Fed. Proc. 29, 697.

Doria, G., M. Martinozzi, G. Agarossi, and S. Di Pietro (1970), In vitro primary immune response resulting from the interaction between bone marrow-derived and thymus cells. Experientia 26, 410.

Doria, G., G. Agarossi, and S. Di Pietro (1971a), In vitro interaction between bone marrow-derived and thymus cells. Adv. Exp. Med. Biol. 12, 63.

Doria, G., G. Agarossi, and S. Di Pietro (1971b), Effect of blocking cell receptors on an immune response resulting from in vitro cooperation between thymocytes and thymus-independent cells. J. Immun. 107, in press.

Doria, G., G. Agarossi, and S. Di Pietro (1971c), Enhancing activity of thymocyte culture cell-free medium on the in vitro immune response of spleen cells from neonatally thymectomized mice to sheep RBC. J. Immun. 107, in press.

Dwyer, J.M., and I.R. Mackay (1970), Antigen-binding lymphocytes in human fetal thymus. Lancet 1, 1199.

Jerne, N.K., and A.A. Nordin (1963), Plaque formation in agar by single antibody-producing cells. Science 140, 405.

Miller, J.F.A.P., and G.F. Mitchell (1968), Cell to cell interaction in the immune response. I. Hemolysin-forming cells in neonatally thymectomized mice reconstituted with thymus or thoracic duct lymphocytes. J. Exp. Med. 128, 801.

Miller, J.F.A.P., and G.F. Mitchell (1969), Interaction between two distinct cell lineages in an immune response. Adv. Exp. Med. Biol. 5, 455.

Mishell, R.I., and R.W. Dutton (1967), Immunization of dissociated spleen cell cultures from normal mice. J. Exp. Med. 126, 423.

Mitchell, G.F., and J.F.A.P. Miller (1968), Cell to cell interaction in the immune response. II. The source of hemolysin-forming cells in irradiated mice given bone marrow and thymus or thoracic duct lymphocytes. J. Exp. Med. 128, 821.

Mitchison, N.A. (1971), The carrier effect in the secondary response to hapten-protein conjugates. II. Cellular cooperation. Eur. J. Immun. 1, 18.

Modabber, F., and A.H. Coons (1970), Presence of antigen binding cells in the thymus of normal mice and their involvement in the primary immune response. Fed. Proc. 29, 697.

Paraskevas, F., S-T Lee, and L.G. Israels (1971), Cell surface associated gamma globulins in lymphocytes. I. Reverse immune cytoadherence: a technique for their detection in mouse and human lymphocytes. J. Immun. 106, 160.

Pernis, B., L. Forni, and L. Amante (1970), Immunoglobulin spots on the surface of rabbit lymphocytes. J. Exp. Med. 132, 1001.

Rabellino, E., S. Colon, H.M. Grey, and E.R. Unanue (1971), Immunoglobulins on the surface of lymphocytes. I. Distribution and quantitation. J. Exp. Med. 133, 156.

Raff, M.C., and J.J.T. Owen (1971), Thymus-derived lymphocytes: their distribution and role in the development of peripheral lymphoid tissues of the mouse. Eur. J. Immun. 1, 27.

Raff, M.C., M. Sternberg, and R.B. Taylor (1970), Immunoglobulin determinants on the surface of mouse lymphoid cells. Nature 225, 553.

TWO STAGES IN DEVELOPMENT OF LYMPHOCYTES

HARVEY CANTOR
Medical Research Council, National Institute for Medical Research, Mill Hill, London NW7 1AA, England

Abstract: The hypothesis is presented that lymphocytes develop through two stages, provisionally termed T_1 and T_2. Supporting evidence comes from experiments performed with adult thymectomy, anti-lymphocyte serum, and anti-θ serum.

1. INTRODUCTION

In a remarkably short time since the demonstration that the lymphocyte plays a critical role in initiating an immune response to antigen (Gowans & McGregor, 1965), it has become evident that an efficient antibody response requires interaction between two classes of lymphocytes (Miller & Mitchell, 1969; Claman,et al 1969; Davies, 1969; Taylor, 1969). However, although evidence for a co-operative interaction between thymus-derived (T) lymphocyte and bone marrow-derived (B) lymphocytes has provided a satisfactory explanation for the crippling effects of neonatal thymectomy and the carrier specificity of secondary responses it has also raised many questions. Immunologists must now re-evaluate basic immunologic phenomena such as tolerance, antigenic recognition and immunologic memory in terms of a two cell model. Moreover, the absence of an obvious mechanism for interaction between T and B cells has stimulated a renewed interest in membrane physiology, membrane-bound antibodies and the definition of chemical factors that might influence cells.

Interest has also focussed on T cells, which have the intriguing dual role of modifying humoral responses and mediating cellular reactions. It is the purpose of this paper to present recent evidence suggesting that peripheral T cells are composed of subpopulations distinguished by different functional and physical properties, probably representing two sequential stages of maturation. Further experiments will be described that suggest the subpopulations of T cells co-operate in a cell mediated response.

2. INDICATIONS OF HETEROGENEITY IN THE T CELL POOL

Ablation of the thymus in neonatal life results in a severe impairment of both cell-mediated and humoral responses in the adult (Miller, 1961, 1962; Arnason et al., 1962). Neonatal thymectomy is also attended by a depletion of lymphocytes in the blood, thoracic duct (Miller, 1961), and periarteriolar areas in the spleen (Parrott et al., 1966). By contrast, removal of the thymus in adult life has been reported to have little effect upon either humoral or cellular responses until very much later in life (Miller, 1965; Taylor, 1965; Metcalf, 1965). These observation and others have supported the idea that peripheral T lymphocytes are long-lived, recirculating cells. We were therefore somewhat surprised to find that adult thymectomy resulted in a 50-60% depletion of θ-bearing cells in the spleen 3-4 weeks later. Moreover, no reduction was seen in the proportions of θ-bearing cells in lymph nodes or blood. This observation suggested the possibility that a subpopulation of T cells might be dependent upon an intact thymus either because it was composed of short-lived cells requiring continuous replacement and/or dependent upon some humoral factor elaborated by the thymus. Separate indications of this cell type, as well as a clue to its functional significance, had also been obtained in several independent systems.

In studies of rosette-forming cells (RFC) to sheep erythrocytes (SRBC) it has been noted that a proportion of immune as well as background RFC can be inhibited by anti-θ and guinea pig complement and there is good evidence that these are T-cells (TRFC) (Greaves & Moller, 1970; Bach et al., 1970).

However, the properties of background TRFC in thymus and spleen are very different from those in peripheral lymph node and blood. Thymus and spleen background TRFC can be inhibited with very low concentration of anti-theta serum, and azathioprine (Bach et al., 1970, 1971), and appear to have only immunoglobulin light-chain on the cell surface (Greaves & Hogg, 1970). Background TRFC in lymph node are far less sensitive to azathioprine and anti-theta (Greaves & Bach, personal communication) and have both light and heavy chains on their surface (Greaves, personal communication). Although background TRFC in spleen disappear soon after adult thymectomy, background TRFC in lymph node are unaffected by this procedure (Bach & Dardenne, 1971; Greaves, personal communication), suggesting that the splenic TRFC are part of a sub-population of T cells that differ from the typical long-lived recirculating T cell which appears not to be affected until very long after thymectomy. Finally, shortly after immunization with SRBC splenic TRFC acquire the characteristics of background lymph node TRFC: they are now relatively resistant to treatment with anti-theta and azathioprine and can be shown to bear both light and heavy chain immunoglobulin on their surface (Greaves & Hogg, 1970; Bach et al., 1971).

Another indication of heterogeneity within T lymphocytes has been obtained in studies of the cells that initiate graft-vs-host response in F_1 neonatal mice. These experiments suggest that two types of parental T cells, which appear to have many similar properties to the two T cell types described above, respond synergistically in F_1 hosts.

For example, a small dose of ALS given to donor mice 3 days prior to testing resulted in a 4-5-fold decrease in the GVH activity of the spleen cells; the addition of peripheral blood lymphocytes (PBL), in numbers far too small to produce any response alone, resulted in substantial recovery of activity (Cantor & Asofsky, 1971b). Similar types of synergy can be demonstrated using mixtures of PBL and thymus cells (Cantor & Asofsky, 1971a) and is seen whether death or splenic enlargement is used as the measure of GVHR. Both types of cells must be genetically capable of recognizing recipient antigens (Cantor & Asofsky, 1970) in contrast to the non-specific effects reported of bone marrow-thymus mixtures in irradiated recipients (Barchilon & Gershon, 1970; Hilgard, 1970). Moreover, since both types of cells are diminished after neonatal thymectomy (Cantor & Asofsky, 1970, 1971b) and are inhibited by anti-θ in the presence of guinea pig complement (Cantor, 1971a) it seems clear that they are both T cells. The two types of T cells differ markedly in their properties: one of the cells ("T_1") is present in higher concentrations in thymus and spleen than in lymph node, blood or thoracic duct (Cantor & Asofsky, 1970, 1971a, and unpublished observations) and is relatively insensitive to ALS *in vivo* (Cantor

& Asofsky, 1971b); the other cell, ("T_2") is present in high concentrations in lymph node, blood and thoracic duct and is very sensitive to ALS *in vivo*. T_1 cells appear to be slightly more sensitive to anti-θ and complement *in vitro* than are T_2 cells, and there is preliminary evidence that T_1 cells decrease earlier than T_2 cells after thymectomy (Cantor, unpublished observations).

Although at present the nature of cooperative interaction between these subpopulations of T cells is unknown, the following information may provide some clues:

(1) An exceedingly small number of T_2 cells is sufficient to produce detectable synergistic responses with T_1 cells (as few as 15×10^3 PBL in combination with 5×10^6 thymocytes produce synergistic responses in F_1 hosts) (Cantor & Asofsky, 1971a).

(2) If C57Bl/6 thymocytes are injected with Balb/c PBL into C57BlxBalb/c F_1 hosts, spleen cells from these F_1 hosts will transfer secondary GVH responses to Balb/c newborn mice but not to C57Bl/6 newborn (Cantor & Asofsky, 1971b).

(3) Although DBA/2 lymph node cells will respond to Balb/c cells in a mixed lymphocyte culture, they will not mount a GVH response in Balb/c hosts (Cantor & Asofsky, 1971c). The addition of small numbers of C57 thymocytes to the inoculum results in a vigorous GVH response (Cantor & Asofsky, 1971c). On the other hand, no response is seen when DBA/2 thymocytes and inactive numbers of C57Bl/6 lymph node cells are injected into Balb/c hosts.

These findings, taken together, are consistent with either of the following two mechanisms:

1. T_2 cells amplify the maturation of T_1 precursor cells, perhaps by increasing proliferation and differentiation of T_1 cells by efficient presentation of alloantigen.

2. T_1 cells, when injected into neonatal hosts, might shed surface receptors, which are adsorbed by activated T_2 cells. This mechanism would allow one to replace the thymocytes with supernatant from an incubated thymocyte suspension. These experiments are currently in progress.

3. AN HYPOTHESIS

These and other data may be interpreted by the idea that there are two, more or less separable subpopulations of T lymphocytes, which, for convenience of discussion, have been provisionally termed T_1 and T_2 (Cantor, 1971b; Raff & Cantor, 1971). The most important differences between them are that (a) T_1 cells are still somehow dependent on the thymus (either because they are short-lived and are continuously being replaced and/or because they are dependent on a thymus humoral factor) while T_2 cells are not; (b) T_2 cells recirculate and are thus sensitive to ALS and thoracic duct drainage while T_1 cells for the most part do not recirculate and are thus relatively insensitive to these procedures; (c) the tissue distribution of T_2 cells conforms to that generally accepted for long-lived recirculating T cells, while T_1 cells are in highest concentrations in thymus and spleen and lowest in blood and thoracic duct.

(ii) These two populations belong to the same cell lineage and represent different stages of maturation, T_1 being less mature than T_2.

(iii) The maturation of T_1 to T_2 is driven by specific antigen. Although this usually occurs in the peripheral lymphoid tissues, it can take place in the thymus prior to migration if the first encounter with antigen occurs there.

(iv) T_2 cells are responsible for the T lymphocyte component of the immune response seen after overt primary or secondary immunization. The only response of T_1 cells to antigen is to become T_2 cells. This implies that the overt primary response of an unimmunized animal is in fact the response of T_2 cells which have matured from T_1 cells under the influence of small amounts of "environmental" antigen, or cross-reacting antigen. The first overt exposure to an antigen stimulates the existing T_2 cells, but also drives T_1 to T_2, increasing the number of T_2 cells which will now give an amnestic response on second exposure to the antigen.

4. THE EFFECTS OF ADULT THYMECTOMY AND ANTILYMPHOCYTE SERA UPON HUMORAL RESPONSES

According to this scheme, adult thymectomy should mainly affect secondary rather than primary responses and small doses of ATS should have the opposite effect. To test this prediction, mice were thymectomized or sham thymectomized at 6 weeks of age, and then immunized with bovine serum albumin (BSA) 3-4 weeks later. Primary responses in intact animals were similar in both groups. (Table 1).

TABLE 1

	Intact Primary (Day 10) Response*	Anti-BSA adoptive secondary response* (spleen) 2×10^7 cells	Anti-DNP ABC ($\times 10^{-8}$M) adoptive secondary (spleen) 2×10^7 cells
Sham	0.64 ± .21	8.44 ± 0.88	10.54 ± 1.20
Tx	0.46 ± .16	1.60 ± 0.41	1.05 ± 1.11

* Anti-BSA ABC mcg/ml serum

However, the ability of spleen cells to transfer secondary responses to lethally irradiated recipients was substantially reduced in the thymectomized group (Table 1). Moreover, spleen cells transferred from unimmunized thymectomized or sham animals to lethally irradiated recipients produced similar early primary responses, suggesting that the loss of secondary responsiveness uncovered by spleen transfer was not simply due to a general unresponsiveness in spleen following thymectomy. This deficit in immunologic memory to BSA resides, at least in part, in anti-BSA helper activity; approximately 8-10 times as many primed spleen cells from thymectomized animals as controls were required to produce the same anti-hapten titers in mice that had also been given DNP-OA primed cells and challenged with DNP-BSA. Representative anti-DNP titers at a single dose of cells are shown in Table 1.

By contrast, although a small dose of anti-lymphocyte serum (0.25 ml) administered three days previously severely impaired the ability of spleen cells to transfer primary responses to BSA in lethally-irradiated, thymectomized recipients, nearly normal secondary responses could be elicited after challenge with BSA in saline three months later. Finally, spleen cells obtained from mice thymectomized three weeks previously and given 0.25 ml ALS 3 days before transfer failed to restore either primary or secondary responsiveness to irradiated thymectomized hosts (Table 2).

TABLE 2

	Adoptive Primary*	Adoptive Secondary*
0.25 ALS 20×10^6 cells	0.15 ± 0.41	4.66 ± 0.55
Thymectomy "	3.41 ± 0.66	2.11 ± 0.66
Control "	4.96 ± 0.40	8.11 ± 0.36
Thymectomy + ALS "	0.14 ± 0.10	0.33 ± 0.16

* Anti BSA ABC mcg/ml serum

5. THE EFFECT OF ADULT THYMECTOMY UPON THE FACTOR OF IMMUNIZATION IN CELLULAR RESPONSES

The disproportionately large number of T cells responding to major histocompatibility antigens (Nisbet et al., 1969; Wilson et al., 1968; Szenberg et al., 1962) might be attributable to relatively large amounts of these antigens in the internal environment leading to a disproportionately large representation in the T_2 cell pool of "unimmunized" animals. This could also account for the ineffectiveness of preimmunization across the major histocompatibility barriers in augmenting a GVHR (Simonsen, 1962) as the T_2 pool for these antigens may already be very large. On the other hand, preimmunizing across minor histocompatibility barriers is known to be far more effective (Simonsen, 1962), possibly because there is a relatively small pool of T_2 cells specific for these antigens. Adult thymectomy should therefore markedly decrease the effect of preimmunization across minor histocompatibility barriers.

To test this, three to four weeks after adult thymectomy or sham thymectomy, CBA ($H\text{-}2^K$) mice were immunized with 2 weekly intraperitoneal injections of C_3H ($H\text{-}2^K$) spleen cells. One week after the last injection, 5×10^6 spleen cells from each group were injected into C_3HxCBA neonatal recipients. Nine days later, recipient spleen indices were determined. Spleen cells from control CBA mice pre-immunized against C_3H cells (which differ at several minor histocompatibility loci) produced vigorous GVH reactions, while cells from adult-thymectomized pre-immunized mice produced barely significant indices (Table 3).

TABLE 3

Donor*	Thymectomy	Recipient	Mean Spleen Index ± SE
CBA	-	C_3HxCBA	1.10 ± .11
CBA preimmunized to C_3H	-	C_3HxCBA	2.31 ± .21
CBA preimmunized to C_3H	+	C_3HxCBA	1.51 ± .16

* 6×10^6 spleen cells from CBA donor mice were injected intraperitoneally into newborn C_3H recipients. Spleen and body weights were obtained nine days later.

6. THE DEVELOPMENT AND REGULATION OF T CELLS

These experiments, and others (Cantor, 1971b), suggest that the peripheral T cell pool is composed of at least two subpopulations of lymphocytes having different functional properties, summarized in Table 4, and probably

TABLE 4

SURFACE MARKERS	T_1	T_2
TL	-	-
θ	++	+
Immunoglobulin light chain	+	+
heavy chain	-	+
IN VIVO PROPERTIES		
Tissues of highest concentration	thymus, spleen	peripheral blood, lymph node, thoracic duct
Recirculation	no	yes
Effect of adult thymectomy	in 2-6 wks	after 40 weeks
Susceptibility to corticosteroids in vivo	+++	+
Effect of immunization	↓ in specifically reactive cells	↑ in specifically reactive cells
FUNCTIONAL PROPERTIES - HUMORAL		
Primary response	-	+
Secondary response	+	+
FUNCTIONAL PROPERTIES - CELLULAR RESPONSE		
Mixed lymphocyte reactivity (Mosier & Cantor, 1971)	-	+
GVHR	+	+

representing different stages of maturation. The observation that T cell memory following immunization with either protein antigens or minor allo-antigens can be substantially depleted by previous adult thymectomy is consistent with the view that T_1 cells can be driven by specific antigen to T_2 cells. In vitro, the high sensitivity of spleen T_1 cells to azathioprine, ALS, and anti-theta is similar to that of thymocytes rather than blood or lymph node T cells, suggesting that T_1 cells maybe less mature than most T cells from lymph nodes or blood. Recently, this subpopulation has also been shown to be extremely sensitive to rapid changes in osmolality , a property also shared by thymocytes.

Recent work has clarified several steps in T cell development. Stem cells migrate from yolk sac and liver in the embryo (Owen & Ritter, 1969) and from bone marrow in the adult (Ford et al., 1966) into the thymus where they proliferate and differentiate into thymocytes (Owen & Raff, 1970), possibly induced by thymus epithelium. In embrionic mice (Schlesinger & Hurwitz, 1969; Owen & Raff, 1970) and probably in the adult (Boyse & Old, 1969), immigrant stem cells do not bear either theta or TL antigens, but probably acquire them during differentiation from stem cell to "immature" thymocyte. These "immature" thymocytes, are characterized by a strong representation of θ and TL alloantigens on their surface (Raff & Owen, 1971a, 1971b) by corticosteroid sensitivity (Levine & Claman, 1970; Blomgren & Anderson, 1970) and by minimal immunocompetence as assayed by their ability to mediate a GVH reaction (Levine & Claman, 1970) or respond to the mitogen phytohemagglutinin (PHA) (Takiguchi et al., 1971). A second, "mature" population of thymocytes is also detectable, primarily in the medulla, and constitutes about 5-10% of the organ's cellularity. This population is characterized by the absence of the TL antigen, and a reduction in the amount of surface θ (Raff & Owen, 1971a), by relative resistance to irradiation and to corticosteroid treatment (Blomgren & Andersson, 1970) and by a lower density (Takiguchi et al., 1971). These cells account for most of the GVH and PHA reactivity in the thymus (Levine & Claman, 1970; Blomgren & Andersson, 1970). It is likely that these 'mature' thymocytes arise from immature thymocytes in situ, and not from peripheral T lymphocytes migrating through the thymus since in ontogeny (Bortin et al., 1969) and after irradiation (Blomgren & Andersson, 1970) GVH activity appears first in the thymus and subsequently in the periphery.

T_1 lymphocytes in the periphery might therefore result from a precocious emigration of immature cortical thymocytes, which would lose TL antigen just before, or during, emigration, having bypassed the thymus medulla. According to this idea, some of these cells would continue to differentiate in the periphery, perhaps in spleen, rather than in the medullary areas of thymus. Alternatively, both T_1 and T_2 cells might arise from TL negative steroid-resistant medullary cells. Although there is some evidence that steroid resistant, TL negative cells are a heterogeneous population composed of immature and mature T cells (Leckband & Boyse, 1971; Mosier & Cantor, 1971; Bach & Dardenne, 1971), at present it is not possible to rule out either possibility.

Once in the periphery, the fragility of T_1 cells seen *in vitro* may provide a clue to the reason for their short-lived nature *in vivo*. There is evidence that T cell proliferation following adrenalectomy is abolished by prior adult thymectomy (Castro & Hamilton, personal communication). This would imply that physiological levels of endogenous corticosteroids normally regulate the size of the immature T_1 cell population, since following adult thymectomy there are amply numbers of long-lived T_2 cells which do not appear to respond to the lowering of steroid levels following adrenalectomy.

Alternatively, it is also possible that normally the thymus and adrenals produce antagonistic factors influencing the T_1 cell population. Adult thymectomy enhances the destructive effects of steroids on T_1 cells, while adrenalectomy amplifies the trophic effects of thymic hormone on this T cell subpopulation.

It should not be necessary to emphasize that at the moment the evidence for this model for T cell maturation is hardly conclusive. Moreover, despite the suggestive nature of these and other experiments, one is still tempted *a priori* to attribute all things to a single type of T cell. However, on the same *a priori* grounds, this maturational scheme provides an efficient mechanism for the development of a compact peripheral T cell pool that is well-suited to a given antigenic environment. If maturation from T_1 cells to T_2 cells can be driven by specific antigen present in the internal environment, the long-lived reculatory T_2 pool will be composed of cells capable of recognizing and reacting to substances commonly encountered in the environment; short lived T_1 cells will die after a period perhaps under the influence of peripheral steroids, in the absence of antigen. In a sense, a constant supply of short-lived, immature T_1 cells from the thymus may be viewed as a continuing immunologic probe that results in flexible protection in a changing antigenic environment.

REFERENCES

Bach, J.F., Muller, J.Y. and Dardenne, M, (1970), *In vivo* specific antigen recognition by rosette forming cells. Nature (Lond.), 227, 1251.

Bach, J.F., and Dardenne, M. (1971), Antigen recognition by T lymphocytes I. Thymus and bone marrow dependence of spontaneous rosette forming cells in the mouse, Cell Immunol., in press.

Bach, J.F. and Dardenne, M. (1971), Antigen recognition by T lymphocytes. III. Evidence for two populations of thymus-derived rosette forming cells in spleen and lymph nodes, submitted for publication.

Barchelon, J. and Gershon, R.K, (1970), Synergism between thymocytes and bone marrow cells in a graft-vs-host reaction, Nature (Lond.) 227, 71.

Blomgren, H. and Andersson, B, (1970), Characteristics of the immunocompetent cells in the mouse thymus: cell population changes during cortisone-induced atrophy and subsequent regeneration, Cellular Immunol., 1, 545.

Boyse, E.A. and Old, L.J, (1969), Some aspects of normal and abnormal cell surface genetics, Ann.Rev.Genet., 3, 269.

Cantor, H., (1971a), Effects of anti-theta antiserum upon GVH activity of spleen and lymph node cells, Cell Immunol., in press.

Cantor, H, (1971b), T cells and the immune response in Progr. in Bioph. and Mol.Biol., v.22, in press.

Cantor, H. and Asofsky, R, (1970), Synergy among lymphoid cells that mediate the graft-vs-host response. II. Synergy in GVH responses produced by Balb/c lymphoid cells of differing anatomic origin, J.exp.Med., 131, 223.
Cantor, H. and Asofsky, R, (1971a), Synergy among lymphoid cells that mediate the graft-vs-host response, submitted for publication.
Cantor, H. and Asofsky, R, (1971b), Synergy among lymphoid cells that mediate the graft-vs-host response. IV. Evidence for interaction between two classes of thymus-derived cells, submitted for publication.
Cantor, H. and Asofsky, R, (1971c), manuscript in preparation.
Claman, H.N. and Chaperon, E.A., (1969), Immunologic complementation between thymus and marrow cells - a model for the two-cell theory of immunocompetence, Transpl.Rev., 1, 92.
Davies, A.J.S., (1969), The thymus and the cellular basis of immunity, Transpl.Rev., 1, 43.
Ford, C.E., Micklem, H.S., Evans, E.P., Gray, J.G. and Ogden, D.A. (1966), The inflow of bone marrow cells to the thymus:studies with part body irradiated mice injected with chromosome marked bone marrow and subjected to antigenic stimulation, Ann.N.Y.Acad.Sci., 129, 283.
Gowans, J.L. and McGregor, D.D., (1965), The immunological activities of lymphocytes, Progr.Allergy, 9, 1.
Greaves, M.F. and Hogg, M.M, (1970), Antigen binding sites on mouse (lymphoid) cells. In:Third Sigrid Juselius Foundation Symposium. Helsinki, EC Cross, A. Academic Press, N.Y.
Greaves, M.F. and Moller, (1970), Studies on antigen binding, I. Origin of reactive cells. Cell Immunol., 1, 372.
Hilgard, H.R, (1970), Synergism of thymus and bone marrow in the production of graft-vs-host splenomegaly in X-irradiated hosts, J.exp.Med., 132, 317.
Leckband, E. and Boyse, E.A, (1971), Immunocompetent cells among mouse thymocytes: a minor population, Science (Washington) 172, 1258.
Levine, M.A. and Claman, H.N, (1970), Bone marrow and spleen: dissociation of immunologic properties by cortisone, Science(Washington), 167, 1515.
Metcalf, D., (1965), Delayed effect of thymectomy in adult life on immunological competence, Nature (Lond.), 208, 1336.
Miller, J.F.A.P., (1961), Immunological function of the thymus, Lancet, 2, 748.
Miller, J.F.A.P., (1962), Effect of neonatal thymectomy on the immunological responsiveness of the mouse, Proc.Roy.Soc.B., 156, 415.
Miller, J.F.A.P., (1965), Effect of thymectomy in adult mice on immunological responsiveness, Nature (Lond.), 208, 1337.
Miller, J.F.A.P.and Mitchell, G.F., (1969), Thymus and antigen-reactive cells. Transpl.Rev., 1, 3.
Mosier, D. and Cantor, H., (1971), Functional maturation of mouse thymic lymphocytes, Eur.J.Imm., in press.
Nisbet, N.W., Simonsen, M. and Zaleski, M, (1969), The frequency of antigen-sensitive cells in tissue transplantation, J.exp.Med., 129, 459.
Owen, J.J.T. and Raff, M.C., (1970), Studies on the differentiation of thymus-derived lymphocytes, J.exp.Med., 132, 1216.
Owen, J.J.T. and Ritter, M.A., (1969), Tissue interaction in the development of thymus lymphocytes, J.exp.Med., 126, 715.

Parrott, D.M.V., de Sousa and East, J, (1966), Thymus-dependent areas in the lymphoid organs of neonatally thymectomized mice, J.exp.Med., 123, 191.
Raff, M.C. and Cantor, H., (1971), Subpopulations of thymus-derived cells. In: Proc. of the First Int.Cong. of Imm., in press.
Raff, M.C. and Owen, J.J.T., (1971), Thymus-derived lymphocytes: their distribution and role in the development of peripheral lymphoid tissues in the mouse, Europ.J.Immunol.. 1, 27.
Schlesinger, M. and Hurvitz, D., (1969), Differentiation of the thymus-leukemia (TL) antigen in the thymus of mouse ambryos, Israel J.Med.Sci., 4, 1211.
Simonsen, M., (1962), Graft-vs-host reactions: their natural history and applicability as tools of research. In: Progr. in Allergy, 6, 349. Editors: Paul Kallos and Byron H.Waksman. Publishers: Basel, N.Y.
Takiguchi, T., Adler, W.H. and Smith, R.T., Identification of mouse thymus antigen recognition function in a minor, low density low θ cell population, Cell Immunol., in press.
Taylor, R.B., (1965), Decay of immunological responsiveness after thymectomy in adult life, Nature, 220, 611.
Taylor, R.B., (1969), Cellular cooperation in the antibody response of mice to two serum albumins: specific function of thymus cells, Transpl.Rev., 1, 114.
Wilson, D.B., Blyth, J.L. and Nowell, P.C., (1968), Quantitative studies on the mixed lymphocyte interaction in rats.III. Kinetics of the response, J.exp.Med., 128, 1157.

DIFFERENTIATION OF THE THYMUS *IN VIVO* AND *IN VITRO*

T. MANDEL, PAMELA J. RUSSELL AND W. BYRD
Walter and Eliza Hall Institute of Medical Research
P.O. Royal Melbourne Hospital
Parkville 3050, AUSTRALIA

Abstract: The differentiation of foetal mouse thymus was studied during normal development, in organ culture and in grafts of cultured thymus placed under the kidney capsule of syngeneic hosts. In all these situations undifferentiated lymphoid stem cells proliferated in the presence of differentiated thymic epithelial cells. The cultured thymus released into the medium a factor which enabled spleen cells from neonatally thymectomized mice to respond *in vitro* to a challenge of sheep red blood cells. It is postulated that an interaction occurs between the thymic epithelial cells and the lymphoid stem cells leading to the production of small lymphocytes and that a competence-inducing humoral factor is produced by the cultures.

1. INTRODUCTION

The precursors of thymus derived lymphocytes or "T" cells enter the thymus early in embryonic development. These large blast-like cells can be identified in the foetal mouse thymus by the 11th or 12th day of gestation by light and electron microscopy (Owen and Ritter, 1969; Mandel, 1970). During the succeeding couple of days the blasts accumulate and begin to proliferate forming a population of small and medium sized lymphocytes. Indeed, in the foetal mouse, a histologically typical thymus is already present by about the 17th day of gestation, well before peripheralization of lymphocytes has occured.

During their intrathymic period of development the stem cells undergo antigenic, functional and morphologic changes. It has been shown that the stem cells do not express either theta (θ) or TL antigens on their surface but by the 15th to 16th day of gestation the small lymphocytes which have differentiated from them are both θ and TL positive (Owen and Raff, 1970).

It has been suggested that the differentiation of T cells involves at least two steps; firstly intrathymic proliferation which leads to the production of θ positive small lymphocytes and secondly, a further differentation into functionally mature cells (Raff, 1971). This second step involves a loss of TL, a decrease in θ (Aoki *et al*., 1969), and an increase in H-2 antigens and may occur outside the thymus. Functional studies have also shown that thymus derived cells are more effective on a cell to cell basis than intrathymic lymphocytes in mediating immunological reactions. However neither stage of differentiation is understood from the point of view of its control (Raff, 1971). It is not known for example, whether full differentiation can only occur within the microenvironment of the thymic epithelium cells or whether some can occur at a distance, perhaps in the peripheral tissues. Thus it is not known whether a direct interaction is necessary between thymic epithelial cells and the stem cells or whether one or more humoral factors are produced by the thymus and act on the target cells at a distance.

Such problems of differentiation are difficult to monitor in the complex environment of the intact animal. It would therefore be of great advantage to have available a simplified system in which many of the variables, uncontrollable in vivo, could be selectively altered. It is possible to isolate and grow the early foetal organ in culture and to study the development of the lymphocytes and the epithelial cytoreticulum. In addition, it is possible to collect media from the cultures for assay for humoral activity. This paper describes some results of studies on long-term organ cultures of the foetal mouse thymus and presents preliminary data which suggest the presence of a humoral factor capable of affecting the function of lymphoid cells.

2. DEVELOPMENT OF THE MOUSE THYMUS IN VIVO

In order to interpret meaningfully the results of a morphologic study of organ cultures, it was first necessary to study the development of the thymus in vivo. The results of this study were used as a basis for comparison with the in vitro studies. Since the observations on normal development of the thymus have been published in full elsewhere (Mandel, 1970), only a few points will be described here.

It was evident that a rapid accumulation of lymphoid stem cells occured in the early thymic anlage even before it was vascularized and while its epithelial cells were still quite undifferentiated. Indeed, on the 12th day of gestation, when the first lymphoid stem cells were identified, the epithelial cells were undifferentiated and could be distinguished from the lymphoid cells mainly by the presence of desmosomes which linked adjacent epithelial cells. Lymphoid cells continued to accumulate in the anlage but showed no evidence of mitotic activity until the 14th day of gestation. By contrast, dividing epithelial cells were frequently seen and could be identified by the presence of desmosomes. Moreover, during this period the epithelial cells showed evidence of differentiation and began to resemble the "cortical" epithelial cells of a fully developed thymus (Mandel, 1968a; 1970). From the 14th day onwards, lymphopoiesis became increasingly prominent and medium and finally small lymphocytes became the predominant cell type of the differentiating thymus.

A thymic medulla did not develop until relatively late in gestation and differentiated "medullary" epithelial cells were first seen on the 17th day. These cells which have a very characteristic morphology have been described in the thymic medulla of many species, including mice (Höshino, 1963; Clark, 1963, 1966, 1968; Gad and Clark, 1968; Mandel, 1970), hamsters (Ito and Hoshino, 1966), guinea pigs (Mandel, 1968b), snakes (Raviola and Raviola, 1967) and amphibia (Klug, 1967). The "medullary" epithelial cells have an ultrastructure suggestive of secretory activity and this role has also been suggested for them from experimental evidence (Clark, 1966,1968). By the time of birth the thymic ultrastructure was essentially fully developed and no further major differences were noted at least up to maturity. It should be noted that at birth the peripheral lymphoid system in the mouse is still poorly developed in marked contrast to the mature appearance of the thymus.

3. DIFFERENTIATION OF FOETAL THYMUS IN ORGAN CULTURE

Intact 13 day foetal mouse thymuses were cultured in order to follow the development of the lymphoid and epithelial cells under conditions where the inflow of new stem cells was abolished and where other extrathymic regulatory factors were absent. The detailed methods and results of this study have been published (Mandel and Russell, 1971) and again only a few points will be discussed.

From the results of a previous study (Mandel, 1970) it was known that lymphoid stem cells were already present in the early foetal thymus. In the absence of new stem cells, it was possible to study the fate of the progeny of the limited number of stem cells present at the start of the culture period. There was a brief period before lymphoid mitotic figures were seen but after the first day of culture numerous dividing lymphocytes were identified both by autoradiography after labelling with H^3 thymidine and by the morphology of the dividing cells. This phase of lymphopoiesis occupied about 10 days during which time the relatively few blasts present initially transformed into numerous small and medium lymphocytes. After this period of rapid proliferation, lymphopoiesis stopped and the number of small lymphocytes gradually decreased. However, even in cultures of 28 days duration, typical intact and apparently viable small lymphocytes were still seen.

By contrast, the proliferation of the epithelial cells occured slowly and dividing epithelial cells were seen at <u>all</u> stages of culture. The absence of lymphopoiesis at the later periods of culture therefore appeared to be due to an exhaustion of the proliferative potential of the stem cells.

This study demonstrated that the differentiation of the foetal thymus <u>in vitro</u> paralleled that observed <u>in vivo</u> with the major difference that no new lymphoid stem cells were allowed to enter the organ. As <u>in vivo</u>, differentiation of epithelial cells occured rapidly, and again lymphopoiesis began once "cortical" epithelial cells had developed.

4. REPOPULATION OF CULTURED THYMUS GRAFTED BENEATH THE KIDNEY CAPSULE

It was necessary to determine whether the cessation of lymphopoiesis was due to an exhaustion of the proliferative capacity of the stem cells or whether it was due to a failure of the epithelial microenvironment to sustain it. In order to do this, thymuses which had been maintained <u>in vitro</u> for 14 days were grafted beneath the kidney capsule of neonatally thymectomized or sham-thymectomized syngeneic mice aged 2 weeks. The morphologic development of the graft was studied by light and electron microscopy (Mandel and Russel, 1971). The cultures were excised at frequent intervals particularly during the first week after grafting, taking care for fixation to include a substantial wedge of underlying kidney. After fixation the graft was carefully separated from the kidney and weighed. Grafts <u>in situ</u> for up to 12 weeks were studied and all neonatally thymectomized mice were examined carefully for evidence of residual thymus by serially sectioning the upper mediastinum.

It will be remembered that at the time of grafting, after 14 days in culture, the thymic rudiment consisted of numerous small and medium lymphocytes and epithelial cells and that at this stage lymphopoiesis had ceased. Within 24 hours of grafting, the lymphocytes disappeared from the graft and it consisted mainly of epithelial cells. There was little evidence of cell death or destruction within the graft in contrast to the situation described for grafts of whole neonatal thymus (Blackburn and Miller, 1967). It is possible that the lymphoid cells migrated out of the graft to peripheral lymphoid tissues, but this point has not as yet been formally studied. During the 2nd and 3rd days the graft consisted of large epithelial cells, many of which were in mitosis. During the 3rd day, large lymphoblast-like cells entered the graft and rapidly began to divide and differentiate so that, by the end of the first week, a lymphoidal though minute thymus was present. Indeed the great majority of the cells at the end of one week were quite typical small and medium lymphocytes and numerous mitotic figures were seen.

The graft continued to grow and usually between 10 and 30 mg of thymic tissue could be recovered from each culture after 8-12 weeks. On one occasion 50 mg of thymus developed from what was originaly a single foetal lobe. However, some grafts developed poorly, and in some cases no evidence of a graft could be found. There appeared to be no difference in the size and rate of growth between grafts in thymectomized or sham-thymectomized hosts, thus confirming the apparently autonomous behaviour of thymic tissue described by Metcalf (1963). When either uncultured foetal thymuses or normal neonatal thymuses were grafted, they too showed growth patterns similar to those of cultured tissue. Possibly the only difference between cultured grafts and normal grafts was the poor development of the medulla and the paucity and poor differentiation of medullary epithelial cells in the former.

This aspect of the study showed that the epithelial cells retained their potential to support the differentiation and proliferation of the lymphoid stem cells when these were again allowed to enter the microenvironment of the thymic epithelial cells. The process of repopulation was similar to that observed in the development of foetal thymic tissue both *in vivo* and *in vitro* suggesting that a similar sequence of events occured. An interesting feature which will be studied in greater detail is the fate of the small lymphocytes developed *in vitro*. It would be interesting to know their degree of immunological maturity and to determine whether they correspond to a fully mature T cell or whether they require a period of extra-thymic development for further differentiation. Their rapid exit from the graft suggests that they may have left it to enter peripheral lymphoid tissues and to join the recirculating pool. Labelling of the cultures with H3-thymidine or the use of a chromosome marker may allow these cells to be traced in the hosts.

5. FUNCTIONAL STUDIES ON CULTURED AND GRAFTED THYMUSES

Morphologic studies showed that the cultured thymus could support lymphopoiesis *in vivo*, but gave no information about the functional activity of the graft. Two parameters of immunological competence were therefore studied, firstly the ability of grafted mice to resist post-thymectomy wasting and secondly their ability to mount a response to a challenge of sheep red blood cells (SRBC). Since these results are incomplete, only preliminary data are given.

Mice in which a graft was found at autopsy did not develop a wasting syndrome in contrast to the high mortality of fully thymectomized ungrafted mice. Thus it appears that the grafts produce sufficent cells to allow the hosts to survive into adult life.

The response to SRBC was tested by challenging the mice with 2×10^8SRBC injected intraperitoneally 8-10 weeks after grafting and 4 days later assaying their spleens for antibody forming cells by the Cunningham modification (Cunningham and Szenberg, 1968) of the Jerne plaque technique (Jerne and Nordin, 1965). In addition to cultured thymic grafts, some mice were grafted with uncultured foetal or neonatal thymuses. In all cases one lobe was grafted into each donor and at autopsy the mediastinal contents were serially sectioned in a search for thymic remnants. Only fully thymectomized mice were included in the study and preliminary results are shown in Table 1. The table shows that cultured grafts could not restore the immune response to levels significantly above those of thymectomized mice. However, even uncultured thymuses, although better than the cultured grafts, were still not capable of fully restoring the response to SRBC. The reason for these poor responses is not clear, and further studies are in progress using multiple grafts in each recipient and also grafts cultured for varying periods of time in order to see whether the response can be improved by a greater mass of tissue or whether it depends on the prior length of culture of the graft.

Table 1
The number of antibody forming cells per spleen making direct plaques to SRBC.

Treatment	No. of mice	Antibody forming cells per spleen Geometric mean	Upper + Lower Limits (S.E.)
Thymectomy (Tx)	27	4,298	5,462 - 3,383
Sham Tx	32	67,480	78,980 - 57,660
Tx and Cultured Graft	17	8,939	11,100 - 7,201
Tx and Foetal Graft	15	16,620	23,340 - 11,820
Tx and Neonatal Graft	6	19,300	26,670 - 13,960
Sham Tx	5	91,680	96,680 - 46,310

6. ASSAY OF TISSUE CULTURE MEDIA

Numerous workers have suggested that the thymus acts as an endocrine gland and secretes one or more humoral factors. It has been suggested that one function of such factors may be the induction of immune competence in lymphoid cells. Trainin, Small and Globerson (1970) exposed spleen cells from neonatally thymectomized mice to an extract of syngeneic thymus *in vitro* and showed that these cells could produce a graft-versus-host reaction. It was therefore of interest to determine whether the thymus organ cultures secreted a factor with competence-inducing activity. Media were collected from the cultures at 3 day intervals, and those from cultures of the same age were pooled, dialyzed against phosphate buffered saline and lyopholized. The dried material was then redissolved in Eagle's Minimal Essential Medium to a concentration five times that of the original. For controls, media were obtained from cultures of foetal mouse liver, lung and kidney and processed identically. The final concentrated media were sterilized by Millipore filtration and were used as a 2½% supplement in the assay. These media are refered to as "conditioned" media. To test for restoration of immune competence, the *in vitro* response of mouse spleen cells to SRBC was used as described by Marbrook (1967). In this system 20×10^6 nucleated spleen cells were immunized *in vitro* with 3×10^7 SRBC and after 4 days the number of antibody forming cells was assayed by a modification (Cunninghan and Szenberg, 1968) of the Jerne plaque assay (Jerne and Nordin, 1963). In this study, spleen cells from neonatally thymectomized mice were exposed for the entire period of culture to a 2½% supplement of conditioned medium. Preliminary results of this study are shown in Figure 1.

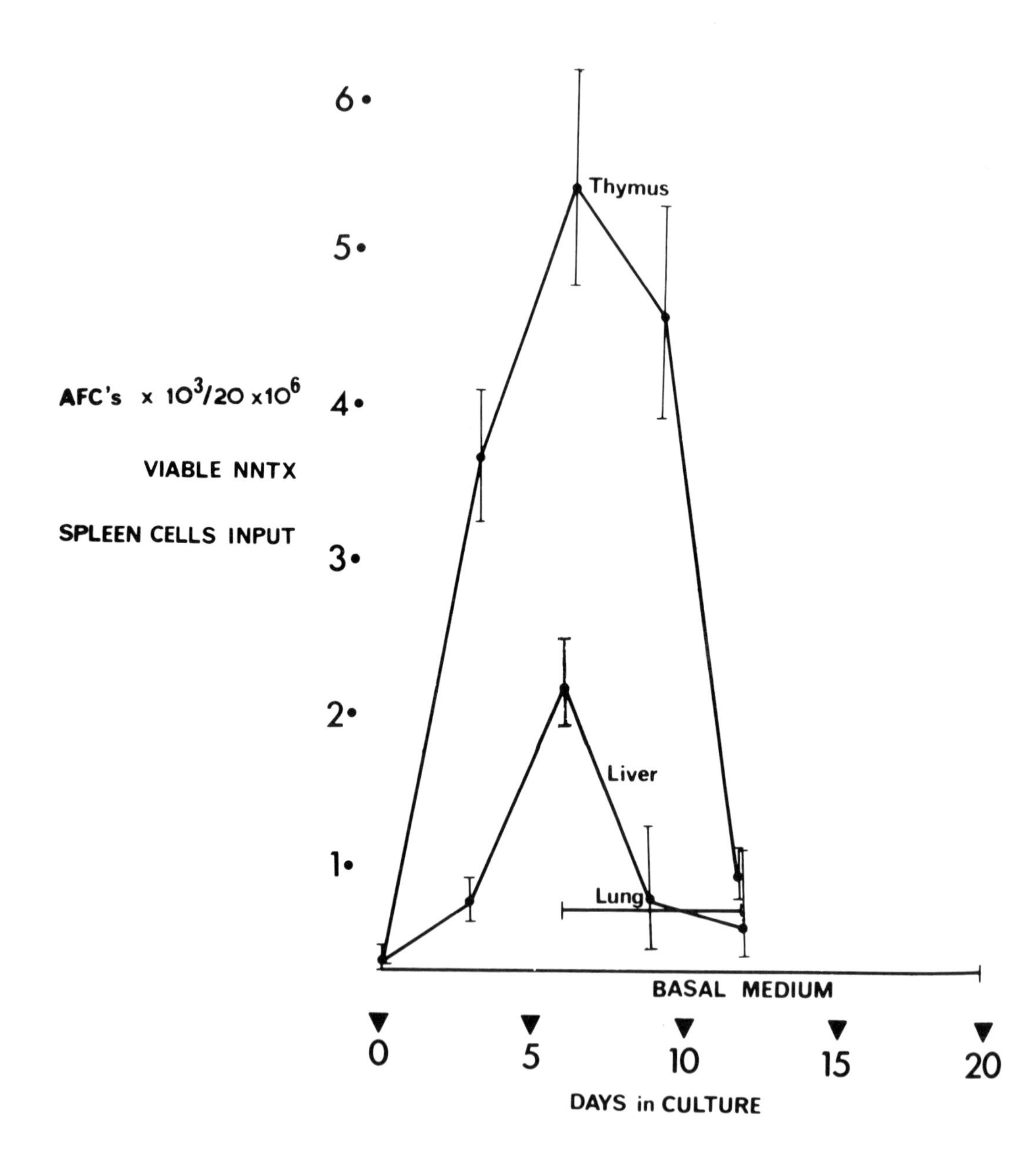

Fig. 1. The number of antibody forming cells (± SE) produced when 20x10^6 viable spleen cells obtained from neonatally thymectomized mice were exposed _in vitro_ to supplements of conditioned media. Cells exposed to basal medium without conditioned medium supplements did not respond. Spleen cells from normal mice exposed to basal medium gave about 2000 AFC/20x10^6 cells.

Spleen cells from normal mice not exposed to conditioned medium supplements normally gave about 2000 plaques per culture, whereas the same number of spleen cells from neonatally thymectomized mice gave only a few hundred plaques. The presence of a thymus conditioned medium supplement restored the ability of incompetent spleen cells to respond. However this restorative effect was only found in media taken from the 3,6, and 9 day organ cultures. The peak response was found with conditioned media obtained in the period between 4-6 days. Conditioned media from lung and kidney cultures were completely ineffective, as was medium not exposed to any organ. A slight restorative effect was obtained with foetal liver conditioned media. It must also be stressed that in this study particular care was taken to ensure that the foetal calf serum used in the assay system was negative for the sheep erythrocyte restorative factor (SERF) described by Byrd (1971).

7. DISCUSSION

Among the major problems of differentiation are the modes of action of inductive microenvironments. Studies on these problems will of necessity have to be performed in vitro since in the intact animal too many uncontrollable factors exist. The thymus presents an ideal system for such studies since it can be isolated at an undifferentiated stage of development and its further development can be followed in culture. Characteristic surface antigenic markers develop on the lymphoid cells populating the thymus (Schlesinger and Hurvitz, 1968) and the functional status of the progeny of these cells can be evaluated. The observations described in this paper show that development of foetal mouse thymus in vitro parallels closely the sequence of events occuring in the intact animal during its normal development. It is not yet known however whether the entire sequence of differentiation, from stem cell to mature T cell, can occur in culture. It has been shown by Owen and Raff (1971) that the acquisition of cell surface antigens marks the first step in the maturation of T cells and it is presumed that the stimulus of this differentiation originates in the thymic epithelial cells (Owen and Ritter, 1969). It has been suggested that a direct cell to cell interaction is required between the lymphoid cells and the epithelial cells (Mandel, 1969) but this does not exclude the possibility of humoral effects which may occur at a distance. Owen and Raff (1971) also suggested that following the first differentiation from stem cell to thymus lymphocyte which occurs in the microenvironment of the thymus, a further differentiation sequence is necessary for the transformation of these cells to the fully functional thymus derived lymphocytes or T cells. This stage also involves surface antigenic modulation which, in TL positive strains results in a loss by the peripheral T cells of the TL alloantigen. Where in the body this second step occurs is far from clear. It has recently been shown that a small pool of immunocompetent, cortisone resistant cells are present in the normal mouse thymus (Andersson and Blomgren, 1970). The origin of these cells is still uncertain, but it is likely that they are the progeny of the cortical thymic lymphocytes. Similar studies, performed with lymphocytes differentiated in vitro should establish whether such cells are really derived from the original stem cells or whether they enter the thymus secondarily. The use of organ cultures will also enable the sequential development of antigenic changes to be studied in a single population undiluted by the progeny of fresh stem cells and by the complex migratory patterns which occur within the intact animal.

The production of one or more humoral factors by the thymus has been frequently suggested but the nature and role of this factor or factors is still unclear. The presence of a factor in thymus-conditioned media reported in this study suggests that it may act on cells which have been seeded from the thymus before birth and which may be at an immature stage of development corresponding to the first stage of differentation suggested by Owen and Raff (1970).

The assay system used in the present study was the development of direct plaques by cells forming antibody to SRBC (Cunningham and Szenberg, 1968). The antibody response to SRBC is a highly thymus dependent system but the cells actually producing the antibody are not T cells (Nossal et al., 1968). Presumably the factor present in the thymus-conditioned medium acts either on immature T cells already present in the spleen of neonatally thymectomized mice and allows these cells to express their potential for collaboration with antibody forming cell precursors, or it in some way replaces entirely the function of T cells. A similar factor has been described by Byrd (1971) in some batches of foetal calf serum, but the relationship between this factor and the substance present in the thymus-conditioned media has not been elucidated. It is also quite uncertain whether either or both of these factors are true thymic hormones or whether they are non-specific adjuvants similar to synthetic polynucleotides (Braun et al., 1971; Johnson et al., 1971).

The use of organ cultures can yield much data on problems of cell interactions in differentation and the thymus-induced differentation steps in the development of T cells from stem cells are particularly suitable for such studies. The developing lymphocytes can be characterized by a number of parameters at various stages of their differentiation by techniques already available.

Acknowledgements: This study was supported by the Jane Coffin Childs Memorial Fund for Medical Research and the National Health and Medical Research Council of Australia; equipment was provided by the Australian Research Grants Committee, J. B. Were and Sons and the Potter Foundation.

REFERENCES

Andersson, B. and H. Blomgren, (1970), Evidence for a small pool of immunocompetent cells in the mouse thymus. Its role in the humoral antibody response against sheep erythrocytes, bovine serum albumin, ovalbumin and the NIP determinant, Cell Immunol. 1, 362.

Aoki, T., U. Hämmerling, E. de Harven, E.A. Boyse and L.J. Old, (1969), Antigenic structure of cell surfaces. An immunoferritin study of the occurence and topography of H-2, θ and TL alloantigens on mouse cells, J. Exp. Med. 130, 979.

Blackburn, W.R. and J.F.A.P. Miller, (1967), Electron microscopic studies of thymus graft regeneration and rejection, I. Syngeneic grafts, Lab. Invest. 16,66.

Braun, W., M. Ishizuka, Y. Yajima, D. Webb and R. Winchurch, (1971) in Biological Effects of Polynucleotides (ed. by R. Beers and W. Braun), Springer Verlag, New York.

Byrd, W., (1971), Restoration of the immune response to sheep erythrocytes by a serum factor, Nature, New Biology 231, 280.

Clark, S. L. Jr., (1963), The thymus in mice of strain 129/J studied with the electron microscope, Amer. J. Anat. 112, 1.

Clark, S.L. Jr., (1966), Cytological evidences of secretion in the thymus. In: Ciba Foundation Symposium, The thymus: Experimental

and clinical studies (ed. by G.E.W. Wolstenholme and R. Porter), London J. and A. Churchill Ltd., p. 3
Clark, S.L. Jr., (1968), Incorporation of sulfate by the mouse thymus: its relation to secretion by medullary epithelial cells and to thymic lymphopoiesis, J. Exp. Med. 128, 927.
Cunninghan, A.J., and A. Szenberg, (1968), Further improvement in the plaque technique for detecting single antibody forming cells, Immunology 14, 599.
Gad, P. and S.L. Clark Jr., (1968), Involution and regeneration of the thymus in mice, induced by bacterial endotoxin and studied by quantitative histology and electron microscopy, Amer. J. Anat. 122, 573.
Hoshino, T., (1963), Electron microscopic studies of the epithelial reticular cells of the mouse thymus, Z. Zellforsch. 59, 513.
Ito, T. and T. Hoshino, (1966), Fine structure of the epithelial reticular cells of the medulla of the thymus in the golden hamster, Z. Zellforsch. 69, 311.
Jerne, N.K. and A.A. Nordin, (1963), Plaque formation in agar by single antibody-producing cells, Science 140, 405.
Johnson, A.G., R.E. Lowe, H.M. Friedman, I.H. Han, H.G. Johnson, J.R. Schmidtke and R.D. Stout, (1971) in Biological Effects of Polynucleotides (ed. by R. Beers and W. Braun), Springer Verlag, New York
Klug, H., (1967), Submikroskopische Zytologie der Thymus von Ambystoma mexicanum, Z. Zellforsch. 78, 379.
Mandel, T., (1968a), Ultrastructure of epithelial cells in the cortex of guinea pig thymus, Z. Zellforsch. 92, 159.
Mandel, T., (1968b), Ultrastructure of epithelial cells in the medulla of guinea pig thymus, Aust. J. Exp. Biol. Med. Sci. 46, 755.
Mandel, T., (1969), Epithelial cells and lymphopoiesis in the cortex of guinea pig thymus, Aust. J. Exp. Biol. Med. Sci. 47, 153.
Mandel, T., (1970), Differentiation of epithelial cells in the mouse thymus, Z. Zellforsch. 106, 498.
Mandel, T. and P.J. Russell, (1971), Differentiation of foetal mouse thymus. Ultrastructure of organ cultures and of subcapsular grafts, Immunology 21, 659.
Marbrook, J., (1967), Primary immune response in cultures of spleen cells, Lancet 2, 1279.
Metcalf, D., (1963), The autonomous behaviour of normal mouse thymus grafts, Aust. J. Exp. Biol. Med. Sci. 41, 437.
Nossal, G.J.V., A. Cunningham, G.F. Mitchell and J.F.A.P. Miller, (1968), Cell to cell interaction in the immune response. III Chromosome marker analysis of single antibody-forming cells in reconstituted, irradiated or thymectomized mice, J. Exp. Med. 128,839.
Owen, J.J.T. and M.C. Raff, (1970), Studies on the differentiation of thymus derived lymphocytes, J. Exp. Med. 132, 1216.
Owen, J.J.T. and M.A. Ritter, (1969), Tissue interaction in the development of thymus lymphocytes, J. Exp. Med. 129, 431.
Raff, M.C., (1971), Surface antigenic markers for distinguishing T and B lymphocytes in mice, Transplant. Review 6, 52.
Raviola, E. and G. Raviola, (1967), Striated muscle cells in the thymus of reptiles and birds: an electron microscopic study, Amer. J. Anat. 121, 623.
Schlesinger, M. and D. Hurvitz, (1968), Serological analysis of thymus and spleen grafts, J. Exp. Med. 127, 1127.
Trainin, N., M. Small and A. Globerson, (1969), Immunocompetence of spleen cells from neonatally thymectomized mice conferred *in vitro* by a syngeneic thymus extract, J. Exp. Med. 130, 765.

IS AN ANTIGEN BRIDGE REQUIRED FOR T AND B CELL COOPERATION IN THE IMMUNE RESPONSE?

G. M. IVERSON

TUMOUR IMMUNOLOGY UNIT, ZOOLOGY DEPT., UNIVERSITY COLLEGE, LONDON

Immunity against both hapten and carrier protein is required to elicite a maximum secondary anti-hapten response (carrier-effect) (Ovary and Benacerraf, 1963). The carrier-effect, in mice, has been shown to be the result of an act of antigen mediated cellular cooperation between at least two types of lymphocytes (Claman et al 1966; Davies et al 1967; Mitchison 1967; Rajewsky 1967; and Mitchell et al 1969). One is derived from the thymus and the other from the bone-marrow (Raff 1970). The thymus derived or T lymphocyte requires an intact thymus for maturation (Miller 1962; Ford et al 1962; Leuchars et al 1965; and Davies et al 1969) and is found in varying proportions in the peripheral lymphoid tissue (Raff 1969; Schlesinger 1969). The T lymphocyte does not secrete any detectable antibody (Harris et al 1948) but is involved in both humoral and cell mediated immune responses (Miller 1961). In the humoral immune response it is referred to as a helper cell. The bone-marrow derived or B lymphocyte develops independently of the thymus. The B lymphocyte is the antibody forming cell precursor (AFCP) (Mitchell and Miller 1969). The thymus derived lymphocyte, or helper cell, binds antigen by means of a carrier determinant and presents the haptenic determinant to a bone-marrow derived lymphocyte, B cell, which is then some how triggered to produce antibody.

Experiments have been designed to show that haptens may also act as the carrier determinant (Iverson 1970; Taylor and Iverson 1970; Mitchison 1970; and Mitchison 1971). The results from such an experiment are given in Table 1.

Table 1
Demonstration of DNP acting as a carrier determinant for BSA

Source of primed cells		ug/ml anti-BSA
Helper	AFCP	
Normal spleen	BSA spleen	0.50 ± 0.25
FDNB lymph node	Normal spleen	0.15 ± 0.10
DNP_8CGG spleen	BSA spleen	4.00 ± 1.51
FDNB lymph node	BSA spleen	6.50 ± 0.75

all mice boosted with DNP_8BSA after the cell transfer

Cells from mice immunized to 2-4-dinitrophenol (DNP) were mixed in irradiated syngenic recipients with cells from mice immunized to bovine serum albumin (BSA). The mice immunized to BSA were treated with anti-thymocyte serum (ATS) shortly before the spleens were harvested. Mitchison (1970) has shown that such treatment with ATS sufficiently depletes the spleen of helper cells, in this case BSA helper cells. Such treatment does not significantly effect the B cell population, in this case BSA antibody forming cells. The recipient mice were then boosted with DNP_8BSA. The groups receiving only BSA primed cells or DNP sensitive cells from mice sensitized to 1-fluro-2-4-dinitrobenzen (FDNB) did not respond. Those groups that received cells from both DNP and BSA primed mice did respond with an anti-BSA response.

Experiments designed to demonstrate that a hapten (DNP) may act as the carrier determinant for yet another hapten (NIP) have given interesting results. Mice were immunized to DNP by painting their abdomen with FDNB. Other mice were immunized with 5-iodo-3-nitrophenacetyl conjugated to chicken gamma globulin ($NIP_{12}CGG$). Lymphocytes from these mice were mixed in irradiated syngenic hosts and boosted with the double conjugate DNP BSA-NIP . One week later the mice were bled and their sera titred for anti-NIP antibody. As can be seen from Table 2 there is no increase in titre of anti-NIP antibody in the groups receiving DNP sensitive helper cells. In fact the reverse is true.

Table 2

Demonstration of the inability of DNP to act as a carrier determinant for NIP

Source of primed cells		Antigen Binding Capacity
Helper	AFCP	(10^{-8}M) Anti-NIP
Normal lymph node	NIP-CGG spleen	153.49 ± 1.81
FDNB lymph node	NIP-CGG spleen	98.61 ± 1.19

all mice boosted with $DNP_6BSA\text{-}NIP_{11}$ after the cell transfer

It was such experimental results that lead Taylor and myself (1970) to propose the homospecific exclusion hypothesis. Simply stated this proposes that when one receptor on the helper cell binds an antigenic determinant then the rest of the similar antigenic determinants on the molecule will also be bound by the same T cell. This then somehow leaves the other haptenic determinants unavailable for the B cell.

This hypothesis predicts that a single DNP group would be able to act as a carrier determinant for NIP. More than one DNP would not be able to do so. To test this hypothesis the following double conjugates were made; $DNP_{SH}BSA\text{-}NIP_6$, $DNP_{SH}BSA\text{-}NIP_{11}DNP_5$ and $DNP_{SH}BSA\text{-}NIP_6Ox_5$. BSA was mildly reduced then alkylated with DNP-malic acid (kindly supplied by Dr. M. Green, NIMR, Mill Hill, London). Spectrophotometry confirmed that there was a single DNP group bound to each BSA molecule. This particular DNP group is covalently linked to an SH group on BSA. For clarity this conjugation is noted by writing DNP before the protein carrier with a subscript $_{SH}$. This $DNP_{SH}BSA$ conjugate was divided into three aliquots and each aliquot was further conjugated with more hapten. NIP was conjugated to one, NIP and more DNP to another and the third was conjugated with both NIP and oxazolone (Ox). These later conjugations were made to the epsilon amino groups of lysine residues. These conjugations are designated by writing the hapten to the right of the protein carrier and the degree of substitution is noted with a subscript.

Each of these compounds was used to boost irradiated hosts that had received cells from mice primed to NIP-CGG plus either normal or DNP sensitive helper cells. The hypothesis predicted that DNP sensitive helper cells would cooperate with NIP sensitive B cells if presented with either $DNP_{SH}BSA\text{-}NIP_6$ or $DNP_{SH}BSA\text{-}NIP_6Ox_5$. On the other hand $DNP_{SH}BSA\text{-}NIP_{11}DNP_5$, a multi DNP conjugate, should not benefit from the DNP sensitive helper cells. As shown in Table 3 these predictions were met.

A further prediction would be that DNP sensitive helper cells but not Ox sensitive helper cells, would be able to cooperate with NIP sensitive B cells if presented with $DNP_{SH}BSA\text{-}NIP_6Ox_5$. As shown in Table 3 this prediction was also satisfied. The difference seen in the anti-NIP response in the three different groups is influenced by that portion of the immunogen being recognized by the helper cell. The helper cell must in some way influence B cell recognition. This means that a B cell must recognize the same immunogenic molecule as the T cell. This observation is substantiated by experiments from Hamaoka et al (1971). Using a double hapten protein conjugate they showed that they could reduce the anti-hapten response to one hapten but not to the other by first mixing the double hapten protein conjugate with antibody directed to one or the other haptens. On the other hand they could deplete the response to both haptens and carrier protein by mixing the double conjugate with antiprotein carrier antibody.

Table 3

Comparison of anti-NIP response with three different double conjugates

Source of primed cells: Helper (lymph node)	AFCP (spleen)	Boosting immunogen	Antigen Binding Capacity (10^{-8}M) Anti-NIP
Normal	NIP-CGG	$DNP_{5,11}$ BSA-NIP_{6}	28.64 ± 1.00
FDNB	NIP-CGG		249.01 ± 2.02
Normal	NIP-CGG	$DNP_{5,11}$ BSA-NIP_{11} DNP_{5}	119.68 ± 2.47
FDNB	NIP-CGG		100.10 ± 4.01
Normal	NIP-CGG		76.23 ± 1.51
FDNB	NIP-CGG	$DNP_{5,11}$ BSA-NIP_{6} Ox_{5}	252.82 ± 1.38
Ox	NIP-CGG		135.03 ± 1.38

It has been shown that for T and B cells to cooperate in the immune response the minimal requirements are (1) the immunogen must have at least two qualitatively different antigenic determinants (Rajewsky et al 1969; Taylor and Iverson 1970), (2) these determinants must be covalently linked (Mitchison 1967; Hamaoka et al 1971) and (3) both the T and B cell must recognize the same immunogenic molecule (Hamaoka et al 1971). The simplest way to account for all of these requirements is to postulate an antigen bridge between the two cells. The T cell binding the immunogen by one antigenic determinant, the carrier, and the B cell binding the same immunogen but to a qualitatively different antigenic determinant, the 'hapten-equivalent'.

Why must there be an antigen bridge? Indications come from experiments in Benacerraf's group. Benacerraf et al (1967 have shown that strain 13 guinea pigs are genetically non-responders to DNP-poly-L-lysine (DNP-PLL) and that this non-responsiveness is at least partially due to a lack of T cells capable of recognizing PLL. Katz et al (1971) have shown that when strain 13 guinea pigs are primed to DNP-ovalbumin and bovine gamma globulin (BGG), given DNP-PLL and boosted with DNP-BGG they are unresponsive to DNP. The interpretation of these results is that when the DNP-PLL was administered to the DNP-ovalbumin and BGG primed strain 13 guinea pigs it could only bind to B or T cells through the DNP epitope but not to both. In this instance there is a genetic lack of T cells capable of reacting with the carrier determinant PLL. Little or no antibody was produced when the animals were challenged with hapten conjugated to a carrier that could be recognized by the T cells. On the other hand when strain 13 guinea pigs primed to DNP-ovalbumin are given both DNP-PLL and allogenic guinea pig lymphocytes (to produce a graft versus host reaction) a secondary response to DNP was observed (allogenic effect). DNP-PLL can act as a tolerogen by binding to B cells alone. On the other hand DNP-PLL can act as an immunogen. The latter requires, as well as binding to B cells, a graft versus host reaction going on at the same time. This suggests that the T cells responsible for the graft versus host reaction are releasing a factor that is stimulating the B cells that have bound the DNP-PLL. This factor must operate only at short range. If the factor could function over a long range then everytime a T cell was stimulated it would in turn stimulate all the B cells. Being operable over a short range requires that for B cells to be stimulated they must be close to the activated T cells. One way of ensuring this is to require an antigen bridge between the two cells.

REFERENCES

Benacerraf, B., Green, I. and Paul, W.E. (1967), The immune response of guinea pigs to hapten-poly-L-lysine conjugates as an example of the genetic control of the recognition of antigenicity, Cold Spring Harbor Symp. Quant. Biol. 32, 569.

Claman, H.N., Chaperon, E.A. and Triplett, R.F. (1966), Thymus-marrow cell combination. Synergism in antibody production, Proc.Soc.Exp.Biol., N.Y., 122, 1167.

Davies, A.J.S., Leuchars, E., Wallis, V., Marchant, R. and Elliott, E.V. (1967), The failure of thymus derived cells to produce antibody, Transplantation, 5, 222.

Ford, C.E. and Micklem, H.S. (1963), The thymus and lymph nodes in radiation chimaeras, Lancet, i, 359.

Hamaoka, T., Takatsu, K. and Kitagawa, M. (1971), Antibody production in mice. IV. The suppressive effect of anti-hapten and anti-carrier on the recognition of hapten-carrier conjugates in the secondary response, Immunology, 21, 259.

Harris, T.N., Rhoads, J. and Stokes, J.A. (1948), A study of the role of the thymus and spleen in the formation of antibodies in the rabbit, J. Immunology, 58, 27.

Iverson, G.M. (1970), The ability of CBA mice to produce anti-idiotype to 5563 myeloma protein, Nature (London), 227, 273.

Katz, D.H., Davies, J.M., Paul, W.E. and Benacerraf, B. (1971), Carrier function in anti-hapten antibody responses. IV. Experimental conditions for the induction of hapten specific tolerance or the stimulation of anti-hapten anamnestic responses by 'nonimmunogenic' hapten-polypeptide conjugates, J.Exp.Med., 134, 201.

Leuchars, E., Cross, A.M. and Dukor, P. (1965), The restoration of immunological function by thymus grafting in thymectomized irradiated mice, Transplantation, 3, 28.

Miller, J.F.A.P. (1961), Immunological function of the thymus, Lancet, i, 748.

Miller, J.F.A.P. (1962), Effect of neonatal thymectomy on the immunological responsiveness of the mouse, Proc.Roy.Soc., B, 156, 415.

Mitchell, G.F. and Miller, J.F.A.P. (1968), Cell to cell interaction in the immune response. II. The source of hemolysin-forming cells in irradiated mice given bone marrow and thymus or thoracic duct lymphocytes, J.Exp.Med., 128, 821.

Mitchison, N.A. (1967), Antigen recognition responsible for the induction in vitro of the secondary response, Cold Spring Harbor Symp. Quant. Biol., 32, 431.

Mitchison, N.A. (1970), Mechanism of action of antilymphocyte serum, Fed.Proc., 29, 222.

Mitchison, N.A. (1971), The carrier effect in the secondary response to hapten-protein conjugates. V. Use of antilymphocyte serum to deplete animals of helper cells, Eur.J.Immunology, 1, 68.

Ovary, Z. and Benacerraf, B. (1963), Immunological specificity of the secondary response with dinitrophenylated proteins, Proc.Soc.Exp.Biol. N.Y., 114, 72.

Raff, M.C. (1969), Theta isoantigen as a marker of thymus-derived lymphocytes in mice, Nature (London), 224, 378.

Raff, M.C. (1970), Role of thymus-derived lymphocytes in the secondary humoral immune response in mice, Nature (London), 226, 1257.

Rajewsky, K. (1967), Tolerance specificity and the immune response to lactic dehydrogenase isoenzymes, Cold Spring Harbor Symp. Quant. Biol., 32, 547.

Rajewsky, K., Schirrmacher, V., Nase, S. and Jerne, N.K. (1969), The requirement of more than one antigenic determinant for immunogenicity, J.Exp.Med., 129, 1131.

Schlesinger, M. and Yron, I. (1969), Antigenic changes in lymph node cells after administration of antiserum to thymus cells, Science, 164, 1412.

Taylor, R.B. and Iverson, G.M. (1970), Hapten competition and the nature of cell-cooperation in the antibody response, Proc.Roy.Soc.Lond., B, 176, 393.

SPECIFICITY AND SUPPRESSION IN THE HELPER SYSTEM

K. RAJEWSKY, C. BRENIG AND I. MELCHERS
Institut für Genetik der Universität Köln
Weyertal 121, D-5 Köln 41, W. Germany.

Abstract: Further evidence is provided supporting the view that the specificity of B-cell receptors represented by humoral antibodies, and the specificity of helper cell receptors are fundamentally similar. The similarities are particularly striking if the specificity of high affinity antibody is taken into account.

An attempt was made to determine the extent and specificity of paralysis induced in helpers and AFCP by low and high doses of antigen (high zone and low zone paralysis). Paralysis was stable and nearly complete in the helpers in both cases, and had the specificity which would be predicted on the basis of antibody specificity. In the AFCP, paralysis was detectable only under high zone conditions, and appeared to affect high affinity cells only.

Differences between T- and B-cell specificity might thus be brought about by differential suppression in paralysis and the possible requirement of a high affinity of the helper cell receptor to the antigen, in order to obtain efficient help. Both mechanisms would increase the selectivity of the helper system in recognising foreign substances as immunogens.

1. INTRODUCTION

The significance of the bicellular mechanism operating in antibody induction is still under debate. In this article we are following a line of thought, which attributes to the helper cells the task of determining and limiting the animal's range of responsiveness. This hypothesis states that the number of epitopes which can combine with helper cell receptors in such a way that cooperative

---

Abbreviations: BSA, bovine serum albumin. SSA, sheep serum albumin. DSA, deer serum albumin. GSA, giraffe serum albumin. CG, chicken gamma globulin. NIP-CAP, N-(4-hydroxy-5-iodo-3-nitro-phenacetyl-)aminocaproic acid. ABC, antigen binding capacity. AFCP, antibody forming cell precursors.

activation of the antibody forming cell precursors (AFCP) can take place is limited, and opposed to an abundance of epitopes that might activate AFCP via cooperation. Indeed, a high degree of selectivity at the T-cell level is suggested by the results of a number of studies (reviewed by Paul, 1970). In addition, studies on the genetic determination of immunological responsiveness suggest that in many instances specific defects might be localised in the helper cell system (reviewed by McDevitt and Benacerraf, 1969).

A number of mechanisms have been proposed which would lead to a limited spectrum of specificities at the level of the helper system. These range from the proposal of a specific helper cell immunoglobulin expressing restricted variability on a genetic basis (Rajewsky, 1969) to the idea that only helper cells with exceedingly high receptor specificities for the antigen could be triggered by antigen (Taylor and Iverson, 1971). Another suggestion is of relevance in this context, namely that the specificity pattern in the T and B line might be different because of differential selection in ontogeny (Paul, 1970; Rajewsky, 1971). This differential selection hypothesis can now be elaborated further in the sense of differential suppression. The argument is based on the recent notion (Chiller et al., 1971; Mitchison, 1971a; Rajewsky, 1971), further extended in the present article, that helper cells can be specifically paralysed by much lower doses of antigen than AFCP. Selective paralysis of helper function has indeed been observed in many instances before (Rajewsky and Rottländer, 1967; Green et al., 1968; Taylor, 1969; Miller and Mitchell, 1970).

In this study an attempt is made to determine the specificity of the helper system under conditions of immunity and tolerance, by comparing it to the specificity at the level of the AFCP. The data support the concept that helper cell receptor specificity reflects the specificity of immunoglobulin combining sites, in particular if the specificity of high affinity antibody is taken into account. Profound paralysis of the expected specificity is induced in the helper population by low antigen doses.

2. MATERIALS AND METHODS

Most of these have been described in full detail in our previous communication (Rajewsky and Pohlit, 1971). We list them here briefly, with a few additional points.

Animals. - Male and female inbred CBA/J mice.

Antigens. - BSA ("reinst", Behringwerke), SSA (Pentex), DSA, GSA and CG (Pentex) were the protein carriers. SSA was further purified by gel filtration through Sephadex G-1oo and chromatography on DEAE Sephadex as described previously. DSA and GSA were prepared in the same way, after precipitation from serum by ammonium sulphate (5o-8o% saturation). The following NIP-carrier-conjugates were used: NIP_{14}-BSA, NIP_{10}-SSA, NIP_{10}-DSA, NIP_6-GSA and NIP_6-CG. In binding inhibition experiments, the NIP-albumin carriers were as active in inhibiting the albumin-anti-albumin reaction as the native proteins, indicating that the coupling procedure had left the native protein determinants intact.

Serology (using a modified Farr assay with iodinated, radioactive antigens), cell transfers and immunizations were essentially as previously described. The standard antigen doses for priming were 150 µg NIP-CG, 400 µg SSA and 800 µg BSA per mouse (alum precipitated plus pertussis), i.e. saturating doses. The doses for secondary stimulation in transfer experiments were 0.1 µg NIP-CG and 10 µg NIP-albumin carrier per host, unless stated otherwise. When cells from paralysed animals were to be tested for helper activity, as many spleen cells as possible were transferred, up to 5×10^7 per recipient. Unpublished extensive experiments have shown that in the conditions of our experiments, the secondary anti-NIP response is directly proportional to the number of helper cells, and the responses could therefore be normalized. For the calculation of helper activity in the paralysis experiments the results were corrected for "overriding" (i.e. the anti-hapten response upon transfer of NIP-CG-primed cells only and stimulation with NIP-albumin carrier), but in most cases overriding was negligible. The dose response curves in the cross stimulation experiments were not corrected for overriding. Overriding was not detectable or minute at doses below 10 µg, and variable above, amounting at the highest antigen dose (100 µg) occasionally to 20% of the cooperative response. NIP-SSA exhibited the strongest tendency to override, NIP-BSA the weakest, NIP-DSA and NIP-GSA were intermediate. The strong variations in the extent of overriding by the various hapten carrier complexes occured in a non-correlated fashion, and there was also no obvious correlation between overriding and the relative positions of the dose-response curves in the cooperating systems.

Paralysis. - Low zone and high zone paralysis were induced following the classical Mitchisonian scheme (Mitchison, 1968). 4-5 month old mice were given 480 r in a 190 kV Siemens X-ray machine (0.5 mm Cu) followed by 3 weekly injections of 10 µg (low zone) or 10 mg (high zone) BSA over a period of 10 wk. 2 wk after the last injection, the animals were primed with SSA or BSA. In all experiments, the response to BSA was $\leq$ 6% in low zone, and $<$ 1.5% in high zone paralysed animals as compared to the controls.

3. EXPERIMENTAL DESIGN

In animals immunized to an antigen we presume an abundance of specific antibody forming cell precursors (AFCP) and of specific helper cells. Since the receptor specificities of the AFCP and the specificity of humoral antibody are presumed identical, AFCP receptor specificity is assessed by immunochemical analysis of the specificity of humoral antibody. Helper cell receptor specificity is more difficult to determine. However, the cooperation hypothesis offers an approach to this problem, since it states that the induction of anti-hapten antibody by a hapten-carrier complex requires the interaction of carrier determinants with the helper cell receptor (Mitchison et al., 1970). Therefore, in the presence of constant amounts of hapten-specific AFCP, the anti-hapten response resulting from the injection of a hapten-carrier complex would reflect the reaction of the helper cells with the carrier molecule.

We use an experimental design which is based on these considerations (Rajewsky and Pohlit, 1971; Fig. 1).

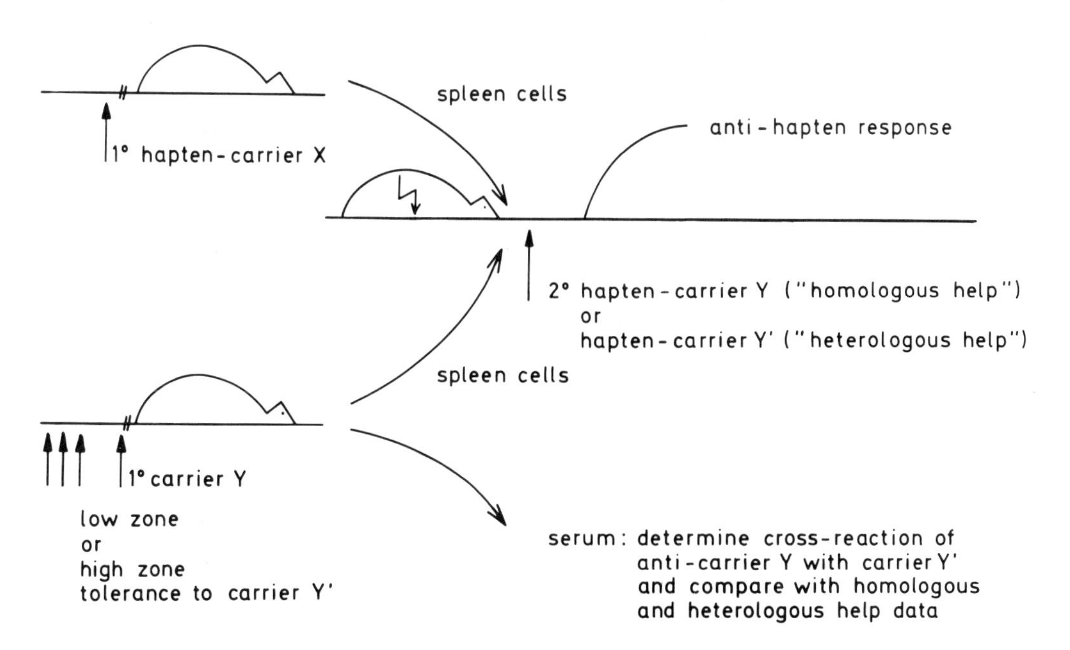

Fig. 1. Experimental Design.

Animals are primed to a suitable carrier molecule Y. 2-4 months later, the animals are bled and their spleens are excised. A spleen cell suspension is prepared and the cells are injected into irradiated syngeneic hosts, together with spleen cells from syngeneic animals, which had been primed with a hapten coupled to a non-cross-reacting carrier X. Groups of hosts are then stimulated with varying doses of hapten coupled to carrier Y or a cross reacting carrier Y', and the secondary anti-hapten responses are determined. Cross stimulation patterns are obtained, which are interpreted to reflect the cross reaction pattern of the carrier molecules at the level of the helper cells. These patterns can be qualitatively compared with the serological cross reactions of the carrier molecules. The serological cross reactions are established by using the antisera obtained from the spleen cell donors that had been immunized with carrier Y. Antigen binding capacities (ABC) for the carriers Y and Y' are determined in these antisera at various antigen concentrations. The ratio of homologous ABC (i.e. ABC for carrier Y) to heterologous ABC (i.e. ABC for carrier Y') is taken as an expression of cross reaction, and its determination over a range of antigen concentrations allows us to compare cross reactions detected by (preferentially) high affinity antibodies to those detected by (preferentially) low affinity antibodies.

The system in addition allows an analysis of the extent and the specificity of paralysis in helper and precursor populations under various conditions. Consider an animal tolerant to carrier Y' and then immunized with carrier Y, the classical Weigle-type tolerance breakdown situation (Weigle, 1961). The extent and the specificity of paralysis in the helper cells can now be determined in the co-operating cell transfer system as described above. In addition, the determination of ABCs for carriers Y and Y' and of their ratios in the animal's serum should enable us to see to what extent the AFCP are affected by the paralysing regimen. Finally, the paralysis experiments are expected to confirm the results of the cross stimulation data and the correctness of their comparison with the cross reaction patterns. If the helper cells reacting with carrier Y' are eliminated by paralysis, the remaining helper activity for carrier Y should correspond to that activity which under normal conditions Y-primed helper cells exhibit for hapten-carrier Y only, i.e. in addition to the cross reacting helper activity for hapten-carrier Y'. Furthermore, if the specificity distribution in helper and precursor cell receptors is similar, then the proportion of helper activity for carrier Y in Y'-tolerant animals should correspond to the proportion of anti-Y antibody, which does not cross react with Y'.

Our intention to measure paralysis in helpers and precursors has largely dictated the choice of suitable carrier molecules. To be able to make use of the careful studies of low zone and high zone paralysis to BSA in mice (Mitchison, 1968), we have purified several cross reacting serum albumins. These offer a number of additional advantages for our system: They can be easily purified in large amounts, iodination or coupling to the NIP-group (used as hapten in these experiments) does not appreciable destroy their native determinants (see MATERIALS AND METHODS), their binding to antibodies can be determined by the Farr assay and their behaviour in the cooperating transfer system is well established (Mitchison, 1971b).

4. CROSS REACTION AND CROSS STIMULATION

The experiments reported in this paper extend previous work, in which the specificity of helpers and AFCP towards BSA and SSA was compared (Rajewsky and Pohlit, 1971).

It was shown in the previous experiments that the two serum albumins exhibited a characteristic cross reaction pattern: Anti-BSA antibodies, over the whole range of antigen concentrations, scarcely distinguished between BSA and SSA. In the case of anti-SSA, this was also true for antibodies of low avidity. However, anti-SSA contained relatively more highly avid antibodies than anti-BSA, and these reacted much better with SSA than with BSA. A higher distinguishing power of high than of low affinity antibodies is expected in a system like ours, where cross reaction is largely mediated by cross reacting antigenic determinants on related protein molecules (see our previous communication). The phenomenon appears again in the present experiments (see Fig. 3).

A-reciprocity was similarly apparent in the cross stimulation pattern, in that NIP-BSA and NIP-SSA could nearly equally well make use of BSA-primed helpers, whereas SSA-primed helper cells distinguished between the two albumin carriers. Again, the latter phenomenon reappears in this study, in the upper section of Table 1. The data suggested that in this system helpers and AFCP exhibit a similar pattern of receptor specificities. The subsequent paralysis experiments, reported below, strongly support this view and confirm the validity of the cross stimulation approach. It is particularly satisfactory to see from these data that the differences in the capacity of the individual albumin carriers to override carrier specificity (see under MATERIALS AND METHODS) and the presence in the animals of carrier -primed B-cells and their products do not appear to grossly affect the cross stimulation results.

Encouraged by the overall consistency of the data, we have extended our analysis to further serum albumin carriers. Fig. 2 shows an experiment, in which three albumin carriers, BSA, SSA and DSA were used, and two types of helper cells, either BSA- or SSA-primed.

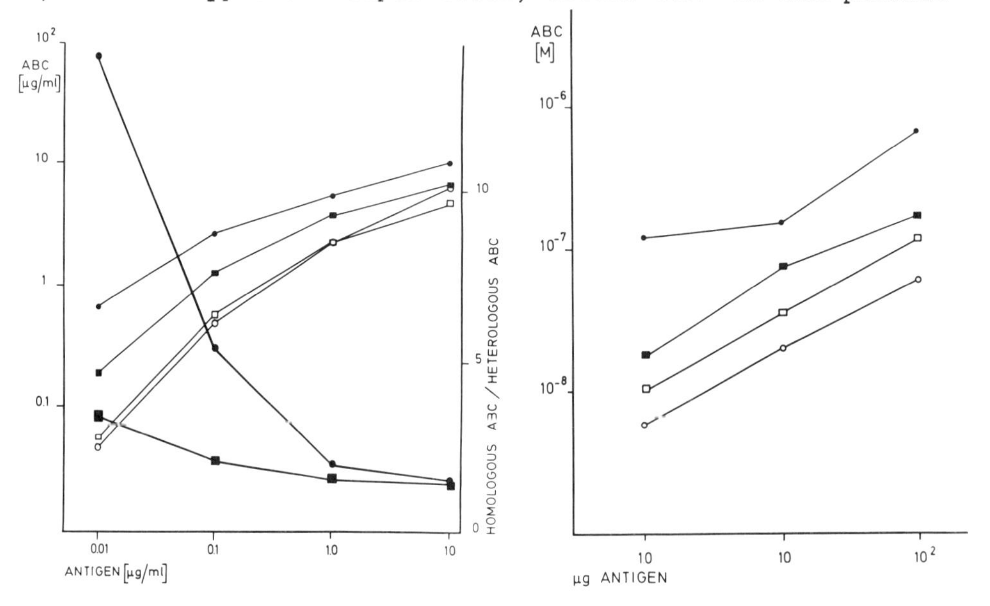

Fig. 2a. Cross reaction of SSA and DSA in anti-SSA (circles) and of BSA and DSA in anti-BSA (squares). Abscissa: antigen concentration at which ABC was determined. Small closed symbols represent homologous, small open symbols heterologous ABCs (left ordinate). Large symbols represent ratio of homologous to heterologous ABC (right ordinate).

2b. Dose response analysis of the anti-NIP response in homologous and heterologous stimulation. Ordinate, molar NIP binding capacity at $1o^{-8}$M NIP-CAP. Abscissa: antigen dose per host. Circles, NIP-CG-primed cells stimulated via SSA-primed helpers, by NIP-SSA (closed circles) and NIP-DSA (open circles). Squares, NIP-CG-primed cells stimulated via BSA-primed helpers, by NIP-BSA (closed squares) and NIP-DSA (open squares). Each point represents geometric mean of a group of 6 animals. Response of the NIP-CG-primed cells to 0.1 μg NIP-CG around $5x10^{-6}$M NIP-binding capacity. Standard deviations of mean log ABC ranged from o.o6 to o.24, corresponding to factors of 1.1 to 1.7. For "overriding" see MATERIALS AND METHODS.

The cross reactions of BSA and DSA in anti-BSA and of SSA and DSA in anti-SSA are represented in Fig. 2a. The patterns are strikingly different in the two antibody systems. In the anti-SSA system, the ratio of homologous to heterologous (i.e. for DSA) ABC increases dramatically with decreasing antigen concentration, indicating again that highly avid antibodies distinguish better between the homologous and heterologous antigen than antibodies of low avidity. In contrast, in anti-BSA serum, binding of BSA and DSA is rather homogenous over the whole range of antigen concentrations. Note in this connection that homologous ABCs decrease relatively more with decreasing antigen concentration in anti-BSA than in anti-SSA. This tendency, not very pronounced in the present experiment, is generally observed in our system and indicates that anti-BSA contains relatively less antibodies of high avidity than anti-SSA, as mentioned above.

Thus, at antigen concentrations in the order of 10^{-7}M, anti-BSA and anti-SSA distinguish between the homologous antigen and the cross reacting albumin equally poorly, whereas the two antisera behave differently if cross reaction is determined at low antigen concentrations. Under these conditions anti-SSA is much more selective than anti-BSA.

We now compare these data with the cross stimulation patterns (represented in Fig. 2b). It can be seen that when BSA-primed helpers are used, the anti-hapten response is almost equally well mediated by NIP-BSA and NIP-DSA, in good agreement with the cross reaction pattern. In contrast, SSA-primed cells help the anti-hapten response much more efficiently via NIP-SSA than via NIP-DSA. This pronounced discriminating power finds its analogy in the cross reaction pattern as determined by antibody of high rather than low avidity.

The experiment represented in Fig. 2 was repeated, with an almost identical result. Furthermore, two analogous experiments, in which our fourth serum albumin carrier, GSA, was used instead of DSA, also led to a very similar result: GSA exhibited a similar cross reaction pattern as DSA when compared with BSA and SSA in their homologous antisera. Again, in cross stimulation, BSA-primed helpers could be used almost equally well by NIP-GSA and NIP-BSA, whereas in the presence of SSA-primed helpers NIP-SSA induced a much higher anti-NIP response than NIP-GSA (In the latter situation, the ratios of homologous to heterologous help in the two experiments were 12.4 and 11.5 at 1.0, 13.4 and 7.1 at 10, and 4.8 and 2.9 at 34 μg antigen per host).

A final comment might be useful. The anti-SSA and anti-BSA sera, which we used to determine the cross reactions as represented in Fig. 2a, were pools of sera not from the donors but from the hosts in the transfer experiment, namely from those groups that had received the highest dose of NIP coupled to the homologous carrier (and therefore had the highest content of anti-carrier antibody). In the anti-SSA serum, we have also determined the cross reaction of SSA and BSA and have found the ratios of homologous to heterologous ABC at low antigen concentrations to be similar to those of SSA and DSA, and thus lower than in primary antisera (see Fig. 3 and our previous communication). This should be kept in mind, when cross stimulation by NIP-BSA and NIP-DSA is compared.

In summary, a satisfactory analogy between cross reaction and cross stimulation patterns is observed, in particular if cross reactions as measured by highly avid antibody are taken into account.

5. PARALYSIS

Mice, low zone or high zone tolerant to BSA, were primed with SSA. Control animals (not paralysed) were always included in the experiments, and also the success of paralysis induction was verified by immunization with BSA. 3-6 weeks later (see Table 1), the animals were bled and their spleens excised. Paralysis in the splenic helper cell population was determined in the cooperating and cross stimulating transfer system, and paralysis in the AFCP was assessed indirectly, by determining SSA/BSA cross reaction in the spleen cell donors'sera, according to the experimental scheme given in Fig.1.

A. Paralysis in helpers

Table 1 shows the results of a series of cell transfer experiments. When SSA-primed cells from BSA low zone or high zone tolerant animals are used as helpers for the anti-NIP response the ratio of homologous help (by NIP-SSA) to heterologous help (by NIP-BSA) is strongly and similarly increased as compared to the controls. This indicates specific paralysis of those helper cells that cross react with BSA. The last two columns of the table show that under conditions of low zone and high zone paralysis to BSA, helper activity for NIP-BSA is indeed strongly and almost completely suppressed, whereas helper activity for NIP-SSA is only slightly reduced. The latter result is in agreement with the cross reaction data, since approximately half of the anti-SSA antibodies present in standard sera bind to BSA at relatively high antigen concentrations($\geq$ 10 μg/ml, Fig. 3) and inhibition experiments have shown that inhibition above 50% of SSA binding to anti-SSA can be achieved only by a very large excess of BSA, in the order of X10^4, and even then does not reach values higher then 70 - 80% (Rajewsky and Pohlit, 1971). It is also satisfying to see that the extent of re-

TABLE 1. PARALYSIS IN THE HELPER SYSTEM

exp. #	paralysis	immunizing antigen	homologous help (by NIP-SSA)[1] / heterologous help (by NIP-BSA)	relative helper activity for NIP-SSA	relative helper activity for NIP-BSA
1	-	SSA	5.0	≙ 1.00	≙ 1.00
2	-	SSA	3.3		
3	-	SSA	2.9		
4	-	SSA	9.3		
5	-	BSA	-		
1	BSA low zone	SSA	323.0	0.49	0.008
2	BSA low zone	SSA	13.1	0.63	0.16
3	BSA low zone	SSA	≥ 36.7	0.21	≤ 0.017
4	BSA low zone	SSA	> 108.0	0.97	< 0.01
5	BSA low zone	BSA	-	-	0.07
1	BSA high zone	SSA	24.2	0.32	0.07
2	BSA high zone	SSA	41.7	0.88	0.08
3	BSA high zone	SSA	≥ 45.0	0.17	≤ 0.01
4	BSA high zone	SSA	> 22.0	0.20	< 0.01
5	BSA high zone	BSA	-	-	0.08

1) ratio of ABCs NIP-CAP

cell transfer was done on day 43 (experiments 1,2), 24 (exp.3), 23 (exp.4) or 21 (exp.5) after immunization

duction of helper activity for NIP-SSA in BSA-tolerant mice is in good agreement with the cross stimulation results in non-paralysed animals (see upper section of table 1 and our previous communication). This, as mentioned above, is a strong argument for the validity of the cross stimulation approach.

The stability of suppression of BSA-specific helper activity upon immunization with a cross reacting antigen is remarkable. It appears to be reflected in the AFCP system in the behaviour of high affinity cells (see below), but our ignorance about the mechanism of T-cell activation and suppression restrains us from drawing a firm analogy here. Systematic studies on the recovery from paralysis in our system are under way.

B. Paralysis in AFCP.

Fig. 3a represents the cross reaction pattern, which is obtained in a typical paralysis experiment, 3 wk after priming with SSA.

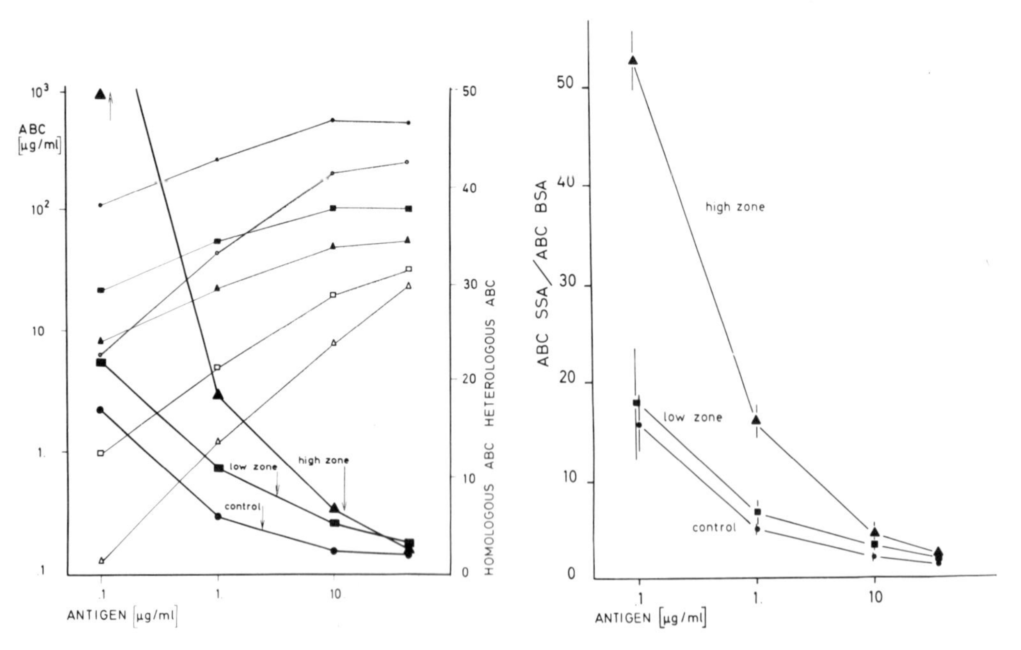

Fig. 3a. Cross reaction of SSA and BSA in sera from BSA-tolerant and control mice, immunized with SSA. Abscissa: antigen concentration, at which ABC was determined. Small open symbols represent ABCs BSA, small closed symbols ABCs SSA (left ordinate). Large symbols represent ratio of ABC SSA to ABC BSA (right ordinate). Circles, analysis in control serum; squares, analysis in serum from low zone tolerant animals, triangles, analysis in serum from high zone tolerant animals.

3b. Arithmetic means of ratio of ABC SSA to ABC BSA in a collection of antisera (see text). Bars represent standard error. Symbols and abscissa as in Fig. 3a.

In this particular experiment (No. 3 in Table 1) helper activity for

SSA was reduced to about 20% in both high and low zone conditions. This compares well with the factor of 5 - 10, by which the ABCs for both SSA and BSA (not regarding, in the latter case, the high zone paralysed animals) are reduced in the paralysed animals, but this symmetry is far from being generally established. However, as a general rule, the high zone tolerant animals differ from the other groups in one particular feature, namely the exceptionally strong decrease of the anti-BSA titers with decreasing antigen concentration. This finds its expression in the ratio of ABCs for SSA to ABCs for BSA over the range of antigen concentrations, which is nearly the same for all groups at 35 μg/ml, but strikingly higher in the high zone tolerant animals than in the low zone tolerant ones and the controls in the range of 1 - 0.1 μg/ml. The average values of five experiments, in which the animals were bled 3 - 5 wk after priming, are represented in Fig. 3b. It was found in a control experiment that normal spleen cells upon transfer into slightly irradiated (200 r) high zone tolerant animals and immunization with SSA produced antibody that cross reacted with BSA even somewhat better than antibody raised in normal animals (ABCs for SSA were 32 μg/ml@1 μg antigen/ml and 12 μg/ml@0.1 μg antigen/ml, ABCs for BSA were 14 μg/ml @1 μg antigen/ml and 3 μg/ml@0.1 μg antigen/ml). This argues against the unlikely (c.f. Mitchison, 1964) possibility that residual antigen might be responsible for the increased ratio of homologous to heterologous ABC in high zone tolerant animals.

The interpretation of these data is as follows: since (1) the quality of the humoral antibody response under high zone tolerance conditions differs from that of a normal response and of a response under low zone tolerance conditions, (2) the quality of humoral antibody in the latter two cases is largely similar, and (3) helper activity is suppressed in both low and high zone paralysis almost to the same extent (Table 1), we conclude that in our system only high zone paralysis detectably affects the AFCP in addition to the helper cells.

Looking at the data, it is tempting to consider them also as suggesting that high zone paralysis is preferentially induced in high affinity AFCP. Furthermore, the data appear to give us a measure of the threshold in terms of affinity, down to which AFCP are paralysed in high zone paralysis. This would be in accord with the earlier findings of other workers (Theis and Siskind, 1968), suggesting that in paralysis high affinity cells are indeed preferentially inactivated, and could also be taken as an argument against T - B interaction in the induction of paralysis. We are still hesitant, however, to accept this interpretation of our data without reservation, in particular because our knowledge of the changes in the cross reaction patterns occurring in the course of the immune response under the various tolerance conditions is still incomplete. It remains to be excluded that the mere fact of having a reduced number of specific AFCP in our animals leads to the observed phenomenon, via some changes in the kinetics of proliferation and selection of high affinity cells.

The antibodies in the sera of the low zone tolerant animals behave in much the same way as those in control sera. It is still puzzling that over the whole antigen concentration range the ratios of homologous to heterologous ABC in the controls are slightly lower than the ones in the low zone tolerant groups, also at high antigen concentrations, where, in terms of quantity of cross reacting antibody, the effect is quite appreciable. There is one striking difference between control and low zone tolerant animals, namely that in the latter animals, as in high zone tolerant ones, a specific part of the helper population has been eliminated. Since indeed the ratios of homologous to heterologous ABCs at high antigen concentrations are almost identical in the case of high and low zone paralysis, we are considering the possibility that the difference between low zone tolerant animals and controls might be due to the difference in antigen presentation.

In summary, the AFCP appear to be affected in this system in high zone, but not low zone paralysis, in accord with our previous conclusion (Rajewsky, 1971). Even in the high zone range, however, the effect is not very impressive in quantitative terms and probably restricted to high affinity cells. That paralysis can be induced in the B-cell population in vivo is supported by the results of a number of workers (Brownstone et al., 1966; Havas, 1969; Playfair, 1969; Henney and Ishisaka, 1969; Chiller et al., 1971; Borel, 1971; Möller et al., 1971; Seppälä and Mäkelä, 1971; Katz et al., 1971), but in many instances paralysis was difficult to achieve or passing. It is noteworthy that Miller et al. (1971) have failed to demonstrate paralysis in AFCP.

6. CONCLUSIONS.

The present data confirm that the helper cells discriminate between cross reacting serum albumins much the same way as do humoral antibodies. The analysis of further pairs of cross reacting carriers reveals the tendency that the discriminating capacity of the helpers reflects that of antibodies of high rather than low avidity. This tendency was not very pronounced in the original study and therefore not recognised at that point.

The data are thus consistent with the view that helpers, as well as AFCP, possess receptors of immunoglobulin nature. This is in line with a number of reports on the inhibition of T-cell function by anti-immunoglobulin antibody (see Mason and Warner, 1970; Lesley and Dutton, 1970; Greaves and Hogg, 1971; Basten et al., 1971), but the subject is still under debate (reviewed by Greaves, 1971).

The results also suggest ways in which differences between the specificity of the helper and the AFCP system might both arise in ontogeny and also be brought about by inborn functional properties of the two cell types.

The paralysis experiments demonstrate efficient elimination of helper cells by low doses of antigen (low zone paralysis regimen), which do not significantly affect the AFCP, in line with the results of the recent experiments of Chiller et al. (1971) and of the joint experimental approach by Mitchison (1971) and ourselves (Rajewsky, 1971). A particular feature of the present experiments is especially relevant in this context, namely that low zone paralysis appears quantitatively almost complete and stable in the helper population in the sense that upon immunization with a cross reacting antigen, helper activity for the latter is reduced to the extent which would be predicted from the extent of cross reactivity on the basis of humoral antibody, whereas it remains almost totally suppressed for the paralytogen.

Therefore, as already discussed by Chiller et al. (1971), the spectrum of receptor specificities must be expected to be much more severely restricted at the T-cell level than at the level of the AFCP, due to the induction of self tolerance and of tolerance to foreign soluble substances that penetrate the organism in low quantities. The argument hinges in part on the assumption that self tolerance and experimental tolerance are induced in the same way, and although this cannot be decided at present, well defined experiments can now be designed to clarify the point. Of fundamental importance in this context is the question whether in terms of the binding energy involved in the interaction of antigen with the receptor, helpers and AFCP have a common threshold, below which they cannot be paralysed by antigen. If this threshold, which we hopefully can approach in our system at least for the AFCP (c.f. fig. 3), would be lower for helpers than for the B cell line, a further severe restriction of the specificity spectrum in the former would be the consequence.

Selectivity in the helper system might also be enhanced by the possible requirement of a high energy of interaction between helper receptor and antigen in order to obtain efficient help, as suggested by the cross stimulation data. The recent work of Stupp et al. (1971), who used an experimental approach similar to ours, appears open to the same interpretation. A similar high affinity requirement has been postulated by Taylor and Iverson (1971) for the process of T-cell activation. It is tempting to speculate that this requirement might be due to a low receptor density on the surface of T-cells as the experiments of Raff et al. (1970) suggest. Tempting as well since in line with the above considerations is the idea that the constancy of T-cell specificity in the course of the immune response (see Paul, 1970) might indicate that only high affinity cells react from the beginning; and that low doses of antigen induce paralysis in the helpers so easily because of the hypothetical high affinity of functionally active helper cells. In the paralysis situation , however, additional mechanisms like the one proposed by Möller (1971) must possibly be invoked to adequatelly explain the differential suppression in helpers and AFCP, since only under high zone paralysis conditions appear high affinity AFCP to be affected. A definite answer to these questions cannot be given at the present stage.

We are aware that differential suppression and, if confirmed, differential affinity requirements may only partly explain how differences in the spectrum of specificities in helpers (and presumably T cells in general) on the one hand and AFCP on the other might be brought about. Thus, the behaviour of cellular antigens in the induction of immunity and tolerance may be quite different from that of soluble antigens of the kind used in this study. Furthermore, the differential suppression process may well be intimately connected with differential selection of specificities in T and B populations (c.f. Jerne, 1971). However, it does not appear necessary to invoke receptors of different nature on T and B cells in order to explain differences of specificity in the two cell populations as they have been observed in many instances (see Paul, 1970; Alkan et al., 1971; Paul et al., 1971 ; Schlossman, this symposium).

Most of the available evidence, including that on the specificity of helper function (Stupp et al., 1971 ; and our cross stimulation data), points to a high degree of selectivity at the level of T cell function. From this and the differential suppression in paralysis it appears that the T-cell population largely limits the animal's range of responsiveness and keeps the crucial balance between immunity and tolerance.

Acknowledgements. - We wish to thank Miss G. v. Hesberg for providing the tolerant mice and Drs. H. Pohlit and H. Seiler for valuable discussions. We are also most grateful to Dr. A. Kalbhen and to the Frankfurter Zoo for providing samples of deer and giraffe serum. This work was supported by the Sonderforschungsberich 74.

BIBLIOGRAPHY.

Alkan, S.S., D. E. Nitecki and J. W. Goodman, (1971), Antigen recognition and the immune response: the capacity of L-tyrosine-azobenzenearsonate to serve as a carrier for a macromolecular hapten, J. Immunol. 1o7, 353.

Basten, A., J.F.A.P. Miller, N.L. Warner and J. Pye, (1971), Specific inactivation of thymus-derived (T) and non-thymus-derived (B) lymphocytes by ^{125}I-labelled antigen, Nature New Biology 231, 1o4.

Borel, Y., (1971), Induction of immunological tolerance by a hapten (DNP) bound to a non-immunogenic protein carrier, Nature New Biology 23o, 18o.

Brownstone, A., N.A. Mitchison, R. Pitt-Rivers, (1966), Biological studies with an iodine-containing synthetic immunological determinant 4-hydroxy-3-iodo-5-nitrophenylacetic acid (NIP) and related compounds, Immunology 1o, 481.

Chiller, J.M., G.S. Habicht and W.O. Weigle, (1971), Kinetic differences in unresponsiveness of thymus and bone marrow cells, Science 171, 813.

Greaves, M.F., (1971), Progr. Imm. 1, in the press.

Greaves, M.F. and N.M. Hogg, (1971), Antigen binding sites on mouse lymphoid cells, in Mäkelä, O. (ed.) Cell interactions and receptor antibodies in immune responses, Academic Press, London, p. 145.

Green, I., W.E. Paul and B. Benacerraf, (1968), Hapten carrier relationshipsin the DNP-PLL foreign albumin complex system: induction of tolerance and stimulation of cells in vitro, J. Exp. Med. 127, 43.

Havas, H.F., (1969), The effect of the carrier protein on the immune response and on the induction of tolerance in mice to the 2,4-dinitrophenyl determinant, Immunology 17, 819.

Henney, C.S. and K. Ishisaka, (1969), A simplified procedure for the preparation of immunoglobulin class specific antisera, J. Immunol. 1o3, 56.

Jerne, N.K., (1971), The somatic generation of immune recognition, Eur. J. Immunol. 1, 1.

Katz, D.H., J.M. Davie, W.E. Paul and B. Benacerraf, (1971), Carrier function in anti-hapten antibody responses. IV. Experimental conditions for the induction of hapten-specific tolerance or for the stimulation of anti-hapten anamnestic responses by "nonimmunogenic" hapten-polypeptide conjugates, J. Exp. Med. 134, 2o1.

Lesley, J. and R.W. Dutton, (1970), Antigen receptor molecules: inhibition by antiserum against kappa light chains. Science 169, 487.

Mason, S. and N.L. Warner, (1970), The immunoglobulin nature of the antigen recognition site on cells mediating transplantation immunity and delayed hypersensitivity, J. Immunol. 1o4, 762.

McDevitt, H.O. and B. Benacerraf, (1969), Genetic control of specific immune responses, Adv. Imm. 11, 31.

Miller, J.F.A.P. and G.F. Mitchell, (1970), Cell to cell interaction in the immune response, J. Exp. Med. 131, 675.

Miller, J.F.A.P., A. Basten and C. Cheers, (1971), Interaction between lymphocytes in immune responses, Cellular Immunology 2, 469.

Mitchison, N.A., (1964), Recovery from immunological paralysis in relation to age and residual antigen, Immunology 9, 129.

Mitchison, N.A., (1968), The dosage requirements of immunological paralysis by soluble protein, Immunology 15, 5o9.

Mitchison, N.A., (1971a), The relative ability of T and B lymphocytes to see protein antigen, in Mäkelä, O. (ed.), Cell interactions and receptor antibodies in immune responses. Academic Press, London, p. 249.

Mitchison, N.A., (1971b), The carrier effect in the secondary response to hapten-protein conjugates. II. Cellular cooperation, Eur. J. Immunol. 1, 18.

Mitchison, N.A., K. Rajewsky and R.B. Taylor, (1970), Cooperation of antigenic determinants and of cells in the induction of antibodies, in Sterzl, J. (ed.) Developmental aspects of antibody formation and structure, Publishing House of the Czechoslovak Academy of Science, Prague, II, p. 547.

Möller, E., O. Sjöberg and O. Mäkelä, (1971), Immunological unresponsiveness against 4-hydroxy-3,5-dinitro-phenacetyl (NNP) hapten in different lymphoid cell populations, Eur. J. Immunol. 3, 218.

Möller, G., (1970), Triggering mechanisms for cellular recognition, in Smith, R.T. and M. Landy (eds.) Immune surveillance, Academic Press, New York, p. 87.

Paul, W.E., (1970), Functional specificity of antigen-binding receptors of lymphocytes, Transplantation Reviews 5, 13o.

Paul, W.E., Y. Stupp, G.W. Siskind and B. Benacerraf, (1971), Structural control of immunogenicity. IV. Relative specificity of elicitation of cellular immune responses and of ligand binding to anti-hapten antibody after immunization with mono-ε-DNP-nona-L-lysine, Immunology 21, 6o5.

Playfair, J.H.L., (1969), Specific tolerance to sheep erythrocytes in mouse bone marrow cells, Nature 222, 882.

Raff, M.C., M. Sternberg and R.B. Taylor, (1970), Immunoglobulin determinants on the surface of mouse lymphoid cells, Nature 225, 553.

Rajewsky, K., (1969), The significance of the carrier effect for the induction of antibodies, in Grundmann, E. (ed.) Current topics in immunology, Springer, Berlin, p. 81.

Rajewsky, K., (1971), The carrier effect and cellular cooperation in the induction of antibodies, Proc. Roy. Soc., Lond. B., 176, 385.

Rajewsky, K. and H. Pohlit, (1971), Specificity of helper function, Progr. Imm. 1, in the press.

Rajewsky, K. and E. Rottländer, (1967), Tolerance specificity and the immune response to lactic dehydrogenase isoenzymes, Cold Spr. Harb. Symp. Quant. Biol. 32, 547.

Schlossman, S. F., (1971), this symposium.

Seppälä, I.J.T. and O. Mäkelä, (1971), Hapten-carrier relationships in immunological unresponsiveness, Eur. J. Immunol. 1, 221.

Stupp, Y., W.E. Paul and B. Benacerraf, (1971), Structural control of immunogenicity. III. Preparation for and elicitation of anamnestic antibody responses by oligo- and poly-lysines and their DNP-derivatives, Immunology 21, 595.

Taylor, R.B., (1969), Cellular cooperation in the antibody response of mice to two serum albumins: Specific function of thymus cells, Transplantation Reviews 1, 114.

Taylor, R.B. and G.M. Iverson, (1971), Hapten competition and the nature of cell-cooperation in the antibody response, Proc. Roy. Soc. Lond. B. 176, 393.

Theis, G.A. and G.W. Siskind, (1968), Selection of cell populations in induction of tolerance: affinity of antibody formed in partially tolerant rabbits, J. Imm. 1oo, 138.

Weigle, W.O., (1961), The immune response of rabbits tolerant to bovine serum albumin to the injection of other heterologous serum albumins, J. Exp. Med. 114, 111.

ANTIGEN BINDING CELLS IN IMMUNE AND TOLERANT MICE

ERNA MÖLLER and O. MÄKELÄ

Division of Immunobiology, Karolinska Institutet, Wallenberglaboratory, Lilla Frescati, 104 05 Stockholm 50, Sweden

and

Department of Serology and Bacteriology, Helsinki University, Helsinki 29, Finland.

Abstract: Antigen binding cells present in mice immune to the hapten NNP are of both T and B cell origin, as judged by their sensitivity to anti-Θ and anti-MBLA sera, respectively. Antigen binding was specifically inhibited by free hapten. The average antigen capturing activity of T and B lymphoid cells was assayed by hapten inhibition at various times after 1° and 2° immunization using hapten coupled on heterologous protein carrier or red blood cells. It was found that antigen binding cells of T origin (T-ABC) were only inhibited by high concentrations of free hapten, irrespective of whether the T-ABC were assessed early or late in the immune response. B-ABC were inhibited by a lower hapten concentration, and the efficiency of inhibition increased with time after immunization.

In mice tolerant to NNP there are no detectable antigen binding T cells. However, significantly elevated numbers of B-ABC are found in such mice. These ABC differ from those of immune mice, because they require a higher concentration of free hapten for inhibition. The significance of these results is discussed.

1. INTRODUCTION

Both thymus-dependent (T) lymphoid cells and thymus-independent or bone-marrow-derived (B) spleen cells of mice immune to sheep red blood cells (SRBC) have the capacity to bind antigen to the cell surface, as judged by the formation of "rosettes", (Greaves and Möller, 1970 a,b, Bach et.al. 1970, Schlesinger, 1970). It has long been discussed whether T lymphoid cells, generally considered to be the mediators of cellular immune reactions, could interact efficiently with haptenic determinants (for ref. and review see Paul, 1970). Anti-hapten reactive helper cells cannot be easily demonstrated (Raff, 1970, Mitchison, 1971 a, Niederhuber et.al., unpublished). Furthermore, antigen binding to B lymphoid cells can be inhibited by free hapten, whereas inhibition of stimulation of DNA synthesis (considered to be a T cell function) is not easily effectuated by free hapten, inhibition requiring that the hapten is bound to the homologous protein carrier (Davie and Paul, 1970, Paul, 1972). However, recent findings by Iverson (pers.comm.) indicate that T cells can act as specific helper cells active towards a haptenic determinant. In the present report we will demonstrate findings indicating that T-ABC can react specifically with haptens present on heterologous red blood cells, and that this reaction can be inhibited by high concentrations of free hapten.

T cells have been considered to carry specific antigen-recognizing receptors on the cell surface, their nature being the matter of intensive studies during recent years. Greaves et.al. (1969), as well as Mason and Warner (1970) demonstrated that biological activities of T cells confronted with antigen could be blocked by anti-light chain antibodies, indicating that T cells were equipped with antigen-recognizing receptors containing immunoglobulin light chain structures. Recent studies by Greaves and Hogg (1971) have shown that immunologically activated T cells expose receptors which contain both light and heavy chain determinants. However, the heavy chain class present on activated T cells seems to be of μ type exclusively.

T antigen binding cells against heterologous red cells appear in increased numbers during both primary and secondary immune responses. However, low dose priming as well as inhibition of B cell proliferation by injection of passive antibodies leads to a selective activation of T-ABC (Greaves et.al. 1970, Möller and Greaves, 1971). Furthermore, tolerance induction to NNP seems to cause a selective inactivation of T-ABC, leaving a proportion of B-ABC intact (Möller et.al. 1971). Thus, T cells seem to have a lower threshold for immune activation as well as for tolerance induction (Rajewsky, 1970, Mitchison, 1971 b, Chiller et.al. 1971). The mechanism of this difference is not known, but is paradoxical, because T cells possess fewer immunoglobulin receptors on the cell surface than B cells (Raff et.al. 1970). However, the geometry of the receptor arrangement as well as the accessability of antigen to these receptors is unknown as yet.

The present study was performed in an attempt to elucidate the mechanism of activation of T and B lymphoid cells in the immune response. Mice were immunized against the hapten NNP coupled to heterologous proteins or red cells. The efficiency by which antigen binding to T and B lymphoid cells could be inhibited by free hapten was studied at various time periods after immunization. Furthermore, we studied the proportion of T and B antigen binding cells present in immune and tolerant mice, respectively and compared the efficiency of the B-ABC present in both immune and tolerant mice to be inhibited in their binding by free hapten.

2. MATERIALS AND METHODS

Mice of the following inbred strains were used in the present study: CBA(H-2^k), (A x CBA) F_1 (H-2^a)(H-2^k) hybrids, B10.5M(H-2^b) and (A x B10.5M)F_1 hybrids.

Antisera: The preparation of AKR anti-C3HΘ serum, as well as of rabbit anti-MBLA (mouse specific B-lymphocyte antigen) has been described before (Greaves and Möller, 1970). Niederhuber's modification (1971) of the method originally described by Raff et.al. (1971) was used for the preparation of anti-MBLA serum. The latter serum has been characterized extensively and the results were published elsewhere (Niederhuber and Möller, in press). Raff (1969) demonstrated that the anti-Θ serum could be used to kill selectively T lymphoid cells, our own studies show that the anti-MBLA serum is a useful tool for the purification of T lymphoid cells. Recently, we have extended our studies with the anti-MBLA reagent and shown that anti-Θ and anti-MBLA react with different cell populations

(Niederhuber et.al., unpubl.). The activity of anti-MBLA is not due to its content of anti-immunoglobulin antibodies (Niederhuber et.al. 1972). The access to reagents selectively cytotoxic for T and B lymphoid cells, makes the use of radiation chimeras unnecessary.

Preparation of purified T and B spleen cells: Cell suspensions from the spleens of 3-5 mice were pooled and divided into three groups. One part was treated with normal mouse serum (or normal heat-inactivated rabbit serum), a second part with anti-Θ serum and a third with anti-MBLA serum. The cells were incubated in serum for 30 min. at 37°C. Thereafter the cells were washed and fresh guinea pig complement, absorbed with agarose was added to all groups and the incubation continued for another 30 min. Thereafter the cells were washed and resuspended to the desired concentration. Anti-Θ as a rule killed 20-40% of the spleen cells, anti-MBLA about 50-80% of the cells. The suspensions were made up to the same volume before use in the assay. The concentration of normal serum treated spleen cells was 25×10^6 trypan blue unstained cells/ml.

Antibody secreting cells: The number of plaque-forming cells (PFC) was estimated with the local hemolysis in gel assay as described by Jerne and Nordin (1965) with modifications for the detection of anti-hapten producing cells described by Pasanen and Mäkelä (1969).

Preparation of hapten-coupled red cells: NNP-azide dissolved in DMFA (dimethylformamide) in a concentration of 40 mg/ml was prepared and stored in the deep freeze. 0.05 ml of this solution was added to 1 ml of carbonate buffer, pH 9.2 (for no. 1 cells). Dilutions of 1/10 and 1/100 of this solution was prepared to give concentrations of NNP-azide to couple no. 2 and 3 cells, respectively. To this solution was added 1 ml of a 20% red cells suspended in carbonate buffer, and the mixture was incubated for 30 min. at room temperature. Thereafter the red cells were washed five times in cold BSS, and finally made up to the desired concentration for immunization, PFC or RFC assay, as described before (Möller et.al. 1971). As a rule, denseley coupled red cells (no. 1) were used for immunizations, no. 2 and 3 cells were used as targets in the PFC and RFC assays.

Hapten inhibition of RFC formation: Free hapten, in the form of NNP-e-aminocaproic acid (NNP-cap.) was dissolved in saline in a concentration of 3×10^{-3} M. 0.1 ml of serial dilutions were prepared in calcium and magnesium free phosphate-buffered saline. 0.1 ml of lymphoid cells were added to each hapten concentration, plus 0.2 ml of buffer and the cells were incubated for 30 min. at room temperature. At this time equilibrium between the free hapten and the cellular receptors was achieved. Thereafter, 0.1 ml of a 1% suspension of hapten coupled red cells were added to each tube. Since NNP-HRBC were used for most immunizations, NNP-SRBC were used for the assay. We have not observed any cross-reactivity between HRBC and SRBC in our tests. The cell mixture was then either centrifuged immediately in the cold for 10 min. and thereafter gently resuspended, or incubated without prior centrifugation at +4°C over night, before the cells were gently resuspended. The number of rosette forming cells (RFC) was enumerated by counts of at least 30.000 cells in each tube. Each lymphoid cell with 5 or more adhering red cells and with a clearly visible central lymphoid cell was scored as a RFC. Buffer treated lymphoid cells gave the optimal response with each test cell suspension. The bond between hapten coupled red cells and lymphoid cells was stable at +4°C, since the addition of free hapten after rosette formation did not influence the number of antigen binding cells. This is interpreted to mean that the hapten coupled red cells probably formed multiple bonds with the cell receptors, which would not be likely to dissociate.

3. RESULTS

Origin of antigen binding cells in the spleens of mice immunized against NNP.

Mice of various inbred strains were immunized with 10^8 no. 1 NNP-HRBC i.p. At various times after immunization groups of 3-5 mice were killed and their spleens removed. Spleens from non-immunized mice served as controls. The cell suspensions were treated with normal serum (N S), anti-Θ serum or with anti-MBLA serum as described above.

Table 1. Origin of antigen binding spleen cells from mice immune to NNP.

Exp.	RFC/10^6 spleen cells treated with normal serum	% inhibition of RFC by treatment with anti-Θ serum	anti-MBLA serum
1.	2620	15	58
2.	3200	31	55
3.	1350	11	44
4.	1640	55	54
5.	2450	41	61
6.	2950	44	57
7.	2500	42	40

On day 7 the percentage of Θ sensitive RFC as a mean is 28 $\pm$ 5 %.

Table 1 shows the number of RFC in antiserum purified spleen cell suspensions. Anti-Θ sensitive RFC are considered to be T-RFC, anti-MBLA sensitive RFC are considered to be B-RFC. As a rule 15-50% of the RFC present in normal serum treated spleens were sensitive to anti-Θ serum, whereas 40-60% of the RFC were sensitive to anti-MBLA serum treatment. In exp. 3, the reduction with anti-Θ serum was only 11% and with anti-MBLA 44%, but in the other experiments the sum of the percent inactivation by the two reagents was more than 70% of the total number of RFC. These results indicate that both T and B lymphoid cells have the capacity to bind hapten-coated red blood cells. The number of background RFC was never higher than

$500/10^6$ nucelated spleen cells in the experiments included in Table 1. Furthermore, the increase of RFC of both T and B origin was found to be specific, since no concommittant increase of RFC against uncoupled SRBC occurred.

Hapten inhibition of T and B antigen binding cells

Free hapten in various concentrations was used to study the inhibitory effect on T and B-RFC formation. It was found that high concentrations of free hapten (6×10^{-4}M) almost completely inhibited both T and B lymphoid cells to form rosettes with hapten coupled red cells. Text figures 1-4 show the inhibition curves for T and B-RFC derived from mice immunized with no. 1 NNP-HRBC at various intervals before the test. It can be seen from these figures that T-RFC without exception required a relatively high concentration of free hapten for 50% reduction of antigen binding. On the other hand, B-RFC seemed to require somewhat less free hapten for inhibition than T cells, this being particularly marked late in the primary response. In some experiments mice were boosted and the number of T and B-RFC was enumerated 10-12 days after the secondary immunization. Text-figure 4 shows one such experiment. Spleen cells from the boosted animals showed less inhibition of B rosettes by high hapten concentration as compared to the B lymphoid cells taken late in the primary response.

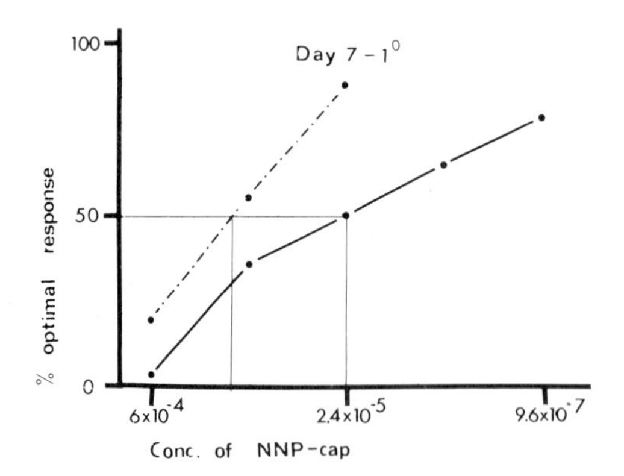

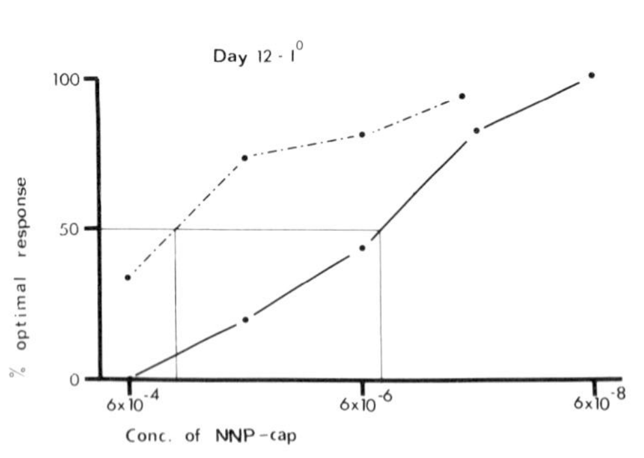

Text-figures 1 and 2. Inhibition of T-RFC, •- . - •, and of B-RFC, • —— • from the spleens of immune mice by various concentrations of NNP-cap. at day 7 and 12, resp. of the primary immune response.

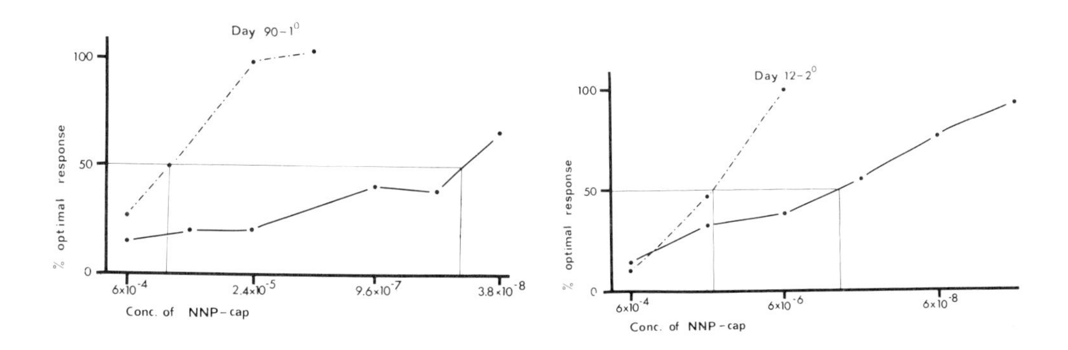

Text-figures 3 and 4. Inhibition of T-RFC, • –.– •, and of B-RFC •—— • from the spleens of immune mice by various concentrations of NNP-cap. at day 90 of the primary and at day 12 of the secondary immune response.

Table 2. Inhibition of T and B-RFC by pretreatment with free hapten.

Exp.	Response	Immunization	Days	Concentration of free hapten needed to cause 50% inhibition of T-RFC	B-RFC
1.	1°	10^8 no. 1 NNP-HRBC	4	6×10^{-4}	2×10^{-5}
2.	1°	"	7	1.7×10^{-4}	2.4×10^{-5}
3.	1°	"	12	1.9×10^{-4}	4.7×10^{-6}
4.	1°	"	90	2.0×10^{-4}	9.3×10^{-8}
5.	2°	"	68	6×10^{-5}	1.2×10^{-6}
6.	2°	"	75	8.5×10^{-5}	7.2×10^{-7}
7.	2°	10^6 no. 1 NNP-HRBC	82	2.5×10^{-5}	6×10^{-8}

In Table 2, the concentration of free hapten which caused a 50% inhibition of the number of RFC was determined by interpolations from the logaritmic plot on the x-axis. As shown, the 50% inhibition concentration of hapten for T-RFC was relatively stable at various time periods after stimulation, whereas the concentration needed to cause 50% inhibition of B-RFC decreased with time after immunization. These differences might reflect changes in average avidity with which the two cell populations interact with antigen at various times after immunization, as will be discussed later.

Induction of tolerance to the hapten NNP

Mice were paralysed against the hapten NNP by three weekly i.p. injections for 10 weeks of 0.3 mg $NNP_{12}BSA$. Thereafter the animals were injected once weekly for another 10 weeks. 7 days after the last tolerance maintaining dose, the mice were challenged with 10^6 - 10^8 no. 1 NNP-HRBC. In exp. no. 4 the mice were instead challenged with 0.4 mg of NNP_{11}-CG (chicken globulin). Previously untreated mice served as controls. As is shown in Table 3, pretreatment with NNP-BSA as described above resulted in a reduced response with regard to antibody secreting cells of the pretreated group tested 7 days after challenge. However, the hapten-conjugate pretreatment resulted only in partial unresponsiveness since non-immune mice had PFC numbers that were always below $10/10^6$ (Möller et.al. 1971). The unresponsive state was specific, since both groups of mice responded equally well to uncoupled HRBC (Möller and Sjöberg, 1971).

Table 3. Immune response of mice pretreated with NNP-BSA.

Exp.	Challenge	Antibody secreting spleen cells in			
		pretreated		normal	
		D.PFC/10^6	ID.PFC/10^6	D.PFC/10^6	ID.PFC/10^6
1	NNP-HRBC	1.67 ± 0.09(47)	1.79 ± 0.10(62)	1.78 ± 0.03(61)	2.29 ± 0.07(196)
2	"	0.75 ± 0.13(6)	0.90 ± 0.23(8)	1.41 ± 0.13(26)	1.86 ± 0.14(72)
3	"	n.t.			
4	NNP-CG	1.33 ± 0.24(21)	1.71 ± 0.01(52)	1.87 ± 0.03(74)	2.41 ± 0.01(256)

PFC/10^6 values are expressed in log 10 values ± s.e. Antilogs in parenthesis.

Studies were also performed on the number of antigen binding cells present in pretreated groups of mice. These studies gave results which were different from those reported above. In all experiments, the numbers of RFC found in the " tolerant " mice was significantly elevated over back-ground values. This findings substantiate our earlier conclusions, where we also found that the percentage of Θ-sensitive RFC was lower in tolerant than in the immune groups (Möller et.al. 1971). Our findings indicated that the T antigen binding cells which can be detected in hapten immune mice, have become inactivated in tolerant mice (Möller, Sjöberg, 1971).

Hapten inhibition of B-RFC in immune and tolerant mice

Spleens from tolerant mice contained considerable numbers of antigen binding cells, which were almost all resistant to treatment with anti-Θ serum plus complement. A majority of these RFC are sensitive to treatment with anti-MBLA serum plus complement, however, which indicates that the antigen binding cells which are present in tolerant mice are mainly B lymphocytes (Möller and Sjöberg, 1971). Hapten inhibition experiments were performed with anti-Θ treated spleen cell suspensions from immune and tolerant mice. Results of two such experiments are shown in text-figures 5 and 6. It can be seen that the B-RFC present in immune mice, gave 50% inhibition values which are of the same order of magnitude as those presented in Table 2.

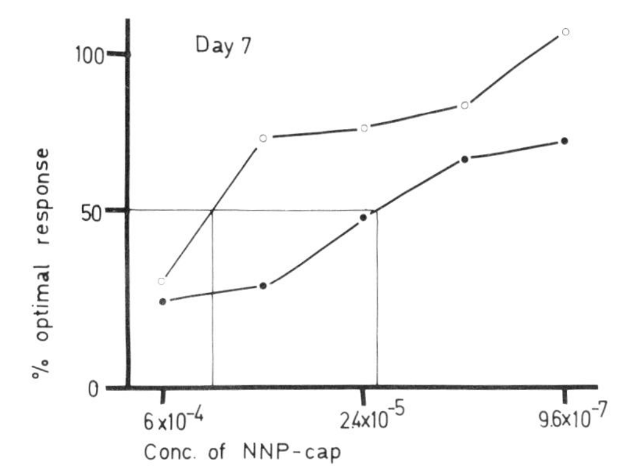

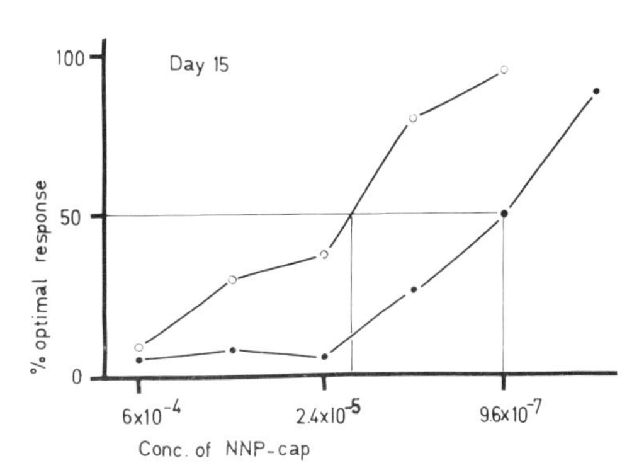

Text-figures 5 and 6. Inhibition of B-RFC from partially tolerant o-o and of B-RFC from immune ●-● animals by various concentrations of free NNP-cap.

However, the B-RFC present in tolerant mice required a higher hapten concentration for 50% inhibition than did B-RFC from immune mice. At day 15, the difference between the two curves was also demonstrable, but at this time the mean concentration of hapten needed for 50% inhibition had decreased in both cell populations. In Table 4, the results of four experiments are summarized.

Table 4. Inhibition of B-RFC in immune and tolerant mice by free hapten.

Exp.	Challenge	Day	Concentration of free hapten needed to cause 50% inhibition of	
			'tolerant' B-RFC	immune B-RFC
1	NNP-HRBC	7	1.4×10^{-4}	4.8×10^{-6}
2	"	7	2.5×10^{-4}	1.9×10^{-5}
3	"	7	3.3×10^{-5}	3.8×10^{-6}
4	NNP-CG	15	1.5×10^{-5}	9.6×10^{-7}

Expts. 1 and 2 were challenged with 10^8 red cells, exp. 3 with 10^6 red cells.

4. DISCUSSION

In the present series of experiments we have found that antigen binding cells in mice immune to the hapten NNP are of both T and B origin. This conclusion is based on the assumption that anti-Θ serum is selectively cytotoxic to T lymphoid cells, anti-MBLA serum to B lymphoid cells. These questions have been discussed at length in other publications (Raff, 1969, Greaves and Möller, 1970, Raff et.al. 1971, Niederhuber and Möller, in press, Niederhuber, 1971, Niederhuber et.al., in press) and will not be dealt with here. Suffice it to say that we consider them useful tools to purify spleen cells into the T and B compartments. In spite of the fact that T cells have been shown to have immunoglobulin-like structures on their surface (Greaves and Hogg, 1971, Basten et.al. 1971), these structures reacting specifically with antigen (Basten et.al. 1971, Roelants and Askonas, 1971) and that the specificity of helper cell activity (a T cell function) is very closely related to B cell receptors (Rajewsky and Pohlit, in press); the point that T cells can recognize hapten might still be controversial. This is mainly due to the fact that Paul et.al. (1971) found that activities of B cells cou'd be blocked with free hapten, whereas activities ascribed to T cells, such as induction of DNA synthesis after antigenic stimulation in vitro, was not inhibited by free hapten, but required the homologous hapten protein complex for complete inhibition (Davie and Paul, 1970). Furthermore, methods like immunofluorescence, which readily demonstrate immunoglobulin on B cell surfaces, fails to detect such on T cells (Raff, 1970 b, Raff et.al. 1970). In systems where helper cells exist which can recognize carrier molecules, very little anti-hapten reactive helper cell activity is demonstrable (Raff, 1970, Mitchison, 1971, Niederhuber et.al., unpubl.).

However, our experiments have shown that a proportion of T antigen binding cells regularly are found in mice immune to the hapten NNP. Recent studies by Iverson (pers.comm.) have also demonstrated that under certain circumstances anti-hapten reactive helper cells exist.

We have also demonstrated that partial tolerance can be induced to the hapten NNP by pretreatment with hapten coated BSA for a prolonged period of time (Möller et.al. 1971). Under these circumstances immunity to the carrier molecule is induced (Seppälä and Mäkelä, 1971). Recently, other authors have found that tolerance can readily be induced against haptenic determinants if pretreatment is performed with the hapten coupled to a "non-immunogenic" carrier molecule (Borek, 1971, Katz et.al. 1971).

Clear evidence exists that antigen acts selectively on cells with specific antigen receptors (Siskind and Benacerraf, 1969, Werblin and Siskind, 1971). Low antigen doses give high affinity antibodies, whereas high antigen doses leads to production of antibodies with lower average affinity. A similar affinity change seems to occur at the cellular level, as was recently described by Andersson (1970). Little is as yet known about the possible changes in reactivity of the T cell receptors with time after immunization. Paul et.al. (1969) found that the initiation of DNA synthesis by antigen in vitro, which is thought to be a T cell property, has certain characteristic features. If priming is performed with low antigen doses, the cells respond in vitro to low antigen doses, whereas high dose priming results in a cell population which reacts only with high antigen doses. However, these authors did not find that the dose which was needed to initiate DNA synthesis in lymphoid cells from high and low dose primed animals changed with time after immunization, a feature which is characteristic of the humoral anamnestic response (Bullock and Rittenberg, 1970).

Bast et.al. (1971) have found that the induction of DNA synthesis in lymph node cells from guinea pigs immunized against HSA was induced by lower antigen doses with increasing time after immunization. This finding would support the notion that the affinity of the cellular receptor for antigen increases with time. Furthermore, the same authors showed that when tolerance to HSA waned, induction of DNA synthesis in vitro was only achieved with very high doses of antigen, characteristic as of cells with low affinity receptors. However, in none of the above mentioned studies was the responding cell population identified as T lymphoid cells. As yet, it has not been conclusively demonstrated that B cells are incompetent to react with increased DNA synthesis after contact with antigen in vitro.

Do our studies give any information as to the avidity changes that might occur in T and B lymphoid cells during the maturation of the immune response? In our test system we are treating purified T and B lymphoid cells with free hapten prior to addition of hapten coated red cells. Equilibrium is allowed to occur before the addition of target cells. It is likely that at a low concentration of free hapten very few receptors will be complexed with hapten, whereas at a higher concentration most of the receptors will have bound hapten. If the affinity of the individual receptors is high, a low concentration of free hapten would suffice to saturate them. When hapten coated red cells are added, the hapten could interact with free receptors on the cell surface, and rosettes could be formed. Since the hapten coat on the red cells is dense, possibilities for multiple bond formation exist, explaining the irreversibility of our test system. Therefore, we could measure pertinent changes occuring in the two cell populations during a maturing immune response. We can not draw any conclusions about absolute values of the interaction forces between receptors and antigen, however, since nothing is known about the exact numbers of receptors or the geometry of receptor arrangement in cells at different stages of differentiation.

We have found that both T and B lymphoid cells are inhibited from binding hapten coupled red cells in the excess of free hapten. Excess hapten does not interfere with formation of rosettes against uncoupled red cells in appropriately immunized mice, however, Therefore, the binding of T and B cells to hapten appears to be a specific phenomenon. Studies on the relative inhibition capacity of hapten at various times after immunization, revealed that anti-hapten reactive B cells undergo a cellular change during immunization. A similar change was not seen in the T lymphoid cell population. This could be explained in different ways.

1. B cells undergo a gradual change and the average avidity of cells for hapten increase with time after immunization.

2. The shift observed in the B cell population reflects a decrease in the number of available receptors.

We believe alternative 1 to be correct, since a) there is a difference in the inhibition curves for B spleen cells derived from tolerant mice compared to immune spleen cells, as expected (Siskind, 1971); b) Andersson (1970) demonstrated an increased avidity at the level of the single antibody secreting cell with time after immunization, and c) no data exist to support the decrease in number of available receptors in B cells.

3. T cells might increase in avidity with time after immunization, but the present test system does not reveal such changes.

a) Demonstration of an increase in avidity of T cells could be blocked by a concommittant increase in the number of available T cell receptors. However, again there are no data to support such a notion. b) T cells do change, but the activated cells are not localized in the spleen. If Raff's (1972) recent findings of T1 cells, being the non-activated T cells residing mainly in the spleen, and the activated T2 cells constituting the recirculating pool of T cells residing mainly in the blood and regional nodes is correct no change might be demonstrated in T spleen cells. This possibility can be tested. However, since active helper cells are often derived from the spleen this argument probably should not carry much weight.

4. T cells do not change in avidity during the maturation of the immune response.

This alternative would argue against Bast et.al.'s (1971) findings, but agree well with those of Paul et.al. (1971) and with hitherto unpublished data by Bullock and Rittenberg (1972). The latter authors found that antibody forming cell precursors increased in effectivity with time after immunization, but the activity of helper cells did not. It is difficult to explain why T cells, if equipped with immunoglobulin-like receptors of a type similar to those present on B cell precursors, would not be sensitive to affinity regulation caused by humoral antibodies. However, it can be argued that cell selection occurs by a hitherto unknown mechanism in the T cell population, since T cells are not ever going to secrete immunoglobulin molecules, but probably secrete other mediators, which lack immunologic specificity.

Finally, it is conceivable that any changes that occur in T cells are not easily visualized, if once activated T cells have a long life span. This notion is supported by the finding that adult thymectomy does not result in immediate deficiencies of T cell activity in the mouse. However, recent studies by Nowell and Wilson (1971) and by Koster and McGregor (1971) indicate that T cells have a relatively rapid turnover. Further experimentation is needed to resolve the present dilemma.

ACKNOWLEDGEMENTS

The skilful technical assistance of Mrs. Lill-Britt Andersson is gratefully acknowledged. The present work was supported by generous grants from the Swedish Medical Research Council, the Knut and Alice Wallenberg Foundation and the Harald Jeansson's Foundation.

REFERENCES

Andersson, B. (1970). Studies on the regulation of avidity at the level of the single antibody forming cell. J. Exp. Med. 132, 77.

Bach, J.F., Muller, J.Y., and Dardenne, M. (1970), Antigen recognition by rosette forming cells. Nature (Lond.) 227, 1251.

Bast, R.C., Manseau, E.J., and Dvorak, H.F. (1971), Heterogeneity of the cellular immune response. I. Kinetics of lymphocyte stimulation during sensitization and recovery from tolerance. J. Exp. Med. 133, 187.

Basten, A., Miller, J.F.A.P., Warner, N.L., and Pye, J. (1971), Specific inactivation of thymus-derived (T) and non-thymus-derived (B) lymphocytes by ^{125}-I-labelled antigen. Nature (Lond.) 231, 104.

Borek, Y. (1971), Induction of immunological tolerance to a hapten (DNP) bound to a non-immunogenic protein carrier. Nature (Lond.) 230, 180.

Bullock, W.W., and Rittenberg, M.B. (1970), In vitro initiated secondary anti-hapten response. II. Increasing cell avidity for antigen. J. Exp. Med. 132, 926.

Bullock, W.W., and Rittenberg, M.B. (1972), Primary in vitro anti-hapten response. Influence of carrier primed cells. Unpublished results.

Chiller, J.M., Habicht, G.S., and Weigle, W.O. (1971), Kinetic differences in unresponsiveness of thymus and bone-marrow cells. Science 171, 813.

Davie, J.M., and Paul, W.E. (1970), Receptors on immunocompetent cells. I. Receptor specificity of cells participating in a cellular immune response. Cell. Immunol., 1, 404.

Davie, J.M., Rosenthal, A.S., and Paul, W.E. (1971), Receptors on immunocompetent cells. III. Specificity and nature of receptors on dinitrophenylated guinea pig albumin-^{125}I binding cells of immunized guinea pigs. J. Exp. Med. 134, 517.

Greaves, M.F., and Hogg, N.M. (1971), Antigen binding sites on mouse lymphoid cells. In "Cell interactions and receptor antibodies in immune responses", Ed. O. Mäkelä, A. Cross and T. Kosunen, Academic Press, London, p. 145.

Greaves, M.F., and Möller, E. (1970 a), Studies on antigen binding cells. I. The origin of reactive cells. Cell. Immunol. 1, 372.

Greaves, M.F., and Möller, E. (1970 b), The origin and significance of rosette-forming cells in the response of mice to sheep erythrocytes. In "Developmental aspects of antibody structure and function", Symposia Prague, Ed. J. Sterzl, p. 627.

Greaves, M.F., Möller, E., and Möller, G. (1970), Studies on antigen binding cells. II. Relationship to antigen sensitive cells. Cell. Immunol. 1, 386.

Greaves, M.F., Torrigiani, G., and Roitt, I.M. (1969), Blocking of the lymphocyte receptor site for cell-mediated hypersensitivity and transplantation reactions by anti-light chain sera. Nature 222, 885.

Jerne, N.K., and Nordin, A.A. (1965), Plaque-formation in agar by single antibody-producing cells. Science 140, 405.

Katz, D.H., Davie, J.M., Paul, W.E., and Benacerraf, B. (1971), Carrier function in anti-hapten antibody responses. IV. Experimental conditions for the induction of hapten-specific tolerance or for the stimulation of anti-hapten anamnestic responses by "non-immunogenic" hapten-polypeptide conjugates. J. Exp. Med. 134, 201.

Koster, F.T., and McGregor, D.D. (1971), The mediator of cellular immunity. III. Lymphocyte traffic from the blood into the inflamed peritoneal cavity. J. Exp. Med. 133, 864.

Mason, S., and Warner, N.L. (1970), The immunoglobulin nature of the antigen recognition site on cells mediating transplantation immunity and delayed hypersensitivity. J. Immunol. 104, 762.

McConnell, I., Munro, A., Gurner, B.W., and Coombs, R.R.A. (1969), Studies on actively allergized cells. I. The cytodynamics and norphology of rosette-forming lymph node cells in mice and inhibition of rosette-formation with antibody to mouse Ig´s. Int. Arch. Allergy 35, 228.

Mitchison, N.A. (1971 a), The carrier effect in the secondary response to hapten-protein conjugates. II. Cellular cooperation. Eur. J. Immunol., 1, 18.

Mitchison, N.A. (1971 b), The relative ability of T and B lymphocytes to see protein antigen. In "Cell interactions and receptor antibodies in immune responses", Ed. O. Mäkelä, A. Cross and T. Kosunen, Academic Press, London, p. 249.

Möller, E., and Greaves, M.F. (1971), On the thymic origin of antigen sensitive cells. In "Cellular interactions and receptor antibodies in immune responses", Eds. O. Mäkelä, A. Cross and T. Kosunen, Academic Press, London, p. 101.

Möller, E., and Sjöberg, O. (1971), Antigen binding cells in immune and tolerant animals. Transpl. Rev. in press.

Möller, E., Sjöberg, O., and Mäkelä, O. (1971), Immunological unresponsiveness against the 4-hydroxy-3,5-dinitrophenacetyl (NNP) hapten in different lymphoid cell populations. Eur. J. Immunol. 1, 218.

Niederhuber, J. (1971), An improved method for anti-B lymphocyte serum preparation. Nature (Lond.) 233, 86.

Niederhuber, J., and Möller, E. (1971), Antigenic markers on mouse lymphoid cells: The presence of MBLA on antibody forming cells and antigen binding cells. Cell. Immunol. In press.

Niederhuber, J., and Möller, E. (1972), Antigenic markers on mouse lymphoid cells: Origin of cells mediating immunologic memory. Submitted for publication.

Niederhuber, J., Britton, S ., and Bergquist, R. (1972), A specific mouse B-lymphocyte antigen (MBLA) demonstrated by double immunofluorescence. Submitted for publ.

Niederhuber, J., Möller, E., and Mäkelä, O. (1972), Unpublished results.

Nowell, P.C., and Wilson, D.B. (1971), Studies on the life history of lymphocytes. I. The life-span of cells responsive in the mixed lymphocyte interaction. J. Exp. Med. 133, 1131.

Pasanen, V.J., and Mäkelä, O. (1969), Effect of the number of haptens coupled to each erythrocyte on hemolytic plaque formation. Immunology 16, 399.

Paul, W.E. (1970), Antigen-binding lymphocyte receptors. Transpl. Rev. 5, 130.

Paul, W.E., Siskind, G.W., and Benacerraf, B. (1968), Specificity of cellular immune responses. Antigen concentration dependence of stimulation of DNA synthesis in vitro by specifically sensitized cells, as an expression of the binding characteristics of cellular antibody. J. Exp. Med. 127, 25.

Paul, W.E. (1972), Proceedings of the First Int. Congress of Immunology. In press.

Raff, M.C. (1969), Theta isoantigen as a marker of thymus-derived lymphocytes in mice. Nature (Lond.) 224, 378.

Raff, M.C. (1970 a), Role of thymus-derived lymphocytes in the secondary humoral immune response in mice. Nature (Lond.) 226, 1257.

Raff, M.C. (1970 b), Two distinct populations of peripheral lymphocytes in mice distinguishable by immunofluorescence. Immunology 19, 637.

Raff, M.C. (1972), Proceedings of the First Int. Congr. of Immunology. In press.

Raff, M.C., Nase, S., and Mitchison, N.A. (1971), Mouse specific bone-marrow-derived lymphocyte antigen as a marker for thymus-independent lymphocytes. Nature (Lond.) 230, 50.

Raff, M.C., Sternberg, M., and Taylor, R.B. (1970), Immunoglobulin determinants on the surface of mouse lymphoid cells. Nature (Lond.) 225, 553.

Rajewsky, K. (1971), The carrier effect and cellular cooperation in the induction of antibodies. Proc. Roy. Soc. Lond. Ser. B, 176, 385.

Rajewsky, K., and Pohlit, H. (1972), Proc. of the First Int. Congr. of Immunology. In press.

Roelants, G.E., and Askonas, B.A. (1971), Cell cooperation in antibody induction. The susceptibility of helper cells to specific lethal radioactive antigens. Eur. J. Immunol. 1, 151.

Schlesinger, M. (1970), Anti-Θ antibodies for detecting thymus-dependent lymphocytes in the immune response of mice to SRBC. Nature (Lond.) 226, 1254.

Seppälä, I.J.T., and Mäkelä, O. (1971), Hapten-carrier relationships in immunological unresponsiveness. Eur. J. Immunol. 1, 221.

Siskind, G.W., and Benacerraf, B. (1969), Cell selection by antigen in the immune response. Adv. Immunol. 10, 1.

Werblin, T.P., and Siskind, G.W. (1971), Effect of tolerance and immunity on antibody affinity. Transpl. Rev., 8, In press.

THE CLUSTER (ROSETTE) ASSAY FOR LYMPHOCYTES BINDING SYNGENEIC AND SHEEP ERYTHROCYTES : SOME IMPORTANT VARIABLES.

H.S. MICKLEM and N. ANDERSON,
Immunobiology Unit, Department of Zoology,
University of Edinburgh, Scotland.

Abstract: Factors such as cell concentration and incubation medium profoundly effect the binding of sheep erythrocytes by normal mouse small lymphocytes at 4° in vitro. Standard techniques greatly underestimate the potential number of cluster-forming cells, especially in the thymus. Similar factors affect the binding of syngeneic or autochthonous erythrocytes in vitro ('syncluster'-formation). Syncluster-forming cells (SFC) are more numerous in the thymus than in other lymphoid tissues under most conditions of assay, but most or all SFC in the lymph nodes appear to be B-cells.

Many authors have demonstrated the existence in normal animals of small lymphocytes which are capable of specifically binding antigens. Experiments with cell populations artificially depleted of specific antigen-binding cells have shown that such cells are involved in the production of an immune response (Ada, 1969; Wigzell, 1970; Bach, 1970; Humphrey, 1971). The binding of antigen by these cells may be impeded by prior incubation of the cells in antisera directed against immunoglobulin antigenic determinants, which suggests that the receptors for antigen are immunoglobulins of some kind (Greaves, 1970). It has also been shown that the capacity of lymphocytes to give the cellular reactions of delayed type hypersensitivity or graft-versus-host reaction can be inhibited by incubation with antisera directed against parts of the immunoglobulin molecule (Greaves, 1969; Mason, 1970). One of the techniques widely used to demonstrate antigen binding has been the rosette or cluster technique. Bach (1970) was able, by depleting a cell population of lymphocytes forming rosettes with sheep or chicken erythrocytes, specifically to inhibit the response to the antigen concerned. This suggests that the cluster-forming lymphocytes (CFC) include what have been termed antigen reactive cells, i.e. cells which are capable of initiating an immune response to a particular antigen. The cluster technique has also been used to provide estimates of the absolute number of antigen reactive cells in a given tissue. The incidence of cells capable of binding sheep erythrocytes has been reported to be far lower in the thymus than in the spleen, lymph nodes and bone marrow (Laskov, 1968; Bach, 1971). A similar paucity of cells binding other antigens in the thymus of normal animals has been reported by Byrt (1969) and Humphrey (1970).

In addition to lymphocytes which bind various heterologous antigens, there is also a substantial population of cells which can be shown to bind autochthonous erythrocytes under suitable conditions in vitro (Micklem, 1971b). These cells, which have been termed syncluster-forming cells (SFC), resemble the cells which bind sheep erythrocytes in having small lymphocyte morphology, in being partially inhibitable by anti-immunoglobulin sera, and in showing immunological specificity. Under the experimental conditions which we have used routinely they are somewhat less numerous than anti-sheep CFC in the spleen, lymph nodes and bone marrow, but much more numerous in the thymus; in fact, the thymus shows the highest incidence of SFC (Micklem, 1971b). Our interest in syncluster-forming cells and their possible importance in relation to immunological self-tolerance and autoimmunity led us to explore the influence of certain variables in the cluster technique. Synclusters and anti-sheep clusters were examined in parallel.

The cluster technique.

The standard method that we used throughout our earlier experiments was based on that of Biozzi (1966) and Zaalberg (1966). Lymphocytic cell suspensions were washed 3 times in Hanks solution. Erythrocytes were prepared by washing heparinized blood cells 3 times in 0.15 molar NaCl. Six million lymphoid cells and 15 x 10^6 erythrocytes were suspended together in 1 ml of phosphate-buffered saline pH 7.2 containing 5% foetal bovine serum, and were incubated overnight in covered siliconised glass tubes at $6^{o}C$. The cells were resuspended by mechanical end-over-end rotation at 20 r.p.m. for 2 minutes. A volume of between 0.9 µl and 200 µl of the suspension was scanned in a haemocytometer or larger chamber, and the number of cluster-forming cells per million nucleated cells, and in the whole tissue, was calculated. Further details are given by Micklem (1970a). Three tightly adherent erythrocytes around a single cell were taken as the minimum criterion for a cluster.

Inclusion of 5% normal syngeneic mouse serum in the incubation medium drastically decreases the number of SFC, and the inhibitory fraction appears to be γG-globulin (Micklem, 1971a).

The variables we wished to investigate in the present experiments were the concentration of lymphocytes and erythrocytes, both absolute and relative to each other. For this purpose it was thought desirable to prevent the suspensions sedimenting. Accordingly, the preparations were made in plastic tubes of exactly 1 ml capacity, care being taken to exclude any air bubble. The tubes were placed on a mechanical roller which gave constant movement round the axis of the tube combined with a gentle up-down movement (Denley Instruments Ltd., Bolney, Sussex, U.K.). The concentration of foetal bovine serum in the incubation medium was lowered to 1%. Preliminary experiments showed that this change did not affect relative SFC number, gave equally good cell viability, and prevented deformation of erythrocytes which was sometimes observed when the concentration of foetal bovine serum was 5%. Very similar results can be obtained with serum free tissue culture medium 199 as the incubation fluid. Constant agitation gave some what fewer clusters under given conditions than the sedimentation-resuspension method, but reduced the incidence of white-cell clumps.

The effect of varying lymphocyte concentration.

Bach (1970) reported a gradual increase in the number of clusters when the number of erythrocytes per lymphocytes was increased from 1 to 100. He suggested that this might be interpreted in terms of heterogeneous affinity of the receptors for antigen, although he pointed out that the situation was much more difficult to interpret than one involving varied concentrations of a soluble antigen. We confirmed Bach's observation concerning the effect of raising the number of sheep erythrocytes while maintaining a constant number of lymphocytes, and also found a similar effect with syncluster formation. However, we found that an equally pronounced effect could be obtained by keeping the concentration of erythrocytes constant and reducing the concentration of lymphoid cells. A remarkably consistent linear increase both in anti-sheep CFC and in SFC was found when the lymphocyte concentration was reduced over a 10^3 fold range. A series of individual cell suspensions from normal CBA mice were tested, and linear regressions of log CFC number on log lymphocyte concentration were estimated for each one. Estimates of mean slope and elevation were made from 4 or more individual regressions for each tissue, statistical analysis having revealed no significant heterogeneity between the individual regressions.

The regressions for anti-sheep erythrocyte clusters in the spleen, lymph nodes, thymus and bone marrow are illustrated in Figure 1. There are no significant differences in slope ($P>0.25$) or elevation ($0.1>P>0.05$) between spleen, lymph nodes and bone marrow. The slope of the thymus regression is significantly steeper than that of the other tissues ($P<0.001$). The effect of this is that at the erythrocyte:lymphocyte ratios most commonly used for enumerating clusters, the thymus appears to have a much lower content of CFC than the other tissues (Bach, 1971). This difference disappears at higher ratios, and the proportion of thymus cells which appear as anti-sheep CFC then approaches 10%.

Figure 2 shows similar data for SFC. The main difference between SFC and anti-sheep CFC is seen in the thymus. The regression for thymus SFC can only be treated as linear up to an erythrocyte:lymphocyte ratio of 50:1. Over that range the two slopes do not differ significantly, but the elevation for SFC is very much higher. Even at low erythrocyte: lymphocyte ratios more SFC are detected in the thymus than in the other tissues, as we have previously reported (Micklem 1971b).

It seemed possible that for some reason only a small proportion of erythrocytes could be incorporated in clusters and that this was a limiting factor in the number of clusters detectable. Two experiments render this explanation unlikely or at least insufficient. Comparisons were made between normal and heat-treated (48°C, 20 mins) erythrocytes in the SFC assay. Heat-treated erythrocytes produced very similar regressions, suggesting that the limiting factor was not a small number of grossly damaged erythrocytes.

Table 1. Anti-sheep CFC in lymph nodes and thymus of normal CBA mice, estimated from preparations made at 10:1 erythrocyte : lymphocyte ratio and various cell concentrations.

Lymphocyte conc./ml.	CFC per thousand nucleated cells	
	Lymph nodes	Thymus
2.5×10^6	2.7	0.4
1.0×10^6	3.9	0.9
2.5×10^5	6.9	14.0
1.0×10^5	23.0	120.0

Each figure represents the geometric mean of 4 observations on two cell suspensions.

Secondly, as shown in Table 1, reduction of total cell concentration while maintaining a 10:1 ratio of erythrocytes to lymphocytes yielded progressively increasing proportions of CFC.

Current experiments are indicating that even when cluster-formation is quickly effected by the technique of McConnell (1969), the number of rosettes can still be increased by lowering the concentration of lymphocytes.

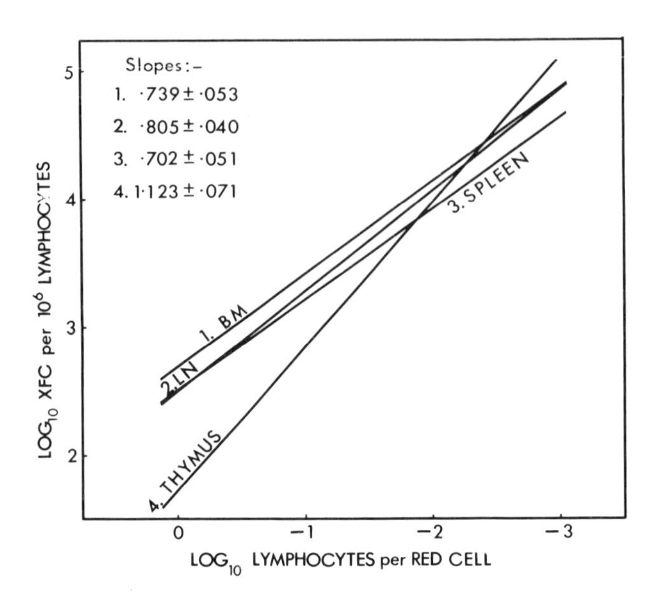

Figure 1. Relationship of anti-sheep CFC ('XFC') concentration to nucleated cell concentration (erythrocytes constant at 25 x 10^6 per ml). Each regression line is calculated from 4 or more individual regressions, each of which represents results obtained with a single cell suspension tested over a range of dilutions.

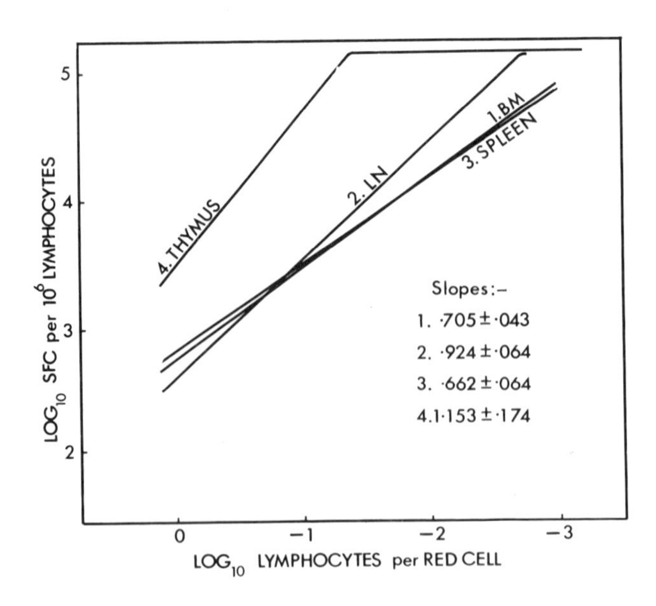

Figure 2. Relationship of syncluster-forming cell (SFC) concentration to nucleated cell concentration (erythrocytes constant at 25 x 10^6 per ml). Each regression line is calculated from 4 or more individual regressions, each of which represents results obtained with a single cell suspension tested over a range of dilutions.

These results can be summed up by saying that most workers, ourselves included, have used ratios (about 3:1 to 10:1) and concentrations (10^6 cells/ml) which yield relatively low numbers of clusters. At this level the test appears to detect antigen-reactive cells (Bach, 1970). The detected numbers of clusters can be increased dramatically by either increasing erythrocyte concentration or decreasing lymphocyte concentration or by decreasing the concentration of erythrocytes and lymphocytes together. Only the first of these procedures would seem to be likely to lead to the detection of receptors of progressively lower binding avidity, as it might be understood according to mass-action principles. The cause of the phenomenon is not understood, nor is it known whether it has any relevance to other systems for studying antigen-binding cells. However, its existence means that estimates of the number of erythrocyte binding cells in a tissue can only be made in the context of rigidly defined experimental conditions. Brain's (1970) results also underline the incompleteness of our understanding of the cluster assay.

The significance of syncluster-forming cells.

Impressed by the high number of SFC in the thymus, we suggested that a function of the thymus might be to generate a population of T-cells carrying receptors which react with 'self' antigens with an energy just below the threshold needed for activation of the cell (Micklem, 1971b, 1971c). The existence of such a population would, it was argued, be advantageous for any 'surveillance' directed against antigenic deviants from the norm, and would also help to account for the large number of cells which seem to partake in graft-versus-host reactions. However, three experimental procedures show that many, and possibly all, of the SFC in the spleen, lymph nodes and bone marrow are B-cells. 14 CBA mice deprived of T-cells by a regime of thymectomy, lethal X-irradiation and bone marrow restoration (Davies, 1969) were kindly provided by Dr. A.J.S. Davies. 74 days after irradiation 6 mice were injected intravenously with 5×10^7 syngeneic thymus cells, and 31 to 39 days later the mice were killed and the synclusters in marrow, spleen, lymph nodes and Peyers patches were enumerated. The results (Table 2) show

Table 2. Mean (geometric) concentration of syncluster-forming cells in tissues of thymectomized, irradiated, bone marrow-reconstituted CBA mice killed 105-113 days post-irradiation. (Erythrocyte:lymphocyte ratio 2.5:1. Lymphocyte concentration 6×10^6/ml).

	SFC per thousand nucleated cells	
	Thymus injected* (6 mice)	No thymus injected (8 mice)
Bone marrow	1.1	1.0
Spleen	1.2	0.7
Lymph nodes	2.9	1.4
Peyers patches	0.4	0.6

* 5×10^7 CBA thymus cells, intravenously, 74 days post-irradiation.

In no tissue was the difference between thymus-injected and non-injected mice statistically significant.

that the T-cell-deprived mice had frequencies of SFC which did not differ significantly from those found in thymus injected and (Micklem, 1971b) in normal CBA mice. Thus it seemed that SFC could develop in normal numbers without the intervention of the thymus or the participation of thymus-processed cells. Our unpublished data on congenitally thymusless Nude mice and their normal littermates also demonstrate that many SFC are B-cells, although the possibility remains open that a minority are T-cells. Table 3 shows

Table 3. Cytotoxic indices (C.I.) and enrichment factors (E.F.) for SFC in CBA lymph node suspensions after incubation for 1 hr. at 37°C in AKR anti-theta serum and rabbit complement. Controls were incubated in normal AKR serum plus complement.
(Erythrocyte:lymphocyte ratio 10:1. Lymphocyte concentration 2.5 x 10^6/ml.)

Expt.	No. of suspensions.	No. of replicate observations/suspension	C.I.*	Expected E.F.**		Observed E.F.
1	4	2	0.716 (mean)	3.52 (mean)	(P<0.001)	10.65 (mean)
2	1	2	0.737	3.80		3.81
3	2	2	0.715 0.627	3.51 2.68		8.73 7.45
4	1	4	0.705	3.38		4.80
5	1	4	0.760	4.17		2.41

* i.e. proportion of total cells killed specifically by anti-theta serum

** i.e. enrichment factor expected if all SFC are theta-negative (calculated as $^1/_{1 - C.I.}$). Lower E.F. would be expected if some SFC are theta-positive.

that lymph node cells are relatively enriched in SFC by prior treatment with anti-theta serum and rabbit complement which destroys most or all T-cells (Raff, 1971). The enrichment factors obtained suggested that all the SFC were B-cells, and indeed that the presence of T-cells might partially inhibit cluster-formation by B-cells. Taken together, all these results imply that no substantial numbers of peripheral T-cells bind syngeneic erythrocytes under the conditions we have investigated (erythrocyte:lymphocyte ratio 2.5:1 - 10:1 and lymphocyte concentration $>10^6$/ml). These conditions produce relatively small numbers of clusters, and we so far know little about the extra SFC which can be revealed by manipulation of the cell concentration.

Similarly, our information concerning specifity and inhibition by anti-Ig sera comes from experiments which yielded low concentrations of SFC (Micklem 1971b). We are also ignorant of the qualities of anti-sheep CFC formed under conditions which yield high numbers. It is clear that at least some anti-sheep CFC in normal mice are immunologically significant. Provisionally it seems reasonable to make the same assumption for the SFC which are found in the peripheral lymphoid tissues, particularly since under standard conditions raised numbers are seen in the lymph nodes of allografted, but not syngrafted mice (Micklem, 1970b). It is not known to what extent SFC and their descendants normally secrete any kind of auto-antibody. Their ability to do so may depend on stimulation of an appropriate T-cell helper population, perhaps by an antigen which cross-reacts with some autochthonous membrane antigen. Conceivably, some of them may produce auto-antibodies significant in the maintenance of 'self' tolerance (Micklem, 1971a). Alternatively, their function may not be primarily immunological, but rather be concerned with some more general somatic cell-cell recognition mechanism. In view of the suppressive effect of mouse serum on SFC-formation, it seems unlikely that clusters as such are formed frequently with autochthonous erythrocytes in vivo.

The significance of thymic SFC is also unknown. They may temporarily be synthesizing and displaying immunoglobulin receptors specific for self patterns, perhaps as a stage in the induction of self-tolerance. But the possibility that they bind syngeneic erythrocytes by means of passively acquired autoantibodies or some other unknown mechanism has not been excluded. The capacity to form synclusters does not seem to be related to the stage of maturity of the thymocyte, since treatment of A-strain thymocytes with anti-TL serum and complement in vitro neither depletes the remaining cell suspension of SFC nor enriches it (unpublished data). Most or all thymic SFC must either lose their receptors or be destroyed before passing to the periphery as T-cells.

This work was supported by the Medical Research Council.

REFERENCES

Ada, G.L. and Byrt, P. (1969), Specific inactivation of antigen reactive cells with ^{125}I-labelled antigen, Nature 222, 1291.

Bach, J.-F., Dardenne, M. and Muller, J.-Y, (1970), In vivo specific antigen recognition by rosette-forming cells, Nature 227, 1251.

Bach, J.F., Reyes, F., Dardenne, M., Fournier, C. and Muller, J.-Y. (1971), Rosette formation, a model for antigen recognition. In Cell Interactions and Receptor Antibodies in Immune Responses (ed. O. Mäkelä, A. Cross and T.U. Kosunen). London, Academic Press, p. 111.

Biozzi, G., Stiffel, C., Mouton, D., Liacopoulos-Briot, M., Decreusefond, C. and Bouthillier, Y. (1966), Etude du phénomène de l'immunocytoadhérence au cours de l'immunisation, Ann. Inst.Pasteur 110, 7.

Brain, P., Gordon, J. and Willetts, W.S. (1970), Rosette formation by peripheral lymphocytes. Clin. exp. Immunol. 6, 681.

Byrt, P. and Ada, G.L. (1969), An in vitro reaction between labelled flagellin or haemocyanin and lymphocyte-like cells from normal animals, Immunol. 17, 503.

Davies, A.J.S., (1969), The thymus and the cellular basis of immunity. Transplant. Revs. 1, 43.

Greaves, M.F. (1970), Biological effects of anti-immunoglobulins: evidence for immunoglobulin receptors on 'T' and 'B' lymphocytes. Transplant. Revs. 5, 45.

Greaves, M.F., Torrigiani, G. and Roitt, I.M. (1969), Blocking of the lymphocyte receptor site for cell-mediated hypersensitivity and transplantation reactions by anti-light chain sera, Nature 222, 885.

Humphrey, J.H. and Keller, H.U. (1970), Some evidence for specific interaction between immunologically competent cells and antigens. In Developmental Aspects of Antibody Formation and Structure (ed. J. Šterzl and I. Řihá), Prague, Academia, Vol. 2, p.485.

Humphrey, J.H., Roelants, G. and Willcox, N. (1971), Specific lethal radioactive antigens. In Cell Interactions and Receptor Antibodies in Immune Responses (ed. O. Mäkelä, A. Cross, and T.U. Kosunen). London, Academic Press, p.123.

Laskov, R. (1968), Rosette-forming cells in non-immunized mice. Nature 219, 973.

Mason, S. and Warner, N.L. (1970), The immunoglobulin nature of the antigen recognition site on cells mediating transplantation immunity and delayed hypersensitivity. J. Immunol. 104, 762.

McConnell, I., Munro, A., Gurner, B.N. and Coombs, R.R.A. (1969), Studies on actively allergized cells. I. The cytodynamics and morphology of rosette-forming lymph node cells in mice, and inhibition of rosette formation with antibody to mouse immunoglobulins. Int. Arch. Allergy 35, 209.

Micklem, H.S. (1971a), The cellular basis of 'self'-tolerance. In Immunological Tolerance to Tissue Antigens (ed. N.W. Nisbet and M.W. Elves). Orthopaedic Hospital, Oswestry, England, p. 237.

Micklem, H.S. and Asfi, C. (1971b), Lymphoid cells reacting against 'self' and their possible role in immune responses. In Morphological and Functional Aspects of Immunity (ed. K. Lindahl-Kiessling, G. Alm and M.G. Hanna) Plenum, New York, p. 57.

Micklem, H.S. and Asfi, C. (1971c), Cells carrying receptors for 'self' constituents: their possible significance in the evolution of the vertebrate immune system. Arch. Zool. exp. gen. 112, 105.

Micklem, H.S., Asfi, C. and Anderson, N. (1970a), Studies of mouse lymphoid cells forming clusters with syngeneic or allogeneic erythrocytes. In Developmental Aspects of Antibody Formation and Structure (ed. J. Šterzl and I. Říhá). Prague, Academia, Vol. 2, p. 701.

Micklem, H.S., Asfi, C., Staines, N.A. and Anderson, N. (1970b), Quantitative study of cells reacting to skin allografts. Nature 227, 947.

Raff, M.C. (1971), The use of surface antigenic markers to define different populations of lymphocytes in the mouse. In Cell Interactions and Receptor Antibodies in Immune Responses (ed. O. Mäkelä, A. Cross and T.U. Kosunen) London, Academic Press, p. 83.

Wigzell, H. and Mäkelä, O. (1970), Separation of normal and immune lymphoid cells by antigen-coated columns. Antigen-binding characteristics of membrane antibodies as analyzed by hapten-protein antigens. J. exp. Med. 132, 110.

Zaalberg, O.B., van der Meul, V.A. and van Twisk, M.J. (1966), Antibody production by single spleen cells: a comparative study of the cluster and agar plaque formation. Nature 210, 544.

ANTIBODY-ANTIGEN REACTIONS AT THE LYMPHOCYTE SURFACE: IMPLICATIONS FOR MEMBRANE STRUCTURE, LYMPHOCYTE ACTIVATION AND TOLERANCE INDUCTION

MARTIN C. RAFF and STEFANELLO DE PETRIS
Medical Research Council Neuroimmunology Project
University College London

Abstract: When multivalent anti-immunoglobulin (Ig) antibodies react with Ig molecules on the lymphocyte surface, they cause the Ig to aggregate into patches and then be actively transported to the tail end of the cell where they are pinocytosed. The implications of this observation for membrane structure and organization, as well as its possible significance for lymphocyte activation and/or tolerance induction is discussed.

Cells interact with one another through direct contact or by means of secreted factors. In either case, the interaction is usually mediated initially by the plasma membrane. Thus, questions about the structure and properties of cell membranes are particularly relevant to this colloquium. One approach to studying plasma membrane structure is to use antibodies directed against cell surface antigens. Unfortunately, for the most part, the chemical nature and function of these antigens are unknown. One notable exception is membrane-bound immunoglobulin (Ig) on the surface of lymphocytes, where we think we know something about the chemistry and function of the molecules that we are studying. In the mouse, for example, most lymphocytes have 7S IgM (i.e. the monomeric subunit of the 19S IgM pentamer) on their surface (Vitetta, Baur and Uhr, 1971) and there is substantial indirect evidence that these Ig molecules serve as specific receptors for antigen (Greaves, 1970). Of the two classes of peripheral lymphocytes (thymus dependent T cells and thymus-independent B cells) only B cells have readily demonstrable Ig on their surface (Raff, 1970) and there is evidence that the Ig present on the plasma membrane is made by the B cell that carries it (Pernis, 1970).

Using immunofluorescence (Taylor, Duffus, Raff and de Petris, 1971) and immunoferritin-electron-microscopy (de Petris and Raff, 1971) to study the distribution of Ig molecules on the surface of mouse spleen lymphocytes, it was found that the interaction of labelled anti-Ig antibody with the membrane Ig molecules induced a dramatic change in the distribution of these molecules on the cell surface, which was followed by pinocytosis of Ig-bearing membrane . Although the observation is simple the available evidence suggests that it has important implications: (1) it indicates that the cell membrane is fluid in nature, which puts limits on the type of organization and patterning that can exist on the cell surface; (2) it implies that previous demonstrations of patchy distributions for a variety of lymphocyte surface antigens (Cerrottini and Brunner, 1967; Aoki, Hämmerling, de Harven, Boyse and Old, 1969) were probably artifactually produced by the interaction with labelled antibody; (3) it suggests a probable mechanism for the phenomenon of antigenic modulation, which is the specific disappearance of a surface antigen induced by antibody directed against the antigen (Old, Stockert, Boyse and Kim, 1968); (4) it points to the possibility that receptor redistribution and/or pinocytosis may be involved in lymphocyte triggering by antigens, antibodies and phytomitogens or in the induction of immunological tolerance; and (5) it makes one wonder if receptor redistribution and/or pinocytosis is important in the interaction of other substances with surface receptors on other cell types.

ANTIBODY INDUCED REDISTRIBUTION AND PINOCYTOSIS OF SURFACE Ig DETERMINANTS

When living lymphocytes are labelled in suspension with fluorescein or ferritin conjugates of anti-Ig antibody, three different labelling patterns can be observed depending on the conditions of the experiment:

(1) Diffuse surface labelling (Fig. 1A) - seen by immunofluorescence when univalent Fab fragments of anti-Ig antibody molecules (i.e. papain digested anti-Ig) are used at any temperature (Taylor *et al.*, 1971) or when multivalent anti-Ig antibody is used strictly at 0°C (including centrifugation and microscopy at 0°C) (Pernis - personal communication). Presumably the same pattern would be seen if the anti;Ig antibodies were all directed at only one or possible two antigenic determinants on the Ig molecule (see later)(Davis, Alstaugh, Stimpfling and Walford, 1971). Although the demonstration of continuously distributed Ig determinants should be possible under appropriate conditions by immunoferritin electron microscopy, this has not yet been attempted.

(2) Patchy labelling (Fig. 1B) - is the main pattern seen when multivalent (i.e. undigested) anti-Ig antibody is used at temperatures somewhere between 4° and 15°C (de Petris and Raff, 1971), or at any temperature above ~4°C in the presence of sodium azide or dinitrophenol (Taylor *et al.*, 1971).

(3) "Cap" labelling (Fig. 1C) - with all of the label (which was initially located diffusely over the surface) accumulated over one pole of the cell (that containing the Golgi apparatus and most of the cell organelles, often elongated as a uropod), is the main pattern seen when multivalent anti-Ig antibody is used at temperatures above ~15°C (Taylor *et al.*, 1971; de Petris and Raff, 1971). Cap formation is almost invariably accompanied by pinocytosis of labelled membrane (de Petris and Raff, 1971) and occurs within minutes if cells treated in the cold and showing diffuse or patchy labelling are warmed to 37° (Taylor *et al.*, 1971). After several hours at 37°C most labelled cells have much of their label within the cytoplasm and many cells no longer have surface Ig determinants demonstrable if attempts are made to relabel them with an anti-Ig-fluorescein conjugate. However, these cells have not lost detectable amounts of H-2 antigens which are still diffusely distributed on the cell surface (Taylor *et al.*, 1971). Under appropriate culture conditions surface Ig determinants reappear after 12-24 hours (Raff and de Petris - unpublished observations).

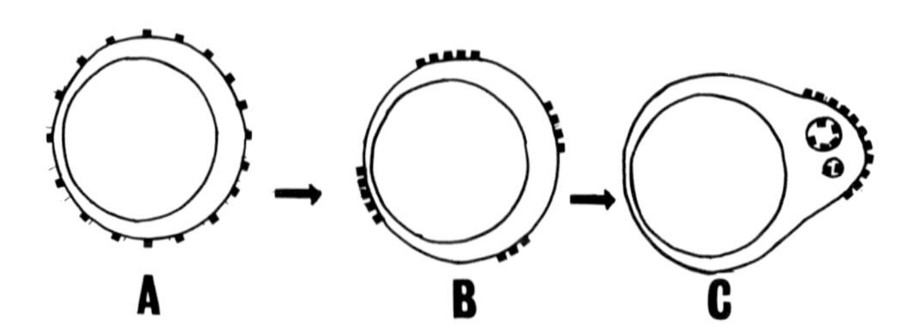

Fig.1. Schematic representations of the three different labelling patterns seen when fluorescein or ferritin conjugated anti-Ig antibody is used to demonstrate Ig on the surface of mouse B lymphocytes. The experimental conditions under which the different patterns are seen are outlined in the text.

The simplest interpretation of these observations is that the normal distribution of Ig is diffuse, either as single Ig molecules or as submicroscopic clusters of molecules relatively free-floating in a liquid matrix (see later). Multivalent anti-Ig antibodies cross-link the mobile surface Ig molecules producing lattices which are microscopically visible as patches. This process of patch formation is probably not an active one and can occur at relatively low temperatures and/or in the presence of azide. At higher temperatures and in the absence of azide, these patches segregate from the unlabelled segments of membrane, moving to the Golgi pole of the cell where pinocytosis of labelled membrane occurs. Cap formation and pinocytosis are metabolically dependent active processes. They do not occur at low temperatures (≥ 4°C) or in the presence of azide, they are independent of extracellular Ca^{++} and Mg^{++} and probably involve contractile microfilament activity to some extent, as cap formation is partially inhibited and pinocytosis is completely inhibited by Cytochalasin B (Taylor et al., 1971). We believe that the mechanism involved in cap formation is the same as that normally involved in cell movement. A detailed discussion of possible mechanisms involved in cap formation is given elsewhere (de Petris and Raff - submitted for publication).

Antibody induced redistribution is not unique for surface Ig. Patchy distributions have been demonstrated for a variety of mouse lymphocyte alloantigens, such as TL, θ, and H-2 (Aoki et al., 1969) and for HL-A on human lymphocytes (F.M. Kourilsky - personal communication). Cap formation can be produced on mouse thymocytes with anti-θ or anti-H-2 antibody but requires in addition a second layer of anti-mouse-Ig antibody, which reacts with the alloantibodies ("Piggy-back" effect)(Taylor et al., 1971). Cap formation has also been observed when phytomitogens such as phytohaemagglutinin (PHA)(Osunkoya, Williams, Adler and Smith, 1970) and Concanavalin A (Smith and Hollers, 1970) react with lymphocytes, and when anti-IgE antibody reacts with IgE-sensitized human basophils (Sullivan, Grimley and Metzger, 1971).

SIGNIFICANCE FOR MEMBRANE STRUCTURE AND ORGANIZATION

The finding that labelled membrane, bearing Ig determinants, can segregate from unlabelled membrane, bearing other determinants (e.g. H-2) implies that there are Ig-bearing units ("mobile units") which can move about in the membrane relative to other membrane components. The finding that this type of mobility is characteristic for many surface components (e.g. Ig, θ, H-2, mitogen receptors) suggests that the lymphocyte plasma membrane has the properties of a two-dimensional fluid (rather than a two-dimensional deformable solid) with mobile units (probably mostly protein) floating in a lipid sea. The same conclusion was reached by Frye and Edidin (1970) studying the diffusion of histocompatibility antigens on newly formed mouse-human heterokaryons, and by Blaisie and Worthington (1969) studying frog retinal receptor disc membranes by X-ray diffraction. The liquid nature of the plasma membrane is lost as the cell is cooled, either because the viscosity of the membrane becomes too high, or the hydrocarbon chains of the membrane phospholipids undergo a liquid-solid phase transition below a critical temperature.

It has been argued that the cell surface must have a highly organized topography with each component in a predetermined position relative to other components in its vicinity, and that this characteristic surface arrangement or grid might play an important role in cell interactions in embryogenesis (Boyse, 1970). The question now arises, how much pattern can there be in a fluid membrane? It is highly likely that all of the studies demonstrating patchy distributions of surface antigens on lymphocytes (suggesting that they are confined to discrete regions of the cell surface) are invalid, since they have all used multivalent antibody or antibody-antigen complexes which are now known to induce patch formation. Thus, when the distribution of H-2 antigens on the surface of mouse lymphocytes was studied by direct immunoferritin electron microscopy (i.e. using anti-H-2-ferritin conjugate) a diffuse distribution of H-2 was observed, while a patchy distribution was observed using the same anti-H-2 antibody in indirect testing (i.e. anti-H-2 followed by anti-mouse-Ig-ferritin conjugate)(Davis et al., 1971). Although the anti-H-2-ferritin conjugate used in the direct technique was multivalent, it was raised in congenic mice differing only at the H-2 locus, and thus may well have been directed against only one or two antigenic determinants, making two-dimensional lattice formation impossible. If this interpretation is correct it implies that the H-2 antigenic determinants are represented only once or at most twice on a mobile unit.

A highly ordered arrangement of the various mobile units does not seem to us to be compatible with the liquid membrane model. However, there may be important organization of individual protein molecules within putative mobile units. For example, it is not clear whether Ig molecules are distributed in the membrane as individual molecules (i.e. mobile unit = 1 Ig molecule) or whether there are a number of Ig or other types of molecules constituting each mobile unit. Although it is suggestive that most H-2 molecules are not located in the same units as Ig (see before)(Taylor et al., 1971), the recent finding that some antilymphocyte sera, which appear not to be contaminated with anti-Ig antibodies, induce Ig cap formation (B. Pernis - personal communication; M. Greaves - personal communication), suggests that Ig mobile units may be composed of more than just Ig molecules. The most compelling evidence for "intraunit" organization comes from the "blocking" studies of Boyse and his colleagues (1968), who studied the effect of the binding of one type of alloantibody to the surface of mouse thymocytes on the subsequent binding of other types of alloantibodies. Assuming that if two antigens are sufficiently close together, absorption of antibody by one will inhibit the subsequent absorption of antibody by the other, they plotted the position of 5 different alloantigenic systems (H-2, θ, TL, LyA and LyB) in relation to one another on the thymocyte surface. Although it is not certain whether all of these antigens are on separate macromolecules (Davies, Boyse, Old and Stockert, 1967) it seems likely that at least some of them are separable. The blocking results strongly suggest that the antigens of these 5 alloantigenic systems are not randomly distributed but are arranged in a specific pattern on the cell surface. Boyse et al. (1968) suggested that the alloantigens of the five systems studied may constitute a single unit which is a repetitive feature of the thymocyte membrane. If that is the case, it would be interesting to see if these alloantigens would segregate as a unit (and thus be equivalent to a mobile unit) when antibody against any one is used to induce cap formation. This type of experiment is difficult to do because in general cap formation does not occur with high concentrations of alloantibody, even with a second layer of anti-mouse-Ig, at least with anti-θ and anti-H-2, the two alloantigens that we have studied. At concentrations of alloantibody where cap formation is induced, all of the alloantigen is usually not moved into the cap.

However, studies of antigenic modulation of mouse thymocyte alloantigens make it clear that TL antigens can be removed from the cell surface by anti-TL antibodies while the antigens determined by H-2(D) are increased on these modulated cells (Boyse et al., 1967). This suggests that although TL antigens and H-2(D) determined antigens map side by side in blocking experiments (Boyse et al., 1968) they can be readily separated from each other by means of antibody. Thus, the forces holding at least some of the molecules together within any one unit must be very weak, unless the interaction between these molecules is altered by the antibody-antigen reaction inducing the modulation. This also suggests that it may be difficult to demonstrate weak associations between different antigens by modulation or segregation (i.e. cap formation) experiments. If two different antigens can be shown to modulate or segregate together when exposed to antibody against either one, it is likely that they are closely associated on the membrane; however, if they modulate or segregate separately, this would not exclude their normally being preferentially associated in the membrane.

How do different antigens become associated in the plasma membrane? In the simplest form, associated antigens may be different determinants on the same macromilecule (e.g. heavy and light chain determinants on surface Ig molecules). In some instances, different proteins that are associated on the membrane may be translated from polycistronic messenger RNA and be inserted into the membrane as preformed units. There is evidence that proximity on the cell surface can parallel genetic linkage (e.g. TL and H-2(D)), but this is not a universal rule (e.g. H-2(D)-determined antigens and Ly-B map side by side in blocking experiments but are determined by unlinked genes (Boyse et al., 1968). Ultimately we must find methods for following the insertion of proteins into membranes and for identifying different antigens on the cell surface simultaneously, and we must learn more about the chemistry and function of the antigens that we are studying.

SIGNIFICANCE FOR THE IMMUNE RESPONSE

Lymphocytes can be triggered to transform into blast cells and divide by a variety of interactions at the cell surface. Thus antibodies directed against surface antigens, phytomitogens reacting with surface carbohydrate determinants, and antigens reacting with their complementary Ig receptors can all activate lymphocytes. For a particular type of lymphocyte, activation by any of these stimuli probably proceeds along a common pathway of response (Greaves, in this colloquium).

A generalization that is emerging from a variety of different lines of experimentation is that multivalent binding of the antibody, mitogen or antigen to the cell surface is required for lymphocyte activation: (1) univalent anti-Ig (Fanger, Hart, Wells and Nissonoff, 1970) or antilymphocyte serum (Riethmüller, Riethmüller, Stein and Hansen, 1968)(i.e. Fab fragments) are not stimulatory whereas whole molecules or $F(ab')_2$ fragments of these antibodies are; (2) monomeric PHA does not activate lymphocytes (Lindahl-Kiessling - personal communication), while polymeric PHA does; (3) polymeric antigens, such as polyvinylpyrolidone (PVP), pneumococcal polysaccharide, E.coli endotoxin, and polymerized flagellin, which can make multivalent attachments to lymphocyte surface receptors, can stimulate B lymphocytes directly to produce IgM antibody without T lymphocyte and possibly without macrophage help. There is no convincing evidence that antigens which cannot bind multivalently to B cells (i.e. which do not have > 2 identical determinants) can activate B cells without some form of help. It seems possible that one of the things that a T cell does when it helps a B cell to make antibody is to present antigen in a multivalent array, either on the surface of the T cell or by cross-linking of antigen molecules by shed T cell receptors.

We would like to think that the requirement for multivalent binding for lymphocyte activation reflects the need for multivalent binding to induce a redistribution of receptor molecules - i.e. patch or cap formation. Monovalent anti-Ig antibody and monovalent antigens do not induce cap formation, while multivalent antibodies and antigens can (Taylor et al., 1971). The finding that mitogens, such as PHA and Concanavalin A, can also induce cap formation (and pinocytosis) strengthens this hypothesis. The correlation between multivalency, capping and lymphocyte activation is so far quite impressive. However, it is likely that receptor redistribution is a necessary but not sufficient condition for activation, since cap formation apparently can be observed under non-stimulatory conditions (Greaves - personal communication).

An important hiatus in the story thus far is the behaviour of antigen-specific receptors on T cells when they interact with antigen. The difficulty in directly demonstrating these receptors (which are almost certainly Ig molecules) has made it difficult to study this problem.

Some forms of immunological tolerance may also be related to cap formation and/or pinocytosis of the Ig receptors. Multivalent antigen has been shown to induce cap formation and pinocytosis (J. Mitchell - personal communication; S. de Petris - unpublished observations), and it is possible that under appropriate conditions, antigen could induce the disappearance of receptors without activating the lymphocyte and thus render the cell tolerant. Small amounts of residual antigen would then be required to remove receptors as they reappeared, and in this way maintain the tolerant state. The examples of immunological tolerance which require, in addition to antigen, small amounts of antibody (possibly to enable multivalent binding of antigen to receptors) (Diener and Feldmann, 1970) would seem to us to be particularly good candidates for this type of mechanism.

ACKNOWLEDGEMENTS

M.C.R. was supported by a Postdoctoral Fellowship from the National Multiple Sclerosis Society of the United States, and S.de P. by a short-term fellowship from the European Molecular Biology Organization.

REFERENCES

Aoki, T., Hämmerling, U., de Harven, E., Boyse, E.A. and Old, L.J., (1969), Antigenic structure of cell surfaces. An immunoferritin study of the occurrence and topography of H-2, θ, and TL alloantigens on mouse cells, J.exp.Med. 130, 979.

Blaisie, J.K. and Worthington, C.R., (1969), Planar liquid-like arrangement of photopigment molecules in frog retinal receptor disk membranes, J.Mol.Biol. 38, 407.

Boyse, E.A., Stockert, E. and Old, L.J., (1967), Modification of the antigenic structure of the cell membrane by thymus-leukemia (TL) antibody, Proc.Nat. Acad.Sci., USA 58, 954.

Boyse, E.A., Old, L.J. and Stockert, E., (1968), An approach to the mapping of antigen on the cell surface, Proc.Nat.Acad.Sci., USA 60, 886.

Boyse, E.A., (1970), Invitation to Surveillance in Immune Surveillance, R.T. Smith and M. Landy, Editors. (Academic Press, New York) p.5.

Cerrottini, J.-C. and Brunner, K.T., (1967), Localization of mouse isoantigens on the cell surface as revealed by immunofluorescence, Immunology 13, 395.

Davies, D.A.L., Boyse, E.A., Old, L.J. and Stockert, E., (1967), Mouse isoantigens: separation of soluble TL (thymus-leukemia) antigen from soluble H-2 histocompatibility antigen by column chromatography, J.exp.Med. 125, 549.

Davies, D.A.L., Atkins, B.J., Boyse, E.A., Old, L.J. and Stockert, E., (1968), Soluble TL and H-2 antigens prepared from a TL-positive leukemia of a TL-negative mouse strain, Immunology 16, 669.

Davis, W.C., Alstaugh, M.A., Stimpfling, J.H. and Walford, R.L., (1971), Cellular surface distribution of transplantation antigens: discrepancy between direct and indirect labelling techniques, Tissue Antigens 1, 89.

de Petris, S. and Raff, M.C., (1971), Distribution of immunoglobulin on the surface of mouse lymphoid cells as determined by immunoferritin electron microscopy. I. Antibody-induced temperature-dependent redistribution and its implications for membrane structure, J.Cell.Sci. - submitted for publication.

Diener, E. and Feldmann, M., (1970), Antibody-mediated suppression of the immune response *in vitro*. II. A new approach to the phenomenon of immunological tolerance, J.exp.Med. 132, 31.

Fänger, M.W., Hart, D.A., Wells, V.J. and Nisonoff, A., (1970), Requirement for cross-linkage in the stimulation of transformation of rabbit peripheral lymphocytes by antiglobulin reagents, J.Immunol. 105, 1484.

Frye, L.D. and Edidin, M., (1970), The rapid intermixing of cell surface antigens after formation of mouse-human heterokaryons, J.Cell Sci. 7, 319.

Greaves, M.F., (1970), Biological effects of anti-immunoglobulins: evidence for immunoglobulin receptors on 'T' and 'B' lymphocytes, Transplant.Rev. 5, 45.

Old, L.J., Stockert, E., Boyse, E.A. and Kim, J.H., (1968), Antigenic modulation. Loss of TL antigen from cells exposed to TL antibody. Study of the phenomenon in vitro, J.exp.Med. 127, 523.

Osunkoya, B.O., Williams, A.I.O., Adler, W.H. and Smith, R.T., (1970), Studies on the interaction of phytomitogens with lymphoid cells. I. Binding of phytohaemagglutinin to cell membrane receptors of cultured Birkitt's lymphoma and infectious mononucleosis cells, Afr.J.Med.Sci. 1, 3.

Pernis, B., Forni, L. and Amante, L., (1970), Immunoglobulin spots on the surface of rabbit lymphocytes, J.exp.Med. 132, 1001.

Raff, M.C., (1970), Two distinct populations of peripheral lymphocytes in mice distinguishable by immunofluorescence, Immunology 19, 637.

Riethmüller, G., Riethmüller, D., Stein, H. and Hansen, P., (1968), In vivo and in vitro properties of intact and pepsin-digested heterologous anti-mouse thymus antibodies, J.Immunol. 100, 969.

Smith, C.W. and Hollers, J.C., (1970), The pattern of binding of fluorscein-labelled Concanavalin A to the motile lymphocyte, J.Reticuloendothel.Soc. 8, 458.

Sullivan, A.L., Grimley, P.M. and Metzger, H., (1971), Electron microscopic localization of immunoglobulin E on the surface of human basophils, J.exp.Med. - in press.

Taylor, R.B., Duffus, W.P.H., Raff, M.C. and de Petris, S., (1971), Redistribution and pinocytosis of lymphocyte surface immunoglobulin molecules induced by anti-immunoglobulin antibody, Nature New Biol. 233, 225.

Vitetta, E.S., Baur, S. and Uhr, J.W., (1971), Cell surface immunoglobulin. II. Isolation and characterization of immunoglobulin from mouse spleen lymphocytes, J.exp.Med. 134, 242.

DIFFERENTIATION AND INTERACTIONS

CELL SURFACE IMMUNOGENETICS IN THE STUDY OF MORPHOGENESIS

DOROTHEA BENNETT, EDWARD A. BOYSE AND LLOYD J. OLD
Division of Immunology, Sloan-Kettering Institute for Cancer Research
and Department of Anatomy, Cornell University Medical College
New York, New York U.S.A.

INTRODUCTION

During ontogeny, the cells of a metazoan, all of them clonal derivates of a single cell, the zygote, develop along divergent pathways, and come to differ profoundly from one another. The various manifestations of this cellular diversification are summed up as differentiation, a general term which has come to mean no more than that two cell types or tissues of one organism consistently differ from one another; it therefore needs qualification or definition for use in particular contexts. For example, it is clear what histological differentiation means. And the definition of Jacob and Monod (1963), that two cells are differentiated from one another when they harbor the same genome but synthesize different sets of proteins, is invaluable. But the latter was not intended to be comprehensive; eg two cells could hardly be more grossly distinguishable from one another than sperm and ovum, yet one of the main reasons for this in many cases is that the ovum contains a large store of nutrients that are not produced by the ovum itself.

The following discussion will be directed to morphogenetic differentiation, whereby the cells of a developing metazoan become divided into separate groups or tissues according to their functions, after which the cells of each tissue become organized into the typical configurations that are its basic microscopic anatomy or cellular organization. These two processes, the emergence of contrasting cell populations and the characteristic assembly of similar cells, may be taken to represent two extremes in the evolution of the organism from the zygote. Intermediary processes involving cooperation among different cell populations and encompassed by the term organogenesis will hardly be touched on here, but it is useful, and we think valid, to use the two extremes outlined as points of reference for discussion of morphogenesis in general. We shall limit discussion further to how the cell surface membrane may be concerned in morphogenesis, first because this particular organelle is likely to provide the immediate key to some important aspects of morphogenesis, and secondly because the main application of immunogenetic methods in the study of morphogenesis, now and in the near future, appears to be in the light they may throw on the genetic specification and organization of cell surfaces. In fact the point of this talk is to suggest that immunogenetics has already established a number of such facts about the surfaces of mammalian cells, and that it promises to yield more; there seems to be a real prospect, even at this early stage, that immunogenetic methods can be used to trace some of the events of ontogenesis from its inception to the evolution of the final cellular phenotypes that mark its completion.

To begin with, it does not seem credible that the entire program of morphogenesis could be contained in the genome in the form of separate genetic instructions for the destination of each individual cell, because there is an obvious limit to what can be expected of selective gene action *per se* in accounting for the more intricate details of tissue architecture. At some point in the ontogenesis of a complex organism therefore it can hardly be doubted that further ramifications of cell division and cooperative assembly are controlled by epigenetic rather than genetic mechanisms, in the sense that differential action of the genome may be involved only in the initial commitment or determination of certain cells to a particular mode of differentiation, after which further division can give rise only to the unique configuration of cells prescribed by that particular state of determination.

For example, it can be envisaged that during a sequence of divisions following the determination of a cell, the generation of new surface membrane is geared to the geometry of cytokinesis in such a way that succeeding cells are imprinted with sequentially modified surface patterns which interlock with those of companion cells only when the cells lie in the right configuration. This would imply individual markings and a unique location for each cell in the assembly, without demanding differential states of the genome. Alternatively the participating cells might be equivalent but have surface characteristics causing them to assume a preferential configuration, in which case reassembly, following experimental disaggregation, could take place with cells in new locations.

Support for either possibility can be cited: The reestablishment of neuronal connections between retina and tectum after surgical section of the optic nerve in the frog (Sperry, 1965) is the most spectacular example of reconstitution which lends itself to interpretation along the lines of separate surface displays for each participating cell, and we shall refer later to instances of cell association suggesting the opposite. Suffice it to say that the question whether cells belonging to a particular assembly are or are not imprinted with individual surface displays is potentially amenable to the sorts of techniques now used in immunogenetics, although the answer is scarcely within our reach at the moment.

Although determination, viewed as the commitment of a cell to one mode of differentiation on the basis of selective gene action, may appear as the likely initiator of morphogenetic events, the opposite is also plausible. In fact, there is no reason to assume that the first switch-points of the morphogenetic program which begins with cleavage of the egg need necessarily be governed by the genome directly; they may be the result of circumstances created by cleavage. Thus for example when a fertilized egg has divided once, the two progeny cells are different from the parent cell in that each now has a proportion of its surface engaged with the corresponding surface of its partner. Presumably this partnership is essential to the ensuing morphogenesis, for if it is broken by separating the two cells, then each may (under certain circumstances) form an entire individual (Tarkowski, 1959) presumably recapitulating in the process the conditioning circumstances of the first division.

To take another example: Some cells in a morula are internal and some are external, the latter differing from the former in having a part of their surfaces exposed. At this stage the first obvious events of cytodifferentiation become evident, the cells of the outer layer flattening and showing specialization in the form of tight junctions (Calarco and Brown, 1969). Nevertheless according to Tarkowski (1961) and others, even at this relatively late stage, embryogenesis may proceed satisfactorily after the cells of the morula have been experimentally reshuffled, (ie the cells are still subject to regulation).

Again, up to the 8-cell stage at least, individual blastomeres can generate entire organisms (Tarkowski, 1967); and according to Gardner and Lyon (1971) single cells of the inner cell mass of one blastocyst, transported to other locations in a second blastocyst, can alter their mode of differentiation accordingly.

On the face of it, this suggests that determination of a cell in early embryogenesis, ie its commitment to a certain mode of cellular differentiation (presumably on the basis of selective gene action), may depend on its position in the cell mass. If this is true, then the signal for genetic determination, in at least some instances, does not come from the genome but from outside the cell, and is related to the position the cell comes to occupy in the clone.

To say the least, it seems one should keep an open mind as to whether morphogenesis procedes genetic determination, or the reverse, in any given context.

Again, during the assembly of cells into definitive configurations (which we have elected to regard as the other extreme of morphogenesis), how much may cellular differentiation be governed by location (morphogenesis) rather than the reverse? Clearly the exterior properties of each cell, together with those of its neighbors, may constitute a microenvironment, possibly a highly specific one depending on particular cell surface appositions, that is essential to the full expression of its phenotype. The question is: Does the information which an individual cell of an organized tissue requires to fulfill its phenotype include not only the determined state of the genome but also details of cellular locale which were decided by the evolution of the clone? To the extent that this is so, it is determination which is heritable and stable whereas cellular differentiation (functional and morphological) is facultative or conditional (see notably Hadorn, 1965). The former, determination, may connote the establishment of a stable genetic program involving selective gene action, whereas morphogenetic differentiation, in the context of cellular organization, suggests epigenetic regulation; ie we entertain as a possibility that the determined state of a single cell is capable of specifying a morphogenetic sequence of cell divisions without further reference to the genome (see below).

The essence of morphogenesis is how cells come to be where they are, but this should perhaps not be divorced from the question of how they stay where they are. From what has been said above, it is possible to take the view that spatial relationships in early ontogeny precede and regulate determination, so deciding the gross organization of the embryo, and that even at the final stages of morphogenesis it is the patterns of cellular organization themselves which initiate and maintain cellular differentiation. There is much in embryology to support this mutual inter-dependence of cellular differentiation and tissue integrity:
(1) Cells of a small piece of an organized tissue placed in culture show a number of abnormalities. Some cells leave the explant and migrate elsewhere. Parenchymal cells, thus apparently released from histotypic constraint, tend to become morphologically (but not necessarily biochemically) "generalized", losing the distinct cellular differentiation they showed previously; they later die, leaving the culture to be taken over by fibroblasts. It is not known why specialized cells are subject to these changes in tissue culture, but high on the list of possible explanations is that maintenance of the proper cellular inter-relations is essential to sustain the differentiated state, and ultimately the viability, of a normal cell, once it has undergone determination and subsequent differentiation.
(2) A fruitful method of answering questions about cell recognition has been to observe the re-association of isolated cells prepared from embryonic rudiments. Do cells of the same type recognize one another and assemble in identifiable histotypic patterns? And do dissimilar cells repel each other? The broad answer to both these questions is that in favorable circumstances they do, but certain conditions must apparently be fulfilled: Successful histotypic reconstitution appears to require that the cells are of comparable embryonic age and are permitted to interact promptly after dissociation; more prolonged maintenance in suspension, or culture as monolayers, causes the cells to lose their ability to form properly organized structures. This suggests that some recognition properties of the cell surface are readily lost if the cells are partially or totally isolated from one another (see eg Moscona, 1968). (3) Much emphasis in the past has been placed on induction during embryogenesis as a cause of differentiation and morphogenesis, "induction" in this context having a strong connotation of "instruction", as if the inducing tissue actually supplies information that initiates a certain new mode of differentiation. More recent evidence tends to discount the idea that specific information is transferred during induction (see especially Holtzer, 1968) and favors a more passive or permissive role for the inducing tissue, which perhaps simply enables already-determined cells to proceed with their program of cellular differentiation. In short, it is not clear to what extent, if at all, we have to take specific induction into account in considering the initiation of morphogenesis.

These considerations lead us to suspect that in multicellular organisms the genome of a cell by itself is insufficient either to initiate selective gene action, or to maintain a fixed state of differentiation once it is achieved.

To say that the genome itself is incapable of initiating differential gene action is to say nothing new; Driesch recognized something of the sort in 1894 when he proposed "specific release mechanisms" acting on the nucleus during embryonic development. To assert that the genome of a cell may by itself be incapable of maintaining a fixed state of differentiation once it is achieved may sound unorthodox, since the impression prevails that once a particular program of selective gene activity and differentiation is established it is a stable one, and is hereditary in that the mitotic descendents of a differentiated cell manifest the same special characteristics. However, as mentioned above, there is much evidence that differentiated cells that are normally fixed in organs or tissues tend to lose morphological individuality, specialized patterns of macromolecular synthesis, and viability, in that order, if they are not supported in their differentiated state by their normal environment.

Accordingly, morphogenesis and differentiation may be thought to proceed by continual and reciprocal sets of interactions between cells and their cellular and acellular environment, which elicit increasingly definitive surface patterns, and simultaneously increase dependence on ordered relationships with similar or complementary cells for the maintenance of those patterns.

IMMUNOGENETICS AND CELL RECOGNITION

Immunogenetics provides methods for answering questions concerning the structural organization of the cell surface, and its genetic and epigenetic control. Some questions relevant to the points raised in the preceding section have already been answered. These are as follows:

(1) Can products clearly specified by identifiable segregating genes be recognized on the cell surface?

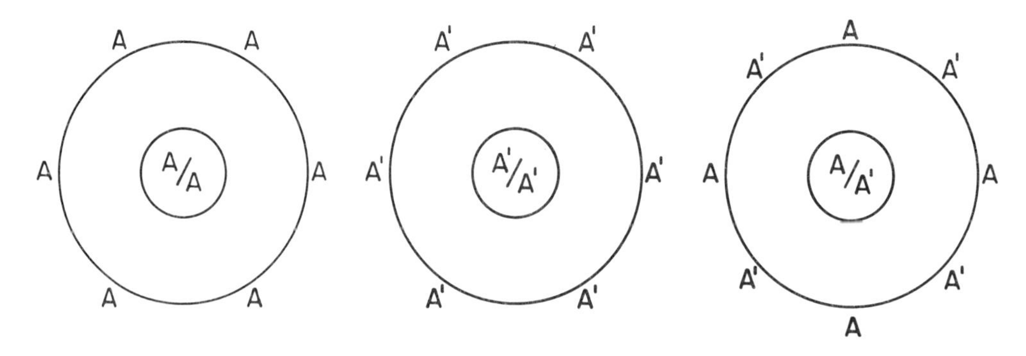

Fig. 1. Components of the cell surface specified by single genes are recognizable serologically.

(2) Do different cell populations in the same individual show phenotypic diversity of genetically specified cell surface structures? In other words, can we discriminate among different cell populations in terms of selective gene action relevant to surface structures?

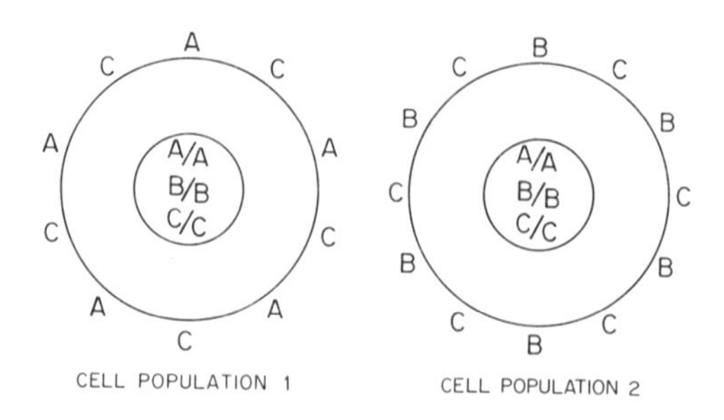

Fig. 2. Different populations of cells in the same individual do not display the same sets of cell surface components.

(3) Are these various constituents arranged indiscriminately throughout the cell surface membrane, or are they organized into characteristic patterns in which each unit bears a defined topographical relation to all other units? In other words, does the information needed to describe fully the phenotype produced by a given genetic locus include not only the nature of the gene product but its topographical relations to other gene products which constitute the cell surface?

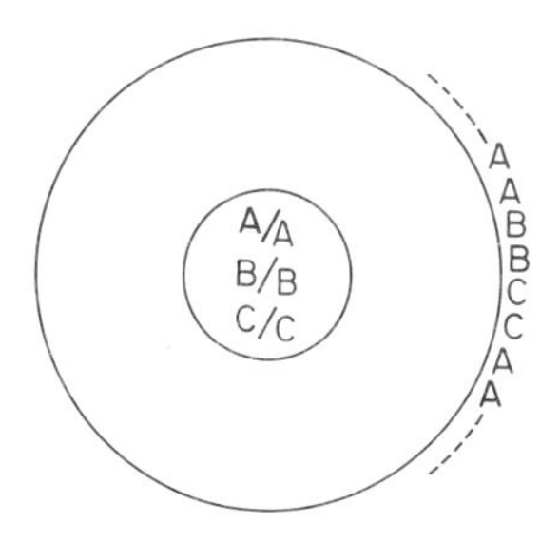

Fig. 3. Cell surface components are organized in characteristic patterns.

Before going on to discuss the data on these points, it is appropriate to consider what in general the methods of immunogenetics have to offer. There is in fact an inherent limitation to serological work directed to answering the questions about morphogenesis which interest us, and it is one which in a sense stems from just the situation we want to analyze. Consider the following: If indeed the evolution of Metazoa depended on the invention of a morphogenetic mechanism involving differentiated cell surface structure, this would imply that genes specifying discriminatory cell surface components existed before the evolution of Metazoa.* Thus we might expect that genes distinguishing the primary germ layers, for example, are very ancient ones. Likewise, as specialized organs evolved, we should expect that genes specifying surface markers for the cells involved accompanied that evolution. However, it is well known that once genes have assumed a particular and necessary function, their potential for further evolutionary variation is thereby restricted, because much of this variation would be incompatible with function. Thus it is reasonable to conclude that all vertebrates, since they share a basic body plan, will also share any sets of genes responsible for surface displays that may be essential to morphogenesis. Similarly, all mammals (eg), being relatively unspecialized animals of comparatively recent origin, may be expected to show even greater homology of cell surface structure, when similar cell types from different species are compared. For example, some hypothetical steps in the evolutionary development of mammals might be diagrammed as follows (Figure 4):

---

*Phylogenetic origin of cell recognition. It is often pointed out that an essential step in the evolution of Metazoa may have been the invention of a device for cell surface recognition and contact maintenance. If so, it seems likely that this was derived from some even older mechanism serving a still more primitive purpose earlier in evolutionary history. At a time when living organisms consisted of unicellular forms it is probable that the invention of sexuality would take evolutionary precedence over the invention of a means for making possible the aggregation of single cells into colonies. Therefore a genetic mechanism for specific cell contact probably arose very early among unicellular forms, and they used it to achieve some primitive process of sexual recombination. The subsequent elaboration and exploitation of this device would involve gene duplication and the development of mechanisms of selective gene action, permitting the expression of alternative phenotypes. No doubt the primitive controls would be environmental ones. Once this was achieved, presumably the development of organisms possessing the triple evolutionary advantages of sexuality, selective gene action, and colony formation was possible.

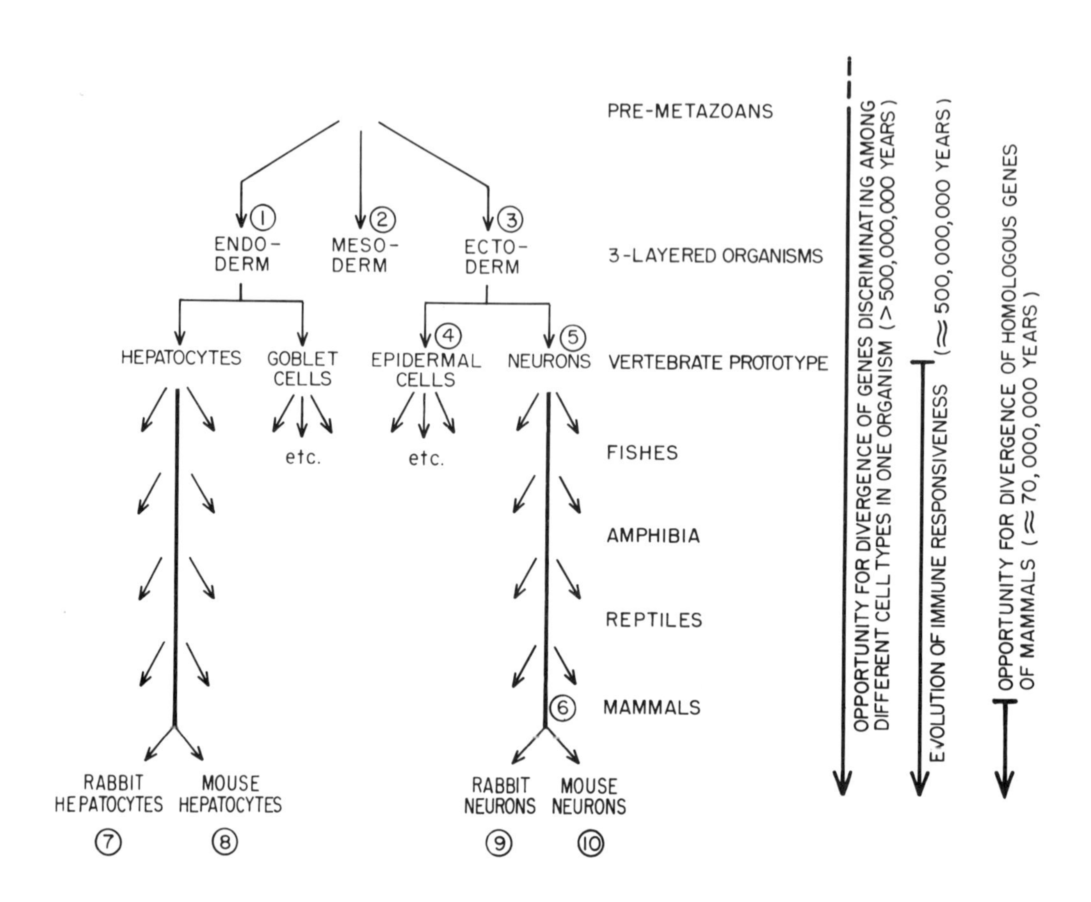

Fig. 4. The divergence of genes discriminating among different cell types in the same species is assumed to be much more ancient than the divergence of homologous genes governing similar cell types. (The numbers refer to cell types schematized in Figure 5).

We might picture that the specialization of cell surfaces in the ectodermal line might have proceeded as follows:

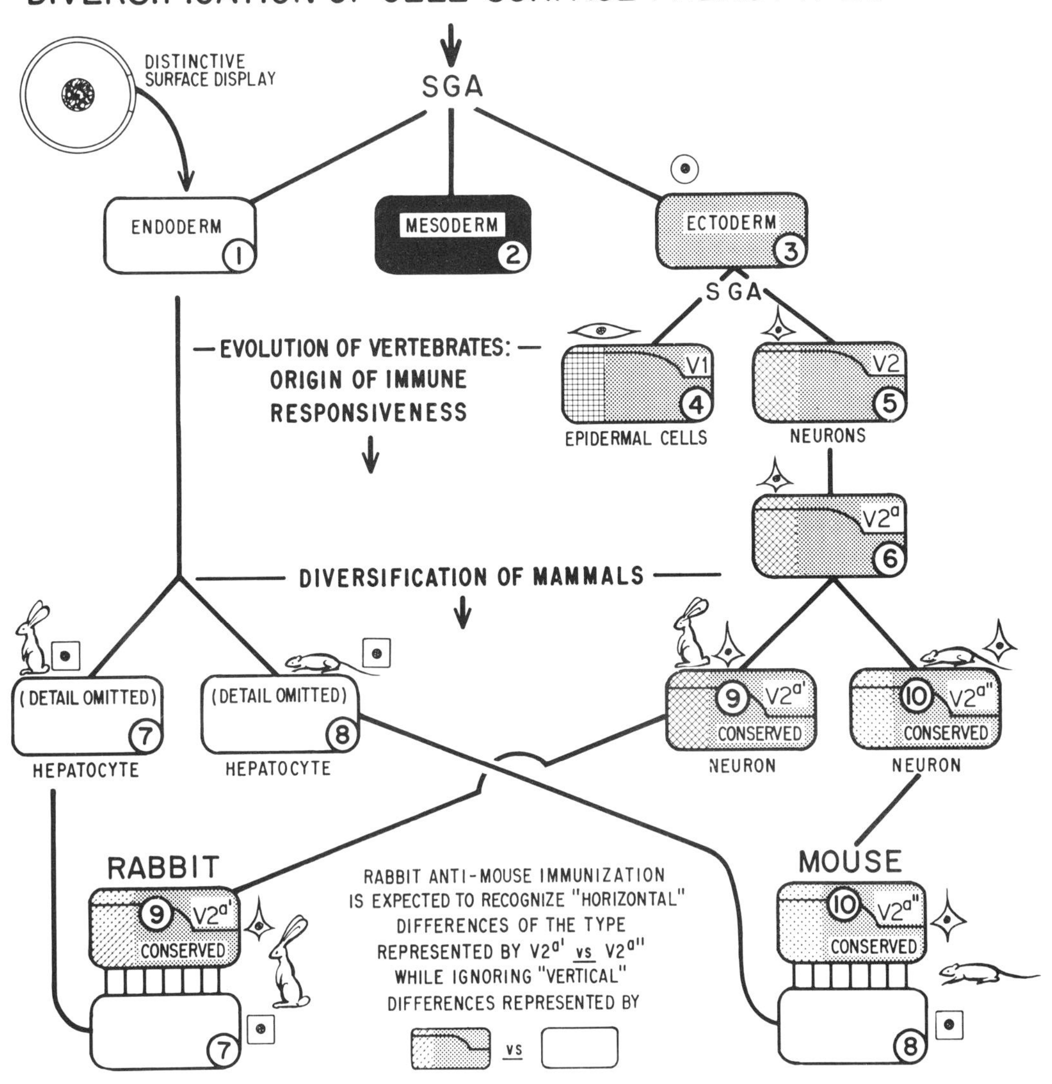

Fig. 5. SGA = Selective Gene Action.

V = Genetic Variation: representing accumulated differences in cell surface structure due to genetic mutation; proportional to the lapse of evolutionary time.

Conserved = Conservation of cell surface structure due to (a) the necessity for conservation of function of gene products, (b) limited lapse of time since divergence of ancestral genes.

Diagrams such as these help us realize that mouse and rabbit will have diverged in respect of (say) their neuron surface components only by the horizontal distance between them, and that they will share, to the large extent required by evolutionary conservatism, the basic genes which originally permitted distinction between neurons and other cells, and which were already in use in their common metazoan ancestors.

The meaning of this for immunological methodology must be considered in relation to the fact that the ability to make an immune response is confined (for practical purposes) to vertebrates, and to the fact that (generally speaking) immune responses do not occur against any element of the organism's own tissues. Thus if we attempt to distinguish mouse neuron cells from another mouse tissue, using antisera prepared in a foreign species such as the rabbit (heteroimmunization) the result will tend to magnify the relatively trivial divergences of structure produced by genic variation, while minimizing the distinctions we wish to look for, namely those arising from cellular differentiation and stemming from genes that parted company in far more ancient times. To digress from the cell surface for a moment, for purpose of illustration, heteroimmunization would completely fail to recognize one major difference between the cells of the Islets of Langerhans and all other cells, namely that the former make a polypeptide called the beta chain of insulin, whereas the latter do not. This is beause this protein has the same primary structure in the rabbit and in the mouse. The same principle must apply in greater or lesser degree to proteins which contribute to the structure of the cell surface. Add to this the fact that immunization with complex biological material like whole cells results in antibody formation against only a few of the more prominent antigens among the many represented in the inoculum, and it can be said that if in fact heteroimmunization can be shown to make any distinction at all between different cell types used for immunization, this would constitute strong evidence that major differences in surface structure are involved in cell surface differentiation. But indeed on the few occasions where heteroantisera have been thoroughly analyzed from this point of view it has been shown that a substantial proportion of the antibody they contain is directed to antigens peculiar to the cell type used for immunization (differentiation antigens, Boyse and Old, 1969); the instances referred to are the antigens MSLA (Shigeno et al, 1968), MBLA (Raff et al, 1971), and MSPCA (Takahashi et al, 1971). In simple terms, although the rabbit recognizes immunologically only the more salient features of the material with which it is immunized, and despite the limitations described above, it reacts strongly to differentiation antigens, providing a very strong hint that selective gene action is a major factor in the construction of cell surfaces.

But heteroimmunization is of limited value in studying cell surface structure because in general the antigens recognized are common to the species donating the cells for immunization. Hence analysis cannot go beyond the simple distinction of one cell type from another with appropriately absorbed heteroantiserum. The antigen called MSLA (above) may in fact be a complex of many different antigens, but it is not possible to sort them out or to find out what genes are specifying them.

Immunization within the species (alloimmunization) is not subject to these limitations, because the antigens recognized are ipso facto polymorphic, so that each one can be traced to a particular gene and recognized as a discrete element of the cell surface. Therefore definitive immunogenetics of cell surface components centers on the study of genetically polymorphic systems (alloantigens). We recognize of course that the analysis of these systems is not directly suited to the special purpose of studying morphogenetically important cell surface properties. This is because, almost by definition, polymorphic variation of antigens represents physiologically trivial differences in structure, of only minor significance to the individual, or else evolution would not have permitted its continued existence, and also because the function of the gene products on which all cell surface alloantigens reside (with the exception of immunoglobulins) is unknown. We shall return to this point later in discussing the selection of systems bearing more direct relevance to morphogenesis. Nevertheless, the study of alloantigenic systems has given some answers to the basic questions posed above, and these answers enable us at least to envisage probable ground-rules applicable to the phenotypic display of cell surface components, and to perceive likely immunological approaches to problems of morphogenesis.

Let us consider briefly what can be said about these questions:

(1) Products specified by identifiable genes can be recognized on the cell surface.

Transplantation genetics and serology, from the time of Landsteiner's demonstration of the ABO blood groups in 1900, abounds with examples of cell surface components that are polymorphic and antigenic, and can be identified as the products of particular genes. Here we shall restrict discussion to serologically demonstrable components on the surface of nucleated cells (because these are more relevant to the questions in hand than are erythrocyte polymorphisms or polymorphisms demonstrable only by such techniques as grafting). A considerable list of such alloantigens can now be drawn up; Figure 6 lists most of them together with a plot of their representation on the cells of various tissues.

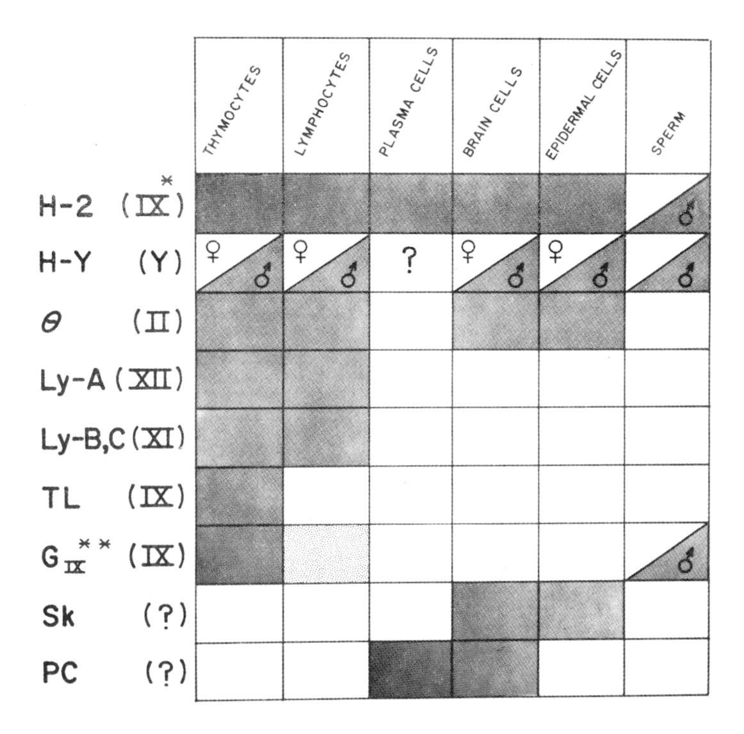

Fig. 6. A graphic representation of some systems of alloantigens on various mouse cells. (After Scheid *et al*, 1971).

*Linkage group

**The expression of antigens controlled by the G_{IX} locus is somehow governed by the genome of murine leukemia virus (Stockert *et al*, 1971, Boyse *et al*, 1971).

(2) Different cell populations in the same individual show surface phenotypic diversity.

The tissue distribution in Figure 6 shows clearly that most alloantigens recognized by serological methods occur on only a restricted category of cell types, and thus fall into the category of "differentiation antigens". That is, the presence of a particular antigen can be used to define a particular class of differentiated cell. Thus allo-immunization, just like hetero-immunization, provides for the recognition of alternate phenotypes. Thus again we have evidence that differentiation products represent major contributions to the cell surface.

Even in the two cases (H-2 and H-Y) where the surface antigen is more widely and perhaps generally represented among tissues there is marked quantitative variation in expression among different cell types. The degree of this variation is such that the amount of antigen present could in many cases be used to distinguish one cell type from another. In other words, both qualitatively and quantitatively the representation of alloantigens on cells is geared to their state of differentiation.

(3) The locations of different gene products on the cell surface show a characteristic topography.

(a) Methods have been devised for ascertaining the positions of two antigens on the cell surface, relative to one another (Boyse et al, 1968, Cresswell and Sanderson, 1968): If two antigens are sufficiently close together, the attachment of specific antibody to one of them blocks the adjacent site, thereby inhibiting uptake of antibody of the second specificity. This can be measured as a decrease of the cells' capacity to absorb the second antibody. For technical reasons this method has so far been applied primarily to antigens recognizable on thymocytes, and as shown in Figure 7, specificities for Ly-A, Ly-B, θ, TL, $H\text{-}2^d$ and $H\text{-}2^k$ seem to occupy fixed or preferential positions relative to one another. That is, from these data we can envisage the cell surface with respect to these components as being a patterned mosaic of repetitive units. The "blocking" method (referred to above) of course gives information only on the smallest complete unit of the mosaic, and is not capable of telling us about the overall distribution of antigens on the cell. Moreover, particularly as such results are obtained with cell populations rather than with single cells, a number of molecular arrangements, of various degrees of complexity, would be consistent with the data. But the important point is that they imply precise molecular patterning of the cell surface and that the topographical location of the gene product on the cell surface is likely to be an inherent property of genes that make cell surface components.

		θ	θ	θ	θ	
	TL.3	TL.2	TL.1	$H\text{-}2^d$	Ly-B	
		θ	θ	θ	θ	
		Ly-A	θ			
		$H\text{-}2^k$				

Fig. 7. Molecular patterning on the cell surface: The "blocking" method of determining the relative positions of antigens on the cell surface indicates that each occupies a defined location in relation to the others. This diagram illustrates the sort of map that would be consistent with the blocking data for alloantigens on mouse thymocytes.

(b) For determining the distribution of antigen on the cell surface as a whole, visualization by immuno-electron microscopy is the method of choice. Such methods are now being extended to the use of more than one visually distinct marker attached to antibody [a small plant virus is now used (Hammerling et al, 1969) in addition to the conventional ferritin] so that it will soon be possible to label two different antigens on one cell simultaneously. Much work of this kind has indicated that each of the antigens so far examined occupies discrete regions of the cell surface and is not uniformly distributed over the whole cell (Aoki, et al, 1969, Stackpole et al, 1971). Three points require emphasis in this connection: (1) At this large scale of organization, no consistent patterning has been evident, in contrast to the regularity of molecular patterns discerned on the fine scale of antibody-blocking, as described above; (2) The use of other procedures for immuno-electronmicroscopy has given a contrary result (uniform distribution) for one of the antigens named above (Davis et al, 1971); (3) All this work has been carried out (for technical reasons) with lymphoid cells, although these are cells which do not participate in any obvious way in morphogenetic events that require cells to assemble. For these three reasons, no safe conclusion can yet be drawn as to the overall representation of gene products on the surfaces of the cells of organized tissues in their native state. But one particularly relevant consideration that does arise from this discussion is the suggestion that the phenotype of an isolated cell, as reflected in the overall distribution of its different surface components, may be variable. Not only may certain components actually be made to disappear from the cell surface as a result of exposure to antibody under certain conditions (antigenic modulation (Old et al, 1968), but also there is evidence of considerable mobility within the cell surface membrane (Frye and Edidin, 1970, Taylor et al, 1971). The point is raised here because it may be relevant to whether or not morphogenetic arrangements are responsible for maintaining the precise differentiated state of the cells of organized tissue, inasmuch as the description of the surface phenotype of such a cell must include information as to the total representation and topography of all its surface components.

(For further detail and discussion see Boyse and Old, 1969, 1971).

SUMMARY: A number of different genetically controlled antigens can be recognized on the surfaces of differentiated cells. Their qualitative or quantitative representation can be used to characterize cell types. They are arranged on the cell surface as an ordered mosaic. Their representation on the membrane is subject to alteration by environmental conditions.

IMPLICATIONS FOR MORPHOGENESIS

We can now return to our original discussion and ask in more definite terms how phenotypic diversity of cell surfaces might be generated during ontogeny, and whether ways can be found to connect it experimentally with the differential cell recognition which must constitute an essential part of morphogenesis and its maintenance. It is intuitively attractive and easy to comprehend that detailed morphogenesis might proceed on the basis of cell recognition mediated by progressively changing patterns of the cell surface. As argued above, it is scarcely possible that all the fine details of morphogenesis could be programmed directly in the genome, and we have suggested: that in early development cell-cell contacts may be essentially non-discriminatory; that the first selective gene activations leading to specific cell surface differentiation may be initiated by environmental differences imposed upon the cells; and that final cellular differentiation may be regulated and maintained by the microenvironment created by the interaction of cell with cell and of cell with environment. Clearly there are many attractions in the hypothesis that detailed morphogenesis requires an orderly diversification of surface structure involving some sort of coding in which only a relatively small set of gene products are used. This might be thought of in terms of sequentially controlled changes in the representation of these few products, the new surface displays of two progeny cells being inherent in the surface display of the parent cell. Surface individuality might then appear as a function of the generation of new surface membrane during cell division, and geared to the geometry of cytokinesis.

As a rider, we add the concept that just as this highly differentiated surface phenotype may depend on the preceding morphogenetic sequence for its origin, it may also depend on the resultant ordered morphology for its maintenance.

The T-locus Mutants

The material presented above only provides evidence that the observed antigenic individuality of cell surfaces, and its genetic and epigenetic control, are compatible with the view that morphogenesis is regulated by mechanisms involving cell surface recognition. It does not however give any information on what these mechanisms may be. For this purpose we can turn, as is so often necessary in embryology, to an analysis of highly abnormal situations, which may yield information that no amount of description of normal processes can provide. Without doubt the richest source of well-defined abnormalities affecting different states of embryogenesis and owing their origin to a single genetic locus is the large series of T-locus mutants.

The T-locus of the mouse occupies a region of linkage group IX near H-2. It is marked by a dominant gene T, which both in heterozygous and in homozygous states interferes with the development and maintenance of the notochord, resulting in the disperson of notochord cells at an early stage in development. In heterozygous embryos (T/+) this causes merely the loss of the distal part of the tail; in homozygous (T/T) embryos the entire notochord is lost, and this produces severe abnormality and death about midway through gestation.

There have also been distinguished at this locus a series of different recessive lethal or semilethal alleles (t-alleles).These variants have all been isolated from wild mice, which show polymorphism at this locus, or obtained by "mutation" from previously known alleles. All these recessive alleles produce morphologically normal heterozygotes (t/+) but interact with T (T/t) to produce taillessness. The recessive alleles can be distinguished from one another by two properties. First, crosses of tailless (T/t) animals carrying identical t-alleles constitute a balanced lethal system. Thus:

T/t x T/t

T/T	T/t	t/t
dies	viable tailless	dies

but matings between tailless animals carrying two different lethal recessive alleles yield a proportion of phenotypically normal progeny (complementation). Thus:

x

T/T	T/t^x T/t^y	t^x/t^y
dies	viable tailless	viable normal

Breeding tests of this kind make it possible to identify the different classes of lethal alleles mentioned above.

The second property by which t-alleles are distinguished is that each of them produces a sharply defined and unique error in ontogenesis, interrupting development in a highly cell-specific or tissue-specific manner at a characteristic stage, ranging from the morula to the late neurula for different t-alleles.

We need not discuss here details of the derangements which these mutants induce (reviewed in Bennett, 1964); it suffices to say that each of them can be interpreted as acting at some switch-point as neuroectodermal development diverges along the following pathway:

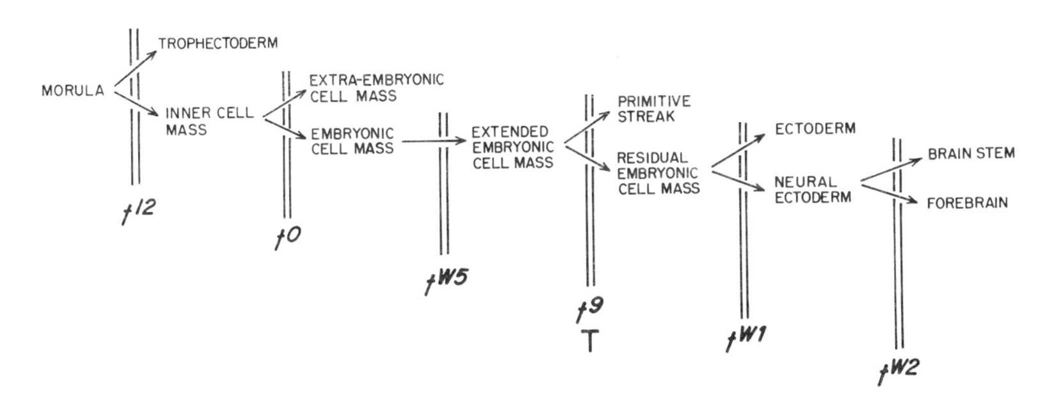

Fig. 8. The pathway of differentiation of ectodermal structures is interrupted at discrete points by alleles of the T-locus.

In addition to these effects on embryonic development, the recessive t-alleles share one other outstanding property, in that they affect the production and probably in some instances the function of spermatozoa. Specifically, they distort the transmission ratio from heterozygous males, but not females, to an extent characteristic of each particular mutant. Males heterozygous (t/+) for a lethal t allele transmit it to an excess of their progeny in proportions ranging from 60-99% depending on the allele concerned. Since their sperm do not show gross deviations from normal in either morphology or number, the most likely explanation rests on a superior fertilizing ability of t-bearing vs +-bearing sperm.

These "alleles" are really pseudoalleleic members of a complex locus. The unusual property of complementation between two lethal "alleles" is one thing that makes this clear. Another factor is their property of generating, with a frequency too high for mutation, new and different alleles by a process accompanied by recombination. Thus it seems very unlikely that these mutants represent conventional point mutations, and implies that the T-locus as a whole may have some structural and functional similarities to the neighboring H-2 locus.

Immunological Study of the T-locus

In the preceding sections we have considered -- (a) evidence linking morphogenesis to the genetic and epigenetic specifications of cell surfaces, and (b) immunogenetic evidence of phenotypic diversity among the cells of single organisms. And it has been pointed out that there is no necessary connection between the two because the function of serologically demonstrable cell surface components is generally unknown; possibly none of them has anything to do with cell recognition. Nonetheless the idea that morphogenesis is mediated by phenotypic diversity of cell surface structure is greatly favored, not only because of its plausibility but also because it appears relatively susceptible to experiment. That is well-known and widely appreciated.

The first step in substantiating the hypothesis is to accumulate evidence that genes that are known to perform critical functions in embryogenesis in fact have products located on the cell surface. If they do not, -- if the locus of action of these gene products is elsewhere, -- then most of what is written above should probably be scrapped; for if cell surface display is the key to morphogenesis, it is highly probable that genes with a key role in morphogenesis will be found to specify components of the cell surface. The corollary would seem to be that if in fact such genes do turn out to be coding for elements of the cell surface, then the hypothesis linking morphogenesis to cell surface specification is to an appreciable extent vindicated.

Viewed in that light, the question whether or not the *T*-locus codes for cell surface structure is a critical one.

These are our reasons for having invested considerable effort in attempting to show that any product of the *T*-locus can be detected serologically on the cell surface.

The fact that different *t* alleles have different transmission rates in *t*/+ males (contrary to Mendelian expectations) was the starting point for our serological study. Sperm production is probably not significantly altered in the heterozygous carriers of different *t* alleles, which suggests that it is sperm *performance* which is affected, and hence (a) that the *T*-locus does code for a product on the cell surface, and (b) that this phenotype is expressed on sperm. This dictated the choice of sperm as the immunizing material for antiserum production. Meanwhile a satisfactory cytotoxicity test had been worked out for demonstrating surface antigens on mouse sperm, and had been applied with success to two known antigenic systems represented on sperm, H-2 and H-Y (Goldberg *et al*, 1970, 1971). For reasons given in the introduction, allo-immunization was judged to be more likely then hetero-immunization to distinguish polymorphic products of the *T*-locus.

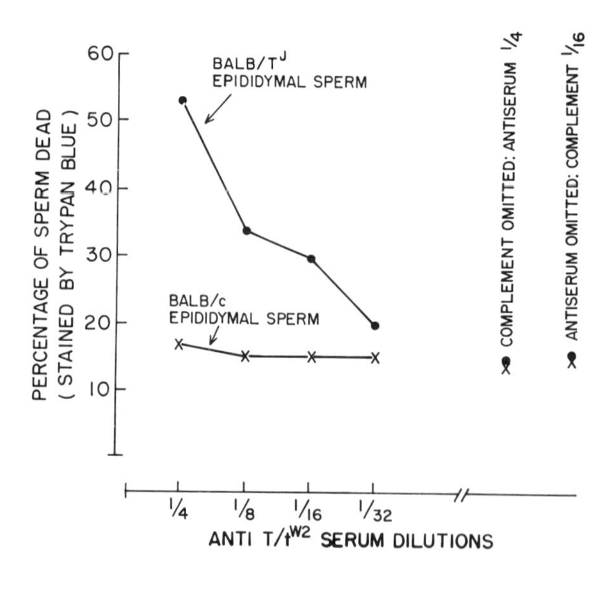

Fig. 9. Detection of antigen T on epididymal sperm of BALB/T^J mice by the cytotoxicity test with serum of + *tf*/+ *tf* male mice immunized with sperm from *T tf*/t^{w2} + mice. The specificity of this reaction for antigen T has been further confirmed by showing that +/+ segregants of backcross populations (from either of two independently derived *T*-mutants) can be distinguished from *T*/+ segregants by blind tests on sperm.

There is one probable success to report; not much perhaps in comparison with the formidable task of attempting an immunogenetic description of the *T*-locus, of the sort available for the neighboring complex locus *H-2*, but enough to go some way to vindicating the proposition that genes controlling embryogenesis do so by coding for elements of the cell surface. The antiserum used for the test illustrated in Figure 9 was +/+ o anti *T*/t^{w2} sperm and was first absorbed with +/+ sperm to take out anti-sperm auto-antibody. At the moment only *T* itself is available in congenic stocks, *ie* in an inbred mouse strain differing from its partner inbred strain at this one locus only. Such strains are invaluable in establishing serological specificity; in Figure 9 the cytotoxicity test is negative on BALB sperm but positive on sperm of the congenic BALB/TJ stock. (The specificity of the cytotoxic antibody is therefore anti-T). According to absorption tests antigen T is not expressed in adult tissues other than sperm, so it meets the criteria of a cell surface differentiation alloantigen. The cytotoxicity method will presumably be of little use in confirming that antigen T is expressed in the embryo, for it is logically to be expected only on certain cells, *eg* the notochord and associated tissues of T-bearing embryos, and probably only transiently in any event. Hopefully it will be possible to apply a more appropriate technique, such as immunofluorescence on tissue sections, for locating antigen T in the tissues of embryos.

Concluding Remarks

As the allele T codes for a cell surface component, it is reasonable to assume that the t-series and + alleles of the T-locus will be found to do so also. It would also be reasonable to conclude that the function of the T-locus is to specify elements of the cell surface that are responsible for steps in morphogenesis. Their expression would presumably be limited to certain cells and certain times during ontogenesis. It will be important to inquire where such antigens are situated on the cell surface, and what alternative antigens may take their place on adult cells. Topography is specially relevant in this connection for in one instance there is evidence of what appears to be substitution of one cell surface component for another that is normally adjacent to it: H-2(D) and TL antigens are coded by linked genes in group IX; they occupy adjacent positions on the surface of thymocytes and can be solubilized and separated from one another by chromatography as fractions of around 50,000 molecular weight. TL is not always expressed on thymocytes, however, and when it is not there is a compensatory increase in H-2(D). This is a so far unique case of an association between linkage of two loci and adjacency of the corresponding gene products on the cell surface. It happens that the T-locus is in the same linkage group, IX, and this invites the surmise that the antigens of the T-locus will map near H-2, perhaps near H-2(K), as the gene order is Tla: H-2(D): H-2(K): T. Another tenuous link between H-2 and T is that the former antigens are weakly if at all expressed in early ontogenesis, whereas the latter, we suspect, are expressed only during ontogenesis (excepting sperm); coupled with the notion of adjacency, this is perhaps sufficient to suggest a possibility that the H-2 and T antigens may have a reciprocal relation to one another, just as H-2 and TL appear to have. A perhaps stronger reason to suggest an intimate association of H-2 with T comes from the necessity to account for the astonishing prevalence of t-alleles in the wild, contrary to what is expected of recessive lethal alleles. One asks therefore whether t-alleles may confer a physiological advantage that outweighs their lethality in the homozygous state and which has made it worthwhile for the species to develop a method of perpetuating them (this mechanism being the preferential transmission of t-alleles by males). It springs to mind that the suppression of crossing over, which all t-alleles found in the wild cause, is the source of this advantage. The effect of this suppression is that alleles at several different loci are inherited *en bloc*, including at least four that code for the cell surface: Tla, H-2(D), H-2(K) and T. If these are members of a cooperative system concerned with cell recognition and morphogenesis, it can be envisaged that certain combinations of alleles are more favorable than others. The advantage conferred by t-alleles might then be explained by their ability to lock these in place and insure their inheritance as single units.

REFERENCES

Aoki, T., U. Hämmerling, E. de Harven, E.A. Boyse, and L.J. Old, (1961), Antigenic structure of cell surfaces. An immunoferritin study of the occurrence and topography of H-2, θ, and TL alloantigens on mouse cells, J. Exp. Med. 130, 979.

Bennett, D., (1964), Abnormalities associated with a chromosome region in the mouse. II. The embryological effects of lethal alleles in the t-region, Science 144, 263.

Boyse, E.A., L.J. Old, and E. Stockert, (1968), An approach to the mapping of antigens on the cell surface, Proc. Nat. Acad. Sci. 60, 886.

Boyse, E.A. and L.J. Old, (1969), Some aspects of normal and abnormal cell surface genetics, Ann. Rev. Genet. 3, 269.

Boyse, E.A., L.J. Old, and E. Stockert, The relation of linkage group IX to leukemogenesis in the mouse, in Proceedings of the Conference on RNA Viruses and Host Genome in Oncogenesis, Amsterdam, May 12-15, 1971, eds. P. Emmelot and P. Bentvelzen (North-Holland Publ. Co., Amsterdam), in press.
Boyse, E.A. and L.J. Old, (1970), The invitation to surveillance, in Immune Surveillance, eds. R.T. Smith and M. Landy (Academic Press, New York) p. 5.
Calarco, P. and E. Brown, (1969), An ulstrastructural and cytological study of preimplantation development in the mouse, J. Exp. Med. 171,253.
Cresswell, P. and A.R. Sanderson, (1968), Spatial arrangement of H-2 specificities. Evidence from antibody absorption and kinetics studies, Trans. 6, 996.
Davis, W.C., M.A. Alspaugh, J.H. Stimpfling, and R.L. Walford, (1971), Cellular distribution of transplantation antigens: discrepancy between direct and indirect labeling techniques, Tissue Antigens 1, 89.
Frye, L.D. and M. Edidin, (1970), The rapid intermixing of cell surface antigens after formation of mouse human heterokaryons, J. Cell. Sci. 7, 319.
Gardner, R. and M. Lyon, (1971), X chromosome inactivation studied by injection of a single cell into the mouse blastocyst, Nature 231, 385.
Goldberg, E., T. Aoki, E.A. Boyse, and D. Bennett, (1970), Detection of H-2 antigens on mouse spermatozoa by the cytotoxicity test, Nature 228, 570.
Goldberg, E., E.A. Boyse, D. Bennett, M. Scheid, and E. Carswell, (1971). Antigens on spermatozoa. Serological demonstration of H-Y (male) antigen on mouse sperm, Nature 232, 441.
Hadorn, E., (1965), Problems of determination and transdetermination, in Genetic Control of Differentiation (Brookhaven Symp. in Biol. #18) p. 148.
Hämmerling, U., T. Aoki, H.A. Wood, L.J. Old, E.A. Boyse, and E. de Harven, (1969), New visual markers of antibody for electron microscopy, Nature 223,1158.
Holtzer, H., (1968), Induction of chondriogenesis: a concept in quest of mechanisms, in Epithelial-Mesenchymal Interactions, eds. R. Fleischmajer and R. Billingham (William and Wilkins, Baltimore) p. 152.
Jacob, F. and J. Monod, (1963), Genetic repression, allosteric inhibition, and cellular differentiation, in Cytodifferentiation and Macromolecular Synthesis, ed. M. Locke (Academic Press, New York) p. 30.
Moscona, A., (1968), Reconstruction of skin from single cells and integumental differentiation in cell aggregates, in Epithelial-Mesenchymal Interactions, eds. R. Fleischmajer and R. Billingham (Williams and Wilkins, Baltimore) p.230.
Old, L.J., E. Stockert, E.A. Boyse and J.H. Kim, (1968), Antigenic modulation: loss of TL antigen from cells exposed to TL antibody. Study of the phenomenon in vitro, J. Exp. Med. 127, 523.
Raff, M.C., S. Nase, and N.A. Mitchison, (1971), Mouse specific bone marrow derived lymphocyte antigen as a marker for thymus-dependent lymphocytes, Nature 230, 50.
Scheid, M., E.A. Boyse, E.A. Carswell, and L.J. Old, Serologically demonstrable alloantigens of mouse epidermal cells, in preparation.
Shigeno, N., U. Hämmerling, C. Arpels, E.A. Boyse, and L.J. Old, (1968), Preparation of lymphocyte specific antibody from anti-lymphocyte serum, Lancet, 320.
Sperry, R.W., (1965), Embryogenesis of behavioral nerve nets, in Organogenesis, eds. R.L. de Haan and H. Ursprung (Holt, Rinehart and Winston, New York) p. 161.
Stackpole, C.W., T. Aoki, E.A. Boyse, L.J. Old, J. Lumley-Frank, and E. de Harven, (1971), Cell surface antigens: serial sectioning of single cells as an approach to topographical analysis, Science 172,472.
Stockert, E, L.J. Old, and E.A. Boyse, (1971), The G_{IX} system. A cell surface allo-antigen associated with murine leukemia virus; implications regarding chromosomal integration of the viral genome, J. Exp. Med. 133, 1334.
Takahashi, T., L.J. Old , C.-J. Hsu, and E.A. Boyse, A new differentiation antigen of plasma cells, Europ. J. Immunol. in press.
Tarkowski, A., (1959), Experiments on the development of isolated blastomeres of mouse eggs, Nature 184,1286.
Tarkowski, A., (1961), Mouse chimaeras developed from fused eggs, Nature 190, 857.
Tarkowski, A., (1967), Development of blastomeres of mouse eggs isolated at the 4- and 8-cell stage, J. Embryol. Exp. Morph. 18, 155.

Taylor, R.B., W.P.H. Duffus, M.C. Raff and S. de Petris, (1971), Redistribution and pinocytosis of lymphocyte surface immunoglobulin molecules induced by anti-immunoglobulin antibody, Nature 233, 225.

ACKNOWLEDGEMENTS

The work on the serology of the T-locus was carried out by Mrs. Ellen Goldberg. We are also indebted to Dr. Donald Michie and Miss Karen Artzt for making some valuable suggestions.

Much of this work was supported by NCI grant CA 08748 (EAB), NIH HD 05482 and TO 1-GM01918 (DB) and NSF GB27259 (DB). DB is a Career Investigator of the Health Research Council of the City of New York.

CELL CONTACT, CONTACT INHIBITION OF GROWTH, AND THE REGULATION OF MACROMOLECULAR METABOLISM

TOM HUMPHREYS
Department of Biology, University of California, San Diego
La Jolla, California, USA

Abstract: Various aspects of the multiple events involved in regulating growth during contact inhibition have been investigated. Serum breaks cell contacts when it releases contact inhibition of growth. The levels of cyclic AMP in chick fibroblasts do not seem to change when contact inhibition is released. During contact inhibition certain cell surface proteins appear to turn over rapidly, ribosomal RNA synthesis is repressed by slowing the rate of precursor chain elongation, and ribosomal RNA turnover increases.

1. INTRODUCTION

Normal cells in culture limit their growth by cellular interactions which depend on cell density. This growth control has most often been termed "contact inhibition of growth" because superficial appearances (Stoker and Rubin, 1967) suggest cell contact is important in the cellular interactions. The critical observation is that cells stop growing at confluent cell densities even though the medium in the culture is capable of supporting growth of cells at lower densities. This density dependent limitation of growth should be distinguished from the many reported examples where growth has stopped because the medium has been depleted of components necessary for growth of cells even at low density (for example see Temin, 1969). The nature of the density dependent interactions which limit growth in culture remain obscure despite the interest generated by considerable evidence indicating that derangement of these cellular interactions is the basis for malignancy (Vogt and Dulbecco, 1960; Pollach *et al*., 1968; etc.). Many lines of evidence suggest that cell contact almost certainly plays a role in growth limitation (Pollach *et al*., 1968; Gurney, 1969; Dulbecco and Stoker, 1970; Burger and Noonan, 1970; Baker and Humphreys, 1971). Components of the medium, especially factors derived from serum, are able to modify these density dependent reactions (Todaro *et al*., 1965; Holley and Kiernan, 1968; Yeh and Fisher, 1969; Paul *et al*., 1971; Ceccarini and Eagle, 1971; Baker and Humphreys, 1971) suggesting to some authors that the supply of growth promoting factors, and not cell contact, plays the major role in growth limitation.

Little is known about the reactions which may or may not involve cell contacts or factors from the medium. They ultimately stop growth and are usually measured by their limitation of DNA synthesis or cell division. But the sequence from the primary density dependent reactions to the limitation of DNA synthesis is mostly conjecture. Some recent studies have suggested cyclic AMP may be involved in contact inhibited cells suggesting that this regulatory nucleotide could be the "second messenger" in the sequence from cell contact. Accumulation of total protein and RNA must be reduced when growth is stopped. This seems to be achieved by both decreasing synthesis and increasing turnover of these components during contact inhibition (Todaro *et al*., 1965; Warren and Glick, 1968; Hodgson and Fisher, 1971; Emerson, 1971; Baker and Humphreys, 1972). The synthesis of certain specific protein and RNA molecules may be repressed while others may be activated (Salas and Green, 1971). Finally DNA synthesis and cell division must be stopped. The roles which cyclic AMP or other regulatory molecules might play in promoting some or all of these changes in cellular metabolism remains to be discovered. The sequence, or possibly even independent parallel sequences, of control events which the cell uses to stop growth cannot even be approached by intelligent guesses at this time.

The various aspects of this web of cellular events involved in regulation of cellular multiplication must be explored and elucidated before a coherent description of growth control is possible. My students and I have begun to examine some of these phenomena in primary cultures of embryo chick skin fibroblasts. I will discuss here experiments to clarify the relationship between cell contact reactions and the alteration of saturation density by serum components (Baker and Humphreys, 1971), cyclic AMP levels in growing and contact inhibited cells, the turnover of certain cell surface components during contact inhibition (Baker and Humphreys, 1972), and regulation of ribosomal RNA increase (Emerson, 1971). In these experiments we have often compared contact inhibited cells with confluent cells released from contact inhibition by adding higher concentrations of serum to the culture medium. Increased serum is necessary for stimulating the changes described, since, contrary to other systems (Todaro _et al_., 1965; Yeh and Fisher, 1969), significant stimulation of these primary chick embryo fibroblasts in the mixed chick and horse serum containing medium is not achieved by changing the medium alone. However, it should be noted that cells at low densities grow faster in medium with higher serum concentrations. Thus, the increased serum used to release contact inhibition may also stimulate the cells to grow faster than they did at subconfluent densities.

2. SERUM AND CELL CONTACT.

The density reached by any given cell type before growth is limited is a function of the characteristics of the cells and the concentration of serum in the medium (Holley and Kiernan, 1967; Clarke _et al_., 1970; Dulbecco, 1970). Although ample evidence suggests that cell contacts and cell surface reactions are involved in growth limitation, the alteration of the limiting density by medium components suggested that a simple understanding of the role of cell contact was untenable. Cells should reach confluence and establish cell contact at the same density irrespective of the serum concentration. Joffre Baker and I ask, "How does serum affect cell contact?" We grew cells in 2% serum and then watched what happened with cell contacts when we added 20% serum (Baker and Humphreys, 1971). Before serum was added the contact inhibited cells covered the entire plate. They were flattened out and the perimeter of each cell was closely applied to the edge of other cells. This changed rapidly after serum was added. There were many cells whose edge was not against the edge of neighboring cells. Instead, cells were often separated by a space where the plate was exposed. It was evident that cell to cell contact was greatly reduced when the serum concentration was increased. This reduction in cell contact by serum could easily explain its release of growth inhibition since at a given cell density cell contact was not as extensive in high serum as in low serum. This provides a completely logical explanation of the serum effect and eliminates any possible paradox which might have been thought to exist between the serum effect and the concept that cell contact determines the level at which growth is inhibited.

We examined the effect of serum more closely using time-lapse photography to speed up the changes in the cells so the reaction would be visually more apparent. When serum was added, the cells immediately began to snap apart and move more rapidly (Table I). It looked as if the cells were able to break away from strong cell to cell adhesions which has stretched them over the plate and slowed down their movement. The serum could have permitted this by reducing the cell adhesiveness which was restraining movement or by stimulating cell movement which allowed the cells to pull away from each other. These two possibilities were tested. If cell adhesions were restraining movement in low serum, cells at low cell density where few cell adhesions should move as rapidly in low serum as in high serum. This proved to be the case (Table I). Adding serum to cells grown at low density in low serum did not stimulate their movement

TABLE I

Effect of serum on cell movement.

CELLS	DENSITY cells/cm^2	% serum added	μm/hr* rate of movement
grown to confluence in 2% serum	9×10^4	-	10
grown to confluence in 2% serum	9×10^4	20	25
grown at low density in 2% serum	$1\text{-}2 \times 10^4$	-	25
grown at low density in 2% serum	$1\text{-}2 \times 10^4$	20	28

*derived from following individual cells in time-lapse films (Baker and Humphreys, 1971).

appreciably (Table I). These results show that cell movement is restricted at higher cell densities and that in the absence of such restrictions serum does not stimulate cell movement. This suggests to us that serum acts by reducing the mutual adhesiveness of cells. This allows previously stable cell contacts to come apart so that the cells start growing and moving more rapidly.

This interpretation relates both release of growth inhibition and increase in cell movement to the same primary action of serum. This suggests that both growth and movement should have similar dose response curves for increasing serum concentration. Cells were grown to confluence in 0.5% serum and stimulated with various concentrations of serum up to 20%. The response curves were quantitatively parallel increasing about linearly to a maximum of 8% serum with no further increase in response up to 20% serum. This agrees with the suggestion that serum stimulates growth and movement by a common mechanism, we believe by reducing mutual cell adhesiveness in the monolayer. This suggests that the cell contact required for the "contact inhibition" to occur is not the mere touching of cells but is rather stable cell adhesions which can limit the movement of the cells. For many years it has been thought that tumor cells are less adhesive (Coman, 1944; Edwards, et al., 1971). These results indicate that the failure of growth regulation in tumor cells may be related to this decreased adhesiveness. Since the increased invasiveness of malignant cells is also related to reduced adhesiveness, both characteristics which define cancer, uncontrolled growth and invasiveness, could be defects in the ability of cells to adhere properly.

3. CYCLIC AMP

Recent suggestions that adenosine 3':5'-cyclic monophosphate (cyclic AMP) may be involved in controlling growth of cells (Johnson et al., 1971; Sheppard, 1971; Otten, 1971) provide a very interesting hypothesis concerning the cellular reaction modified by the stable cell contacts. Cyclic AMP serves as a "second messenger" which transmits a signal reaching the cell surface into the cell itself. Adenylate cyclase, the enzyme which forms cyclic AMP, is a plasma membrane enzyme associated with a cell surface receptor site which activates the enzyme upon the appropriate stimulus at the cell surface. In the case of contact inhibition, the adhesion of two cells could stimulate adenylate cyclase to produce cyclic AMP which inhibits growth.

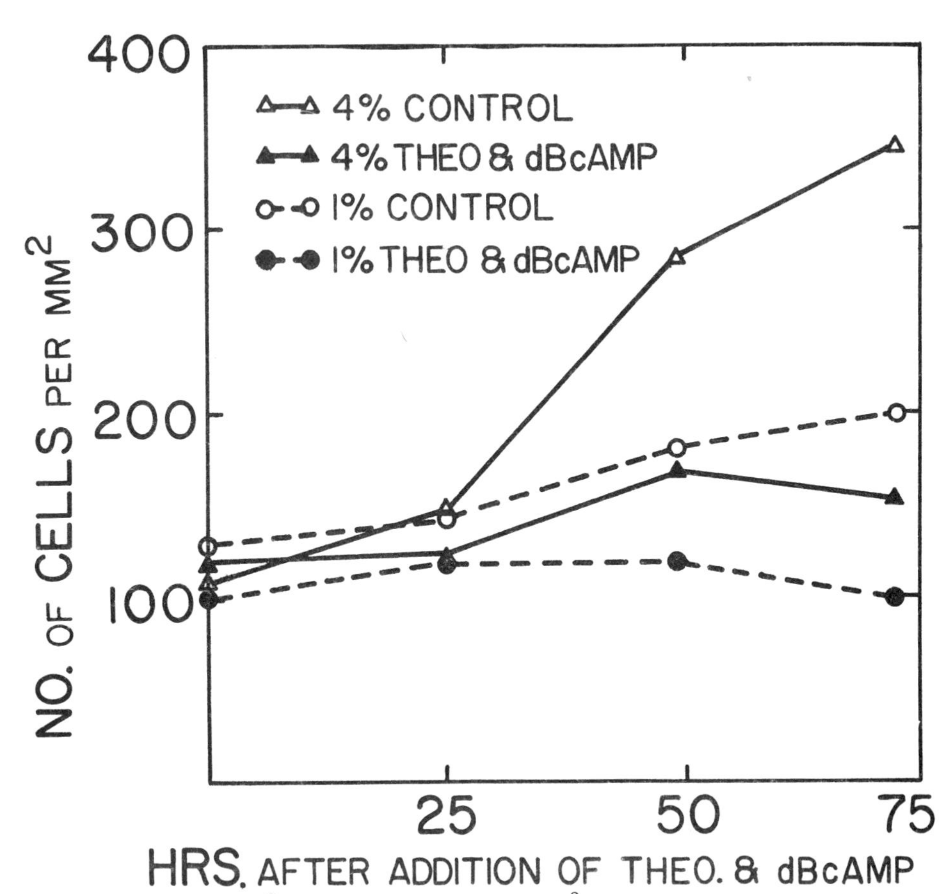

FIGURE 1. The effect of 10^{-2} M theophylline and 10^{-3} M dibutyryl cyclic AMP on the growth of subconfluent chick embryo fibroblasts in 4% and 1% serum. Cells in a measured field were counted under phase microscopy at daily intervals.

Cyclic AMP reactions were examined in chick fibroblasts to evaluate the possibility that it played a role in contact inhibition of growth. I confirmed observations on other cells (Johnson et al., 1971; Sheppard, 1971) that exogenous dibutyryl cyclic AMP and theophyllin change morphology of the cells and slow growth. The effect of these compounds depended somewhat on the growth conditions of the cells. If cells were growing at subconfluent densities in 1.0% serum with a doubling time of about 2 days, addition of 10^{-2} M theophyllin and 10^{-3} M dibutyryl cyclic AMP stopped growth completely (Fig. 1). Growth of cells in 4% serum with a doubling time of about 24 hours was only slowed by the same concentrations of the drugs (Fig. 1). There was no evidence in our experiments that cyclic AMP made cells more sensitive to contact or that contact was required for cyclic AMP to have its effect as described for tumor cells (Sheppard, 1971). Cyclic AMP also decreased the number of cells dividing in confluent cultures and reduced the serum stimulation of these cells. About 5 to 10% of the cells grown to confluence in 0.5% serum normally synthesize DNA during a 20 hour period. Addition of theophyllin and dibutyryl cyclic AMP reduced this number considerably (Table II). If untreated confluent cultures are stimulated with serum most cells synthesize DNA within 20 hours. Addition of the drugs reduced this reaction also (Table II). These inhibitory effects of theophyllin and dibutyryl cyclic AMP could be due to normal physiological action or could equally well be the result of toxic reactions produced by the

TABLE II

Effect of 10^{-3} M dibutyryl cyclic AMP and 10^{-2} theophyllin on cells synthesizing DNA.

CELLS	DRUGS	% Cells synthesizing DNA in 20 hour period*
Confluent culture 0.5% serum	-	9.4
Confluent culture 0.5% serum	+	2.3
Confluent culture 20% serum added 0 time	-	90
Confluent culture 20% serum added 0 time	+	61

*measured as described in Baker and Humphreys (1971).

very high non-physiological concentrations which had to be used to obtain any effect with these compounds.

A critical question is whether the cell regulates cyclic AMP levels during contact inhibition (Otten et al., 1971). I measured cyclic AMP under various growth and inhibited conditions using both incorporation of H^3-adenosine into cyclic AMP assuming its specific activity was equal to ATP specific activity and by the method of Gilman (1970). Cyclic AMP levels varied according to the growth conditions. Cyclic AMP was higher in nonconfluent cells growing with a 2-day doubling time in 0.5% serum than cells at a similar density in 4% serum with 1-day doubling time (Table III). Cyclic AMP, however, failed to change when cells grew to confluence and became contact inhibited (Table III). If cells were stimulated with serum the level of cyclic AMP remained constant from one-half minute to 8 hours (Table III).

TABLE III

Cyclic AMP in cells.

CELLS	CYCLIC AMP* 10^{-18} moles/cell
Nonconfluent in 0.5% serum-growing	17.6
Nonconfluent in 4.0% serum-growing	13.8
Confluent in 0.5% serum-contact inhibited	17.1
Confluent with 20% serum added 2 hrs-growing	16.6

*assayed according to Gilman (1970).

It has been suggested that growing cells are larger than contact inhibited cells (Otten et al., 1971) and, thus, that the actual concentration of cyclic AMP is greater in the latter cells than the absolute amount per cell would indicate. The much larger content of ribosomal RNA in growing chick cells would seem to agree with this idea (Emerson, 1971). However, growing hamster embryo fibroblasts also have more ribosomes but are no larger than stationary cells (Becker et al., 1971). We compared relative volumes of chick fibroblasts using a coulter counter and although growing cell size was slightly more heterogenous they were not significantly different from contact inhibited cells in medium with 0.5% serum (Figure 2).

Cyclic AMP concentrations do seem to vary inversely with the rate of growth in growing chick fibroblasts. However, in the light of my results it is

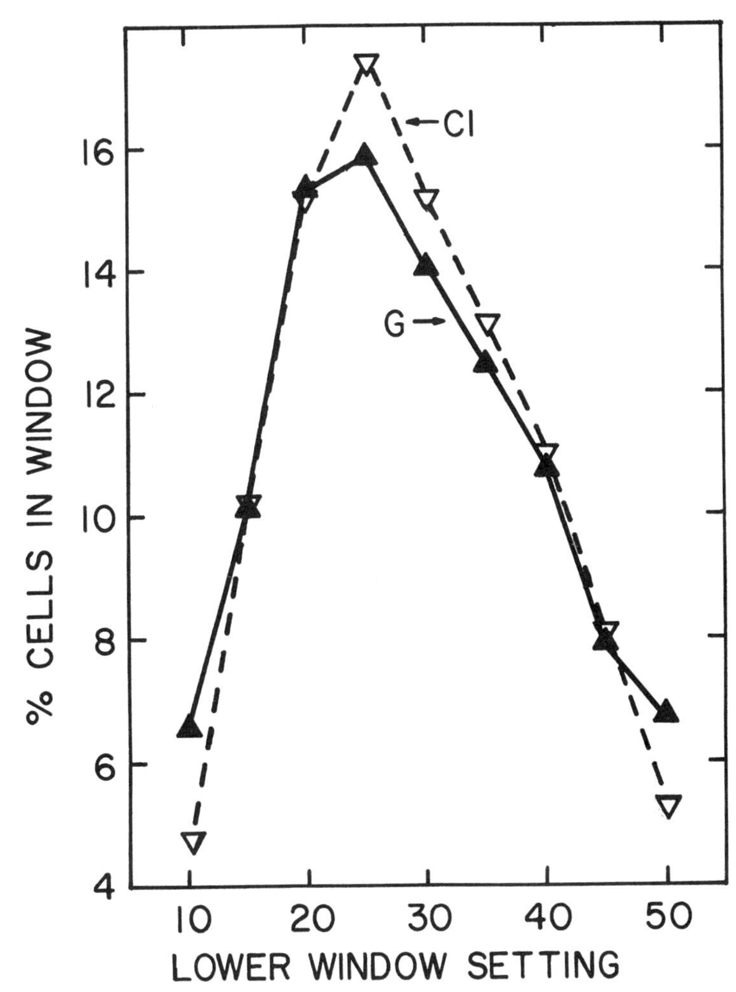

FIGURE 2. The size distribution of growing (solid triangles) and contact inhibited (open triangles) chick embryo fibroblasts. The cells after removal from the plate with 0.02% EDTA in calcium-and magnesium-free medium were sized with a Model B Coulter counter. Cells were counted in 5 unit windows with settings of 1/amplification of 4 and 1/ampeture current of 1.

---

difficult for me to believe that cyclic AMP is very significant for controlling contact inhibition. The idea is certainly attractive and since other data seem to contradict mine (Otten *et al*., 1971), it must be explored further.

4. CELL SURFACE TURNOVER

Joffre Baker has made some interesting observations which suggest that certain specific cell surface components may begin to turn over when contact inhibition occurs (Baker and Humphreys, 1972). During experiments on the necessity of protein synthesis for the serum effect, he noticed that inhibition of protein synthesis with cycloheximide accentuated the serum effect on cell contacts. At the same time he was examining the possibility that cells became agglutinable with the plant lectin, concanavilin A(ConA), when stimulated by serum and tested cycloheximide treated cells which had been serum stimulated. It turned out that serum had no effect on agglutinability but that cycloheximide treated cells were agglutinable with or without serum.

The agglutinability of cells by plant lectins has been correlated with an alteration of the cell surface characteristic of cells which no longer exhibit contact inhibition (Pollack and Burger, 1969). The surfaces of normal, non-agglutinable cells can be converted to the agglutinable state by trypsin treatment (Burger, 1969) suggesting that the sites reacting with lectins are present on normal cells but are made unavailable by protein containing molecules. Thus the alteration produced in the surfaces by inhibition of protein synthesis of normal cells indicates the loss of the surfaces components normally making the sites unavailable and suggests they are continually degraded or lost from the cell surface and require protein synthesis for replacement.

These studies were begun using the subjective depression-slide assay for agglutinability of cells removed from the culture plate with EDTA (Burger and Goldberg, 1967; Burger, 1969; Pollack and Burger, 1969; Inbar and Sachs, 1969). As is apparently characteristic, results were eratic and he sought a better assay. He performed the 10-minute agglutination in shell vials on a shaker with a 4" diameter of rotation designed for cell aggregation in 200 μl volumes (Henkart and Humphreys, 1970). Under these more vigorous shaking conditions, ConA did not produce large clumps and the result was not obvious to the unaided eye. However, examination of the cells with the compound microscope showed most of the normal cells to be single while most of the agglutinable cells were in small clumps of 2 to 10 cells. Since all cells even in clumps were individually countable, a quantitative estimate of agglutinability could be achieved simply by counting the percent of the cells in clumps of 2 or more cells in any random field. The percent of cells agglutinated was very insensitive to ConA concentration between 60 and 2000 μg/ml. This assay still showed a disturbing unpredictable high agglutination of control cells in 10 to 20% of the experiments. Such experiments were discarded. In general, however, the assay was more reproducible and certainly more satisfying than the subjective estimate of agglutination on a slide.

Using this assay, Baker measured the agglutinability of chick embryo fibroblasts from confluent cultures which had been treated with 2×10^{-5} M cycloheximide, a concentration which reduced H^3-leucine incorporation to 5% of control levels. ConA agglutinated about 30% of the untreated cells (Table IV). This control level of agglutination may be associated with the embryonic origin of these cells (Moscona, 1971). Background aggregation in the absence of ConA was about 10%. As the cells were incubated with cycloheximide, agglutination increased linearly for 4 to 6 hours when a maximum agglutination of 70% to 80% was reached (Table IV). This level of agglutination was equal to that produced by a standard trypsinization (Burger, 1969). The increased agglutination was the result of inhibition of protein synthesis, since two other inhibitors of protein synthesis, emetine and pactamycin, used respectively at 10^{-6} M and 10^{-7} M to inhibit more than 95%, also caused the cells to become agglutinable (Table IV). The change to agglutinability cannot be associated with extensive damage to the cells since agglutinability was rapidly reversed when the inhibitor was removed (Table IV).

These results indicate that continued protein synthesis is required to maintain the normal cell surface and suggests that specific cell surface materials are being either rapidly degrading or lost from the cell surface and must be replaced by new protein synthesis. Warren and Glick (1968) examined the turnover of cell surface glycoproteins in growing and stationary cells and found that turnover of these components was more extensive in stationary cells. We therefore examined the effects of cycloheximide on growing cells and found no evidence of increased agglutinability even after 20 hours of treatment (Table IV). The alteration of the cell surface when protein synthesis is inhibited is specific to cells which are contact inhibited and suggests that cell surfaces are relatively more stable in growing cells.

TABLE IV

The agglutination of cells with Concanavilin A after inhibition of protein synthesis.

CELLS	TREATMENT	% AGGLUTINATION*
Confluent	-	32
Confluent	2 hours cycloheximide	43
Confluent	4 hours cycloheximide	70
Confluent	20 hours cycloheximide	78
Confluent	4 hours emetime	76
Confluent	5 hours pactamycin	80
Confluent	4 hours cycloheximide; 6 hours recovery	23
Growing	6 hours cycloheximide	29
Growing	-	29
Growing	20 hours cycloheximide	32

*percent of cells in clumps of 2 or more after shaking 10 minutes in 250 μg/ml ConA on a large diameter of rotation shaker (Henkart and Humphreys, 1970). In all cases about 10% of the cells clumped in the absence of ConA.

During contact inhibition, accumulation of total protein must be stopped. Since protein synthesis is reduced but still remains 20 to 30% of growing cells (Hodgson and Fisher, 1971; Stannens and Becker, 1971) considerable general protein turnover must occur. The relationship of this turnover to the turnover of cell surface proteins also needs to be examined. We do not know what portion of the proteins turn over during contact inhibition or how rapidly. For example, assuming no growth and 20% to 30% of growing cell synthesis in contact inhibited cells all proteins could turn over with a half life of 3 to 5 days or 2-4% could turn over with a half life of about 3 hours. Because the turnover of the cell surface suggest half lives on the order of the latter possibility, I labeled cells with H^3-leucine for short periods of time and then followed decay of label in pactamycin. The label showed no significant decay over an 8 hour period suggesting that general turnover of protein is much slower than the cell surface components.

This alteration from a normal to a malignant-like cell surface which can be experimentally induced and reversed in all cells at the same time might prove to be very useful for studying the nature of the cell surface alterations involved. It might be possible to characterize the cell surface proteins which begin to turn over during contact inhibition. This system could also prove useful for study of the cell surface reactions involved in contact inhibition. Why should cell surface molecules which appear to have a unique function in contact inhibition begin to turn over rapidly when cells become contact inhibited? Do the primary interactions involved in contact inhibition reverse as the cells change to the agglutinable state during inhibition of protein synthesis? The secondary results of growth which are most easily measured require protein synthesis and thus cannot be used to determine if the cells are unable to continue appropriate mutual interactions as they become agglutinable. Our observations show that the cells do not break mutual contact in cycloheximide. It would be interesting, for example, to determine if the cells become more permeable (Cunningham and Pardee, 1968).

5. RIBOSOMAL RNA SYNTHESIS

Ribosomal RNA is among the many specific macromolecules accumulated in growing cells whose increase must be limited during contact inhibition. Study of the molecular details of regulation of ribosomal RNA synthesis and accumulation during contact inhibition seemed to be especially promising because these molecules are well characterized and extensive information on the molecular biology of ribosomal RNA synthesis was available. The synthesis and processing of ribosomal RNA has been well worked out (Penman _et al._, 1966). The ribosomal RNA polymerase has been isolated and identified (Roeder and Rutter, 1970). Ribosomal genes have been isolated (Wallace and Birnstrel, 1966). In addition, Charles Emerson had worked out quantitative methods for measuring _in vivo_ rates of ribosomal RNA synthesis using radioactive RNA precursors and had considerable experience studying regulation of ribosomal RNA synthesis (Emerson and Humphreys, 1970; 1971a; 1971b). He, therefore, initiated a series of experiments on the regulation of ribosomal RNA synthesis during contact inhibition. I will review his findings which are published in in detail (Emerson, 1971).

The rate of synthesis of ribosomal RNA was measured in growing, contact inhibited, and serum stimulated cells using rates of entry of radioactive adenosine into ATP pools and into RNA (Emerson and Humphreys, 1971b). ATP was determined using the luciferin-luciferase assay. Incorporation into ATP was determined after chromatographicly purifying it. Radioactivity in ribosomal RNA was determined by the size of radioactive 18S, 28-32S, and 45S peaks in sucrose gradients. The rate of synthesis of ribosomal RNA was 5.2×10^{-15} gm per 2N DNA per min in growing cells; 2.1×10^{-15} gm in contact inhibited cells and 6.9×10^{-15} gm in cells stimulated by increased serum for 4 hours. These results indicate that the rates of ribosomal RNA synthesis is reduced about 3 fold during contact inhibition and that serum rapidly reverses this repression of synthesis. Present data indicates that this serum reversal occurs between 1 and 2 hours after serum addition.

Although contact inhibited cells had a reduced rate of synthesis, they were still synthesizing ribosomal RNA much faster than they were growing. The actual rates of accumulation of ribosomal RNA was measured chemically in the three kinds of cultures. Growing cells accumulated 5.0×10^{-15} gm per 2N DNA per min, contact inhibited cells 0.6×10^{-15} gm and serum stimulated 6.9×10^{-15} gm. The rates of synthesis equal the rates of accumulation in growing and serum stimulated cells but are 3 fold less in contact inhibited cells. This indicates that 18S and 28S ribosomal RNA is turning over with a half life of about 40 hours in contact inhibited cells. Serum stimulation of cells not only activates ribosomal RNA synthesis but it also stops the turnover of the ribosomal RNA.

The measurements of synthesis above were based on the entry of radioactivity into 18S and 28S RNA. However, ribosomal RNA is synthesized as a 45S precursor which is converted by a series of steps that include degradation of portions of the 45S molecule. It seemed possible that regulation of formation of 18 and 28S RNA could be controlled by regulating the proportion of the 45S converted to 18 and 28S without actually regulating gene activity. Emerson, therefore, tried to measure the rate of formation of 45S precursor molecules. He obtained very peculiar kinetics of increase in absolute amounts of radioactive 45S RNA which could not be easily interpreted as a defined rate of synthesis. The curves showed an accelerating rate of synthesis for about 5 to 10 minutes in growing cells and about 20 to 30 minutes in contact inhibited. One interpretation of these results was some unexpected effect of equilibration of isotope with the radioactive precursor pools. This seems very unlikely because the synthesis of radioactive heterogenous RNA does not show these anomolus kinetics.

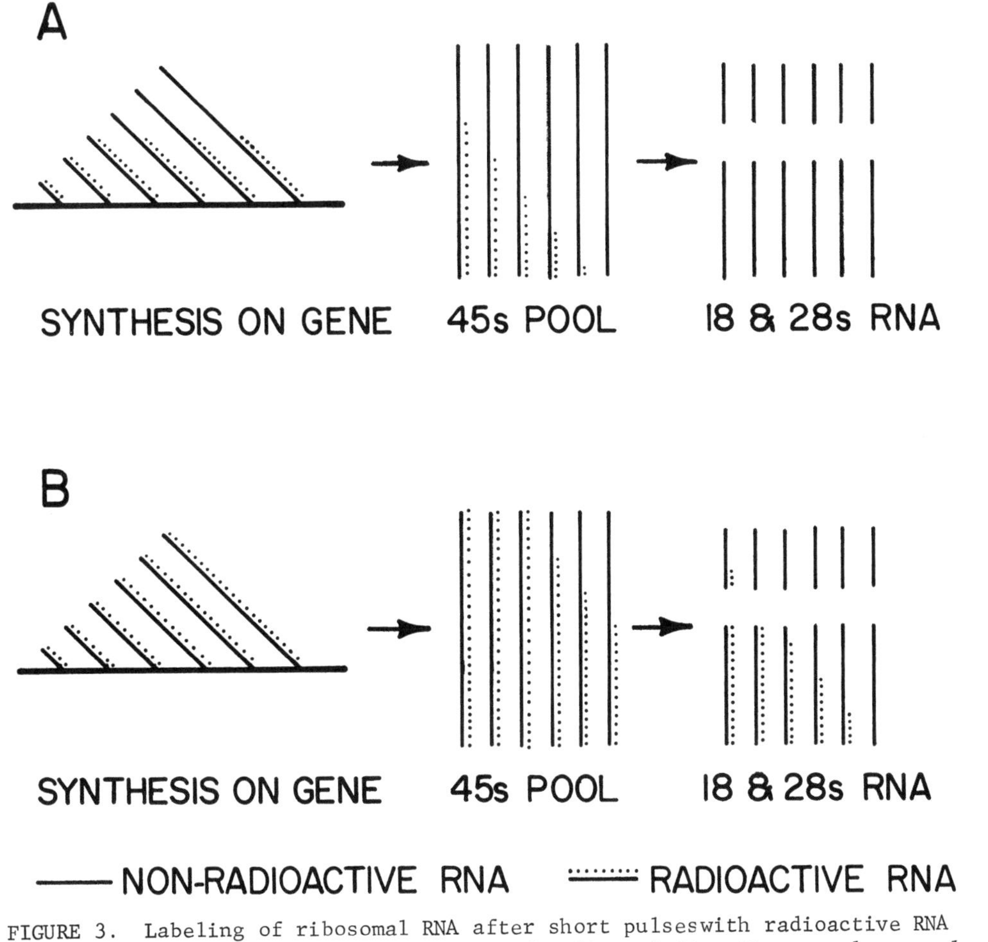

FIGURE 3. Labeling of ribosomal RNA after short pulseswith radioactive RNA precursor. A, pattern of incorporation of radioactivity after a pulse equal to one-half the time required to synthesize a 45S precursor molecule. B, pattern of incorporation after a pulse equal to the processing time of the 45S RNA precursor plus one-half synthesis time.

Upon recognizing what he believed to be the correct interpretation for the apparently anomolus kinetics of synthesis of radioactive 45S RNA, it became obvious that they should have been expected, although not to the extent observed in contact inhibited cells. Fig. 3 contains a schematic to explain this conclusion. Synthesis of 45S RNA molecules probably occurs as many molecules in tandem on one gene (Miller and Beatty, 1969) and require a finite length of time. During a labeling period shorter than the time required to synthesize a complete 45S molecule (Fig. 3A) each completed 45S molecule is only partially radioactive and thus does not add as much radioactive RNA to the 45S pool as a whole molecule. The longer the labeling period the larger the portion of 45S molecule labeled and the more radioactive RNA each completed 45S molecule added to the 45S pool. When one synthesis time has passed each new molecule is fully labeled and adds a constant amount. The length of time required to reach linear synthesis is a measure of the synthesis time of a 45S molecule.

The period of accelerating labeling rate in growing cells indicates a 5 to 10 minute synthesis time and in contact inhibited cells a 20 to 30 minute synthesis time. This means that the rate of chain polymerization of 45S molecules is reduced about 3 fold in contact inhibited cells. If it is assumed that the number of growing RNA chains on the ribosomal genes does not increase upon contact inhibition, this reduction in rate of chain polymerization fully accounts for the regulation of ribosomal RNA synthesis during contact inhibition.

Independent evidence was sought for the idea that the time for synthesis of 45S RNA molecules was longer in contact inhibited cells than in growing cells. It seemed especially important that this evidence be independent of any effects which could be attributed to kinetics of entry of radioactivity into RNA precursor pools. Emerson reasoned that if it took longer to make a 45S RNA molecule it would take longer for label to enter 18S RNA than 28S RNA since 18S is toward the beginning, 5' end of the 45S RNA molecule (Fig. 3B). The ratio in radioactivity in 18S RNA to 28S RNA was measured after various pulse lengths. Upon short pulses the 28S RNA was substantially more radioactive than the 18S RNA but after longer labeling periods they became equally radioactive in both growing and contact inhibited cells. However, for any given pulse length the assymetry of labeling of 18S and 28S RNA was greater in contact inhibited cells than in growing cells confirming the idea that 45S chain formation takes longer in contact inhibited cells. These results suggest that ribosomal RNA synthesis is regulated during contact inhibition by limiting the rate of movement of the RNA polymerase along the gene. This could be achieved by limiting chain termination such that the polymerases are piled up along the gene and move forward each time a chain is terminated. It is also possible that the actual rate of polymerization is controlled.

6. SUMMARY

Contact inhibition is a multifaceted set of events, which may be sequential or parallel, all involved in reducing growth of cells. It apparently begins with cell contact which produces a signal or signals that ultimately stops growth. Cyclic AMP may be major candidate for the primary signal but our data is not consistent with this idea. Our measurements in chick fibroblasts, while confirming the idea that cyclic AMP varies with growth rate of growing cells show no change during contact inhibition. There are a number of changes in macromolecular metabolism which occur during contact inhibition. We have investigated the turnover of cell surface proteins as well as synthesis and turnover of ribosomal RNA. The data suggests that ribosomal RNA synthesis is regulated by controlling the rate of movement or termination of the ribosomal RNA polymerase on the ribosomal gene. Other changes in macromolecular metabolism have been described (for example, Hodgson and Fisher, 1971; Salas and Green, 1970) and, of course, the central one involves DNA synthesis. No doubt many others exist.

At this time it is not clear which of the changes in metabolism are sequential and causitive and which ones are parallel responses to the same or similar signals. For example, it is conceivable that one signal, such as cyclic AMP, could regulate protein synthesis, protein turnover, ribosomal RNA synthesis and ribosomal RNA turnover and many other changes. Alternatively, the cell contact might stimulate the production of a number of regulatory molecules. Such molecules could include guanosine tetraphosphate (Cashel and Kalbacher, 1970) which could specifically regulate the rate of chain elongation of ribosomal RNA precursor molecules and other molecules which regulate other changes in metabolism. The web of events involved in contact inhibition must be explored much more fully before these possibilities can be sorted out.

REFERENCES

Baker, J. B. and T. Humphreys (1971). Serum-stimulated release of cell contacts and the initiation of growth in contact inhibited chick fibroblasts. Proc. Natl. Acad. Sci. U.S.A. 68, 2161.

Baker, J. B. and T. Humphreys (1972). Turnover of molecules which maintain the normal surfaces of contact inhibited cells (mss submitted).

Becker, H., C. P. Stannens, and J. E. Kudlow (1971). Control of macromolecular synthesis in proliferating and resting syrian hamster cells in monolayer culture. II. Ribosome complement in resting and early G_1 cells. J. Cell Physiol. 77, 43.

Burger, M. M. (1969). A difference in the architecture of the surface membrane of normal and virally transformed cells. Proc. Natl. Acad. Sci., U.S.A. 62, 994.

Burger, M. M. and A. R. Goldberg (1967). Identification of a tumor-specific determinant on neoplastic cell surfaces, Proc. Natl. Acad. Sci., U.S.A. 57, 359.

Burger, M. M. and K. D. Noonan (1970). Restoration of normal growth by covering of agglutinin sites on tumor cell surface, Nature 228, 512.

Cashel, M. and B. Kalbacher (1970). The control of ribonucleic acid synthesis in E. coli. V. Characterization of a nucleotide associated with the stringent response. J. Biol. Chem. 245, 2309.

Ceccarini, C. and H. Eagle (1971). pH as a determinant of cellular growth and contact inhibition, Proc. Natl. Acad. Sci., U.S.A. 68, 229.

Clarke, G. D., M. G. P. Stoker, A. Ludlow and M. Thornton (1970). Requirement of serum for DNA synthesis in BHK 21 cells: effects of density, suspension and virus transformation, Nature 227, 798.

Coman, D. R. (1944). Decreased mutual adhesiveness, a property of cells from squamons cell carcinomas, Cancer Res. 4, 625.

Cunningham, D. D. and A. B. Pandee (1969). Transport changes rapidly initiated by serum addition to contact inhibited 3T3 cells, Proc. Natl. Acad. Sci., U.S.A. 64, 1049.

Dulbecco, R. (1970). Topoinhibition and serum requirement of transformed and untransformed cells, Nature 227, 802.

Dulbecco, R. and M. G. P. Stoker (1970). Conditions determining initiation of DNA synthesis in 3T3, Proc. Natl. Acad. Sci., U.S.A. 66, 204.

Edwards, J. G., J. A. Campbell and J. F. Williams (1971). Transformation by polyoma virus affects adhesion of fibroblasts, Nature New Biology 231, 147.

Emerson, C. P., (1971). Regulation of the synthesis and the stability of ribosomal RNA during contact inhibition of growth, Nature New Biology 232,101.

Emerson, C. P. and T. Humphreys (1970). Regulation of DNA-like RNA and the apparent activation of ribosomal RNA synthesis in sea urchin embryos. Quantitative measurements of newly-synthesized RNA. Develop. Biol. 23, 86.

Emerson, B. P. and T. Humphreys (1971a). A simple and sensitive method for quantitative measurement of cellular RNA synthesis, Anal. Biochem. 40, 254.

Emerson, C. P. and T. Humphreys (1971B). Ribosomal RNA synthesis and the multiple, a typical nucleoli in cleaving embryos, Science 171, 898.

Gilman, A. G. (1970). A protein binding assay for adenosine 3':5'-cyclic monophosphate, Proc. Natl. Acad. Sci., U.S.A. 67, 305.

Gurney, T. (1969). Local stimulation of growth in primary cultures of chick embryo fibroblasts. Proc. Natl. Acad. Sci., U.S.A. 62, 906.

Henkart, P. and T. Humphreys (1970). Cell aggregation in small volumes on a gyratory shaker, Exptl. Cell Res. 63, 224.

Hodgson, J. R. and H. W. Fisher (1971). Formation of polyribosomes during recovery from contact inhibition of replication, J. Cell Biol. 49, 945.

Holley, R. W. and J. A. Kiernan (1968). Contact inhibition of cell division in 3T3 cell lines. Proc. Natl. Acad. Sci., U.S.A. 60, 300.

Inbar, M. and L. Sachs (1969). Structural difference in sites on the surface membrane of normal and transformed cells, Nature 223, 710.

Johnson, G. S., R. M. Friedman and I. Pastan (1971). Restoration of several morphological characteristics of normal fibroblasts in sarcoma cells treated with adenosine 3':5' cyclic monophosphate and its derivatives. Proc. Natl. acad. Sci., U.S.A. 68, 425.
Miller, O. L. and B. R. Beatty (1969). Visualization of nucleolar genes, Science 164, 955.
Paul, D., A. Lipton and I. Klinger (1971). Serum factor requirements of normal and simian virus 40-transformed 3T3 mouse fibroblasts. Proc. Natl. Acad. Sci., U.S.A. 68, 645.
Penman, S, I. Smith and E. Holtzman (1966). Ribosomal RNA synthesis and processing in a particulate site in the HeLa cell nucleus, Science 154, 786.
Pollack, R. E. and M. M. Burger (1969). Surface-specific characteristics of a contact inhibited cell line containing the SV40 viral genome, Proc. Natl. Acad. Sci., U.S.A. 62, 1074.
Pollack, R. E., H. Green and G. J. Todaro (1968). Growth control in cultured cells: selection of sublines with increased sensitivity to contact inhibition and decreased tumor-producing ability, Proc. Natl. Acad. Sci., U.S.A. 60, 126.
Roeder, R. C. and W. J. Rutter (1970). Specific nucleolar and nucleoplasmic RNA polymerases, Proc. Natl. Acad. Sci., U.S.A. 65, 675.
Salas, J. and H. Green (1971). Proteins binding to DNA and their relation to growth in cultured mammalian cells. Nature New Biology 229, 165.
Sheppard, J. R. (1971). Restoration of contact-inhibited growth to transformed cells by dibutyryl adenosine 3':5"-cyclic monophosphate, Proc. Natl. Acad. Sci., U.S.A. 68, 1316.
Stanners, C. P. and H. Becker (1971). Control of macromolecular synthesis in proliferating and resting syrian hamster cells in monolayer culture. I. Ribosome function, J. Cell.Physiol. 77, 31.
Stoker, M. G. P. and H. Rubin (1967). Density dependent inhibition of growth in culture, Nature 215, 172.
Temin, H. M. (1969). Control of cell multiplication in uninfected chicken cells and chicken cells converted by avian sarcoma viruses, J. Cell.Physiol. 74, 9.
Todaro, G. J., G. K. Lazer and H. Green (1965). The initiation of cell division in a contact inhibited mammalian cell line, J. Cell.Comp. Physiol. 66, 325.
Vogt, M. and R. Dulbecco (1960). Virus cell interaction with a tumor producing virus. Proc. Natl. Acad. Sci., U.S.A. 46, 365.
Wallace, H. and M. L. Birnstrel (1966). Ribosomal cistrons and the nucleolar organizer, Biochim. Biophys. Acta 114, 296.
Warren, L. and M. C. Glick (1968). Membranes of animal cells. II. The metabolism and turnover of the surface membrane, J. Cell Biol. 37, 729.
Yeh, J. and H. W. Fisher (1969). A diffusible factor which sustains contact inhibition of replication, J. Cell Biol. 40, 382.

DIRECT INTERACTION BETWEEN ANIMAL CELLS

JOHN D. PITTS

Department of Biochemistry, University of Glasgow, Scotland

Abstract: A form of direct interaction between animal cells in tissue culture permits intercellular nucleotide exchange. The interaction is common to many cell types, but not all. Some cell types interact with reduced efficiency, and some do not interact. The inability to interact is dominant in mixed cultures of cells which can interact and cells which cannot. Hybrid cells formed by fusion of cells which can and cells which cannot interact, can interact. Cells interact through a 100-fold range of cell density; at low cell densities, cytoplasmic extensions allow interaction. Cellular interactions of this type can be detected *in vivo*. Such interactions may play an important role in growth control in tissue culture and in developmental processes *in vivo*.

1. INTRODUCTION

An animal may be considered as a population of interacting cells, the growth and function of each cell being coordinated to the varying requirements of the total population. These cell-cell interactions are of two types, direct and indirect. Direct cell-cell interactions require contact between the interacting cells and signals are transmitted either by transport of signal substances through intercellular junctions or by the modification of the cellular membranes at the points of contact, such modifications, in turn, leading to control of cellular properties. Indirect interactions are mediated by signal substances (hormones) which are transported through the extracellular fluids to act on target cells.

The mechanisms of hormone action have been the subject of intensive study and, at least in the case of some hormones, a general picture is beginning to emerge. Direct cell-cell interactions, on the other hand, have received much less attention and little is known about their nature. However, the importance of direct interactions between cells has been realised for some time, in the field of growth control in animal cell tissue culture (Stoker, 1967) and in the field of development and differentiation (Wolpert, 1969; Crick, 1970; Newell, 1971). Most theories of development require the flow, from cell to cell, of morphogens which ellicit different developmental responses at different distances from the morphogen source. But how does the morphogen get from cell to cell? To quote Crick (1970), 'One is led to postulate a special mechanism which allows a relatively quick passage of the morphogen from one cell to another in the tissue of interest'.

A form of cell-cell interaction is described in this paper which allows rapid transfer of nucleotides from one cell to another. This may be particularly pertinent to the problem outlined by Crick, as a nucleotide, cyclic-AMP, has been invoked in at least one system (Bonner, 1970) as the mediator of a developmental process.

2. INTERCELLULAR JUNCTIONS

Many forms of intercellular junctions have been described. These range from the sophisticated junctions between the highly specialised cells of nervous tissue, to simpler junctions which appear to be formed between many cell types, *in vivo* and in tissue culture (Furschpan and Potter, 1968). Such junctions have been investigated in two ways, by direct examination with the electron microscope, and by the study of the movement of substances, which cannot normally cross the cellular membrane, between cells in contact.

Low resistance junctions, junctions between cells with a much lower electrical resistance than the cytoplasmic membrane, are formed between a variety of animal cells *in vivo* (Loewenstein and Kanno, 1964, 1966) and in tissue culture (Potter, et al, 1966).

Another form of direct cell-cell interaction in tissue culture, which depends on the formation of intercellular junctions which are permeable (unlike the cell membrane) to nucleotides, has been described by Subak-Sharpe et al (1966, 1969) and by Pitts (1971).

3. INTERACTION BETWEEN CELLS IN CULTURE

Mutant hamster BHK21/13 cells which lack the enzymes inosinic pyrophosphorylase (BHK-*IPP*⁻ cells) or thymidine kinase (BHK-*TK*⁻ cells) are unable to incorporate hypoxanthine or thymidine, respectively, into cellular nucleic acid, except when they are cultured in contact with wild-type BHK cells. This is clearly illustrated by confluent cultures of 1:1 mixtures of BHK-*IPP*⁻ and BHK-*TK*⁻ cells in which all cells incorporate both hypoxanthine and thymidine.

The phenotypic modification of mutant BHK cells in contact with wild-type cells could be explained either by the transfer of (1) and appropriate enzyme (or mRNA or gene) or (2) the appropriate nucleotides which are beyond the enzyme blocks. These two categories can be distinguished on the basis of separation experiments. If the enzyme is transferred, the mutant cells should behave for a time after separation from wild-type cells, as phenotypic wild-types, and incorporate hypoxanthine or thymidine. On the other hand, if nucleotides only are transferred, the mutant cells should regain their mutant phenotype immediately after separation. Such separation experiments show that the enzymes are not transferred to a significant extent (Pitts, 1971). Similar results with the same conclusions were obtained by Cox et al (1970) using human fibroblasts derived from normal and Lesch-Nyhan (*IPP*⁻) patients.

Not all cells in tissue culture interact in this way. Mutants derived from the mouse cell line L929, L-*IPP*⁻and L-*TK*⁻ cells are interaction negative (*int*⁻) and maintain the mutant phenotype in mixed culture (Pitts, 1971; C. Favre, personal communication). This inability of L929 cells to interact with each other is dominant in mixed cultures of L929 and BHK cells (Table 1), i.e. both cells of an interacting pair must be *int*⁺ for nucleotide transfer to occur.

Table 1
Incorporation of ^{3}H-hypoxanthine by BHK and L cells in mixed cultures

Culture	0-10 grains/cell	10-50 grains/cell	> 50 grains/cell
	Number of cells		
BHK-IPP⁻	523	0	0
L	0	0	499
BHK-IPP⁻ + L	207	13	262
L-IPP⁻	467	0	0
BHK	0	2	509
L-IPP⁻ + BHK	259	27	227

Cells were plated (10^6/dish) in 5 cm dishes and incubated overnight at 37°C. ^{3}H-hypoxanthine was added and the cells fixed 8 h later. After washing with cold 5% TCA, the cells processed for autoradiography.

This dominance of the int⁻ phenotype in mixed int⁻:int⁺ populations, makes it possible to examine the interaction properties of a large number of cell types without first isolating mutants of each type. Fig. 1 shows the grain counts in the cells of 1:1 confluent mixtures of BHK-IPP⁻ cells with different wild-type cells. In the BHK-IPP⁻:BHK mixture all the cells incorporate hypoxanthine, but in the BHK-IPP⁻:L929 cell mixture, only the wild-type L929 cells incorporate the base.

Most cell types examined (see Table 2) interact with the BHK-IPP⁻ cells with 100% efficiency (i.e. the two populations cannot be distinguished) but two cell types, BSC-1 and HeLa, appear to interact with intermediate efficiency (Fig. 1). In both these cases, two cell populations can be distinguished, the mutant cell population incorporating 55% (BHK-IPP⁻:BSC-1 cell mixture) and 20% (BHK-IPP⁻:HeLa cell mixture) of the label incorporated by the wild-type population.

Apart from the L929 cells, the only other cell type which has been shown to be int⁻ is a mast cell line (C. Favre, personal communication).

4. INTERACTION PROPERTIES OF HYBRID CELLS FORMED BY FUSION OF int⁺ AND int⁻ CELLS

Hybrid cells formed between BHK-int⁺ and L-int⁻ cells by fusion with inactivated Sendai virus, interact with BHK cells (table 3). This is consistent with the idea that L929 cells have lost some gene function necessary for interaction and the corresponding BHK gene function is dominant in the hybrid cell. Examination of the hybrid cell surface with antisera prepared against hamster cells and mouse cells, showed that antigens derived from both cell types were present (McCargow and Pitts, 1971).

Figure 1
Efficiency of interaction of different cell types

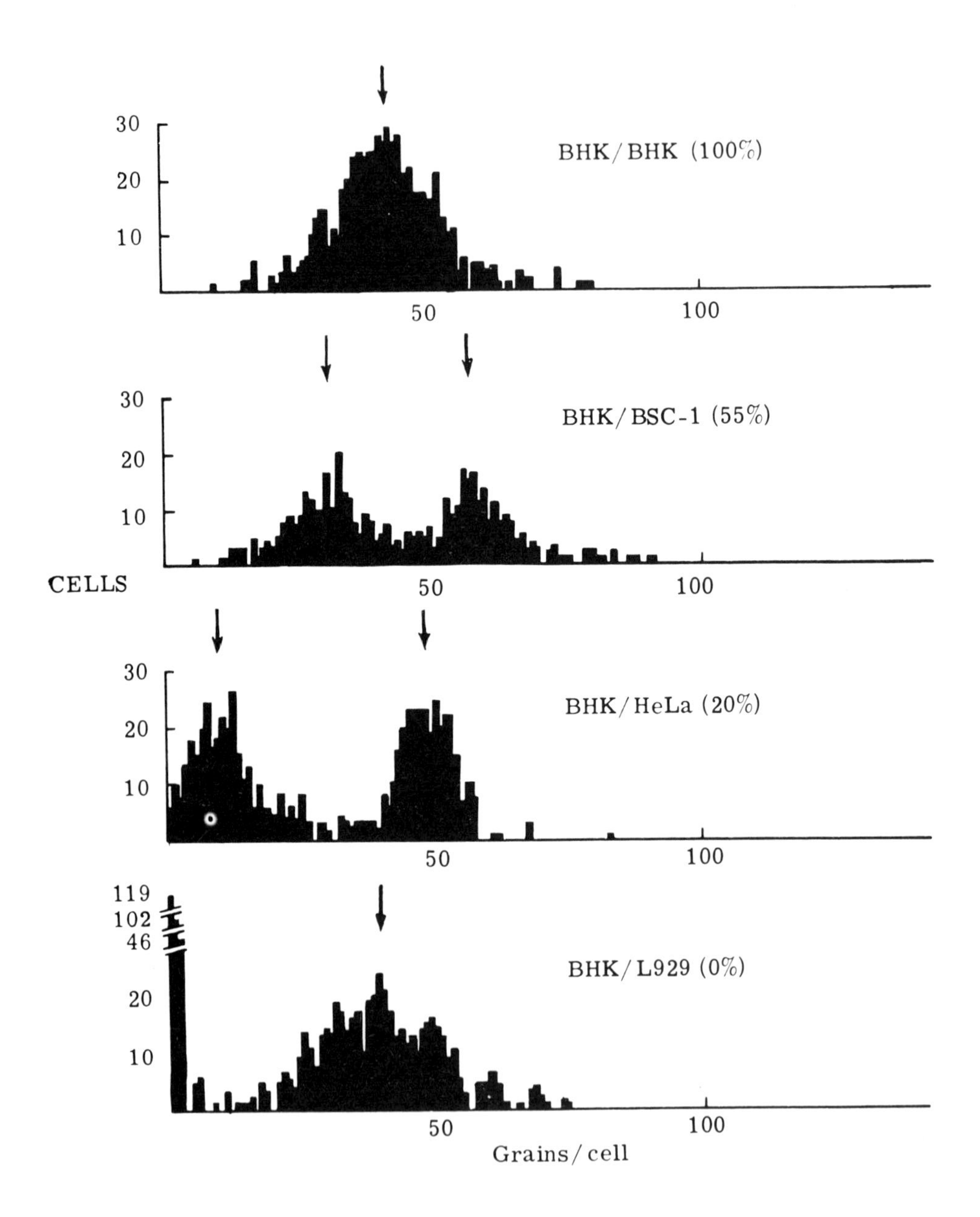

Cells were plated at 5 x 10^5/dish. For other details see legend to Table 1.

Table 2
Interaction properties of different cell types

Donor cells	Recipient cells	Complete Interaction	Intermediate interaction	No interaction
BHK	BHK	+		
BHK	3T3	+		
RSV-BHK	BHK	+		
3T3	BHK	+		
SV-3T3	BHK	+		
Mouse embryo	BHK	+		
Human embryonic lung	BHK	+		
Hel 2000	BHK	+		
Chick embryo	Chick embryo	+		
RSV-chick embryo	RSV-chick embryo	+		
BHK-L hybrid	BHK	+		
Py-BHK	Py-BHK		+	
BSC-1	BHK		+	
HeLa	BHK		+	
L	L			+
BHK	L			+
L	BHK			+
Mast cell	Mast cell			+

BHK = a hamster fibroblast line BHK21/13 (Macpherson and Stoker, 1962). RSV-BHK = Rous sarcoma virus transformed BHK. 3T3 = a mouse fibroblast line (Todaro and Green, 1963). SV-3T3 = SV40 transformed 3T3. Hel 2000 = a human fibroblast line. RSV-chick embryo = Rous sarcoma virus transformed chick embryo cells. BHK-L hybrid = a hybrid cell line formed by fusion with Sendai virus of BHK and L929 cells. Py-BHK = polyoma virus transformed BHK. BSC-1 = an African green monkey kidney cell line (Hopps, et al 1963). HeLa = a human tumour cell line (Gey, et al 1952). L = a mouse cell line L929 (Sanford et al, 1948).

Table 3
Interaction properties of hybrid BHK/L cells

Cells	0-10 grains/cell	10-50 grains/cell	> 50 grains/cell
	Number of cells		
BHK-IPP⁻	512	0	0
BHK/L hybrid	0	1	499
BHK-IPP⁻ + BHK/L hybrid (1:1 mixture)	2	3	505

Cells were plated at 5 x 10^5/dish. For other details see legend to Table 1.

5. INTERACTION AT LOW CELL DENSITIES

Under confluent conditions (about 5 x 10^6 cells/5 cm petri dish for BHK cells) all the cells in the population interact, and at very low cell densities (less than 10^4 cells/5 cm dish) most of the cells attach at widely separated sites and do not interact. The interaction of BHK wild-type and BHK-IPP⁻ cells has been investigated (Table 4) at cell densities between these two extremes and it is clear that as long as the cells are able to spread and touch, even only at the extremeties of long cytoplasmic extensions, interaction takes place efficiently. Even after a 100-fold dilution of the confluent 1:1 mixture, 93% of the cells are phenotypically wild-type (Table 4). The percentage of substrate covered by cellular material at this concentration (5 x 10^4 cells/dish plated; 7.9 x 10^4 cells/dish at the time of observation, 18 h after plating) was estimated by visual observation of stained preparations using an eye-piece grid, at 13%.

6. INTERACTION *IN VIVO*

Recent experiments (Pitts, unpublished) have shown that tumours formed from BHK-IPP⁻ cells in hamsters can incorporate ^{3}H-hypoxanthine which shows that nucleotide transfer occurs *in vivo* and is not limited to tissue culture. Such interaction of cells *in vivo* may explain why heterozygote mothers of Lesch-Nyhan children are clinically normal, despite the fact that such heterozygotes have two genotypically distinct cell populations, one IPP⁻ and the other wild-type (Migeon, 1968).

The ability of individual cells of the cellular slime moulds to act synergistically during development may be another example of this type of cellular interaction. If only 10% of wild-type cells of *D. discoideum* are mixed with mutant cells which alone are blocked in development, apparently normal fruiting bodies can be formed and the spores produced are of both parental types (Ennis and Sussman, 1958).

Table 4
Interaction of BHK cells at different cell densities

Cells plated/5 cm dish (1:1 mixture BHK-IPP⁻:BHK)	0-10 grains/cell	10-50 grains/cell	> 50 grains/cell
	Number of cells		
5×10^6	0	0	379
10^6	1	0	417
2×10^5	2	4	407
10^5	0	13	381
5×10^4	21	6	340
2×10^4	159	9	227
10^4	172	2	206

Mixtures (1:1) of BHK-$\underline{IPP}^-$ and BHK wild-type cells were plated at different densities, as indicated, in 5 cm dishes. After 15 h, ^{3}H-hypoxanthine was added and the cells were fixed 3 h later. After washing with 5% TCA the cells were processed for autoradiography.

A similar interacting system has been described for BHK cells in tissue culture which allows the growth of a population of two cell mutants under selective conditions where neither cell mutant can grow alone (Pitts, 1971). BHK-$\underline{IPP}^-$ and BHK-$\underline{TK}^-$ cells fail to grow when cultured separately in HAT medium (medium containing aminopterin, hypoxanthine and thymidine) because they lack IMP and TMP respectively. In mixed confluent culture, however, both mutants grow at the wild-type rate, presumably by nucleotide exchange.

7. EVOLUTION OF CELLULAR INTERACTION

The maintenance of metabolites within the cell is one of the basic problems of unicellular organisms and the solution to the problem has evolved as a cellular membrane which is impermeable to most metabolites. Accordingly, metabolites found commonly outside the cell are transported across the membrane by specific transport mechanisms. Intermediate metabolites, such as nucleotides, not usually found outside the cell, are maintained in the cell by the permeability barrier of the membrane. Molecules mediating control must also be kept in the cell and some control molecules may have evolved from intermediate metabolites (e.g. cyclic-AMP).

The evolution of multicellular organisms however, modified the problem. Any organised unit of cells must still maintain its metabolites but if each cell has a similar concentration, there is then no necessary restriction on free exchange between the cells. Control molecules, like cyclic-AMP, would also be free to move from cell to cell, and this would be an advantage to any group of cells which needed to respond in the same way to, for example, environmental changes. Further evolution has most certainly imposed certain restrictions on this free movement, but circulating hormones in higher animals act, in some cases, by liberating cyclic-AMP. The involvement of such a second messenger probably guarantees, in some cases, the coordinate response of an interacting complex of target cells.

In view of this, it is perhaps not surprising that junctions are found between animal cells which allow the free passage of nucleotides. It seems quite likely that such junctions are also freely permeable to other types of molecule, of metabolic importance and of importance in the control of gene function.

REFERENCES

Bonner, J.T. (1970), Induction of stalk cell differentiation by cyclic-AMP in the cellular slime mould D. discoideum. Proc. natn. Acad. Sci., U.S.A., 65, 110.

Cox, R.P., Krauss, M.R., Balis, M.E. and Dancis, J. (1970), Evidence for transfer of enzyme product as the basis of metabolic cooperation between tissue culture fibroblasts of Lesch-Nyhan disease and normal cells. Proc. natn. Acad. Sci., U.S.A., 67, 1573.

Crick, F. (1970), Diffusion in embryogenesis. Nature, 225, 420.

Ennis, H.L. and Sussman, M. (1958), Synergistic morphogenesis by mixtures of D. discoideum wild-type and aggregateless mutants. J. gen. Microbiol., 10, 110.

Furschpan, E.J. and Potter, D.D. (1968), Low resistance junctions between cells in embryos and tissue culture. In 'Current Topics in Developmental Biology' (Eds. A. Monroy and A.A. Moscana), 3, 95.

Gey, G.O., Coffman, W.O. and Kubicek, M.T. (1952), Tissue culture studies of the proliferative capacity or cervical carcinoma and normal epithelium. Cancer. Res., 12, 264.

Hopps, H.E., Bernheim, B.C., Nisalak, A., Tijo, J.H. and Smadel, J.E. (1963), Biologic characteristics of a continuous kidney cell line. J. Immunol., 91, 416.

Loewenstein, W.R. and Kanno, Y. (1964), Studies on an epithelial (gland) cell junction. Modifications of surface membrane permeability. J. Cell Biol., 22, 565.

Loewenstein, W.R. and Kanno, Y. (1966), Intercellular communication and the control of tissue growth: lack of communication between cancer cells. Nature, Lond., 209, 1248.

McCargow, J. and Pitts, J.D. (1971), Interaction properties of cell hybrids formed by fusion of interacting and non-interacting mammalian cells. Biochem. J., 124, 48P.

Macpherson, I. and Stoker, M. (1962), Polyoma transformation of hamster cell clones - an investigation of genetic factors affecting cell competence. Virology, 16, 147.

Migeon, B.R., Kaloustian, V.M., Nyhan, W.L., Young, W.J. and Childs, B. (1968), X-linked hypoxanthine-guanine phosphoribosyl transferase deficiency: heterozygote has two clonal populations. Science, 160, 425.

Newell, P.C. (1971). The development of the cellular slime mould D. discoideum: a model system for the study of cellular differentiation. In 'Essays in Biochemistry' (Eds. P.N. Campbell and F. Dickens), 7, 87.

Pitts, J.D. (1971), Molecular exchange and growth control in tissue culture. In the Ciba Foundation Symposium 'Growth Control in cell cultures' (Eds. G.E.W. Wolstenholme and J. Knight), Livingstone: London. p. 89.

Potter, D.D., Furschpan, E.J. and Lennox, E.S. (1966), Connections between cells of the developing squid as revealed by electrophysiological methods. Proc. natn. Acad. Sci., U.S.A., 55, 328.

Sanford, K.K., Earle, W.R. and Likely, G.D. (1948), The growth in vitro of single isolated tissue cells. J. natn. Cancer Inst., 9, 229.

Stoker, M. (1967), Contact and short range interactions affecting growth of animal cells in culture. In 'Current Topics in Developmental Biology' (Eds. A. Monroy and A.A. Moscana), 2, 107.

Subak-Sharpe, H., Burk, R.R. and Pitts, J.D. (1966), Metabolic cooperation by cell to cell transfer between genetically marked mammalian cells in tissue culture. Heredity, 21, 342.

Subak-Sharpe, H., Burk, R.R. and Pitts, J.D. (1969). Metabolic cooperation between biochemically marked mammalian cells in tissue culture. J. Cell Sci., 4, 353.

Todaro, G.J. and Green, H. (1963), Quantitative studies of the growth of mouse embryo cells in culture and their development into established lines. J. Cell Biol., 17, 299.

Wolpert, L. (1969), Positional information and the spatial pattern of cellular differentiation. J. theoret. Biol., 25, 1.

CELL INTERACTION IN NEURAL CREST DEVELOPMENT

JAMES A. WESTON
Department of Biology, University of Oregon
Eugene, Oregon

Abstract: The neural crest has special attributes as a system to investigate the role of cell interactions in the control of morphogenetic movements, the formation of specific cell contacts, and the regulation of cellular phenotypic expression.

Work is summarized that suggests that the crest is composed of a pluripotent cell population that migrates extensively through the vertebrate embryo and localizes precisely. The action of non-specific environmental factors is sufficient to account for this precise pattern of migration and localization. Nerve Growth Factor (NGF) for example, may act as a general morphogenetic factor in this system, causing changes in the contact interactions between crest cells, and hence their morphogenetic behavior.

Certain phenotypes expressed by neural crest cells may be regulated by the degree of cell interaction and by local environmental factors that impinge on the cells during their migration and after they localize. Some of these parameters can be experimentally controlled *in vitro*.

Finally, the neural crest system provides some promising opportunities to use genetic methods to analyse the regulation of crest morphogenetic and differentiative behavior.

1. INTRODUCTION

The neural crest of vertebrate embryos produces populations of cells that exhibit dramatically different morphogenetic and differentiative behavior. The crest may be identified initially as part of the lateral neuroepithelium of the embryonic medullary plate. As the edges of the medullary plate fold and fuse to form the neural tube, crest cells begin an extensive migration through the developing embryo. The emigration of these cells follows definite pathways, and the onset, direction and orientation of cell migration appears to be closely regulated. When migration ceases, the pattern of localization of these cells is exquisitely precise, and is ultimately correlated with the various phenotypes that crest cells express.

Thus, after the crest cells have undergone their major migration, some may be found dispersed uniformly in or on epithelial surfaces. These cells generally will differentiate as pigment cells. After migrating within the somite other crest cells aggregate as compact clusters embedded in a mesenchyme matrix. Such cells ultimately undergo nervous differentiation to form sensory or autonomic ganglia. The neural crest exhibits other modes of distribution and cell specialization, as well (e.g. parts of the cranial skeleton, and teeth), but these will not be considered further here (cf. Hörstadius, 1950; Weston, 1970, 1971).

There is, at least in the case of neurons and pigment cells, a general correlation between spatial arrangement of crest cells, and their phenotypic expression. Neuronal differentiation is normally manifest only when crest cells interact closely in cohesive clusters. Melanogenesis proceeds, on the contrary, when crest cells are initially dispersed (cf. Cowell and Weston, 1970). The crucial question is whether phenotypic differentiation of crest cells precedes, and is necessary for, the precise spatial patterns of cellular distribution, or *vice versa*. The answer to this question will bear directly not only on our understanding of mechanisms controlling the precise morphogenetic behavior of crest cells, but also on the role of the embryonic environment in regulating the course of cell specialization.

Two hypotheses may be formulated to account for crest cell migration and localization. The first hypothesis suggests that precociously determined cells undergo a "directed migration" following highly specific and directional environmental cues. This hypothesis assumes that phenotypically different populations of crest cells (e.g. melanoblasts, autonomic neuroblasts, and medio-dorsal and ventrolateral sensory neuroblasts) exist at the onset of migration, and exhibit differential susceptibility to specific directional and locational cues in the environment. It further assumes that the embryonic environment itself can provide such information to the migrating crest cells.

An alternative hypothesis--that crest cell migration begins randomly and that the pattern of the subsequent cellular migration and localization is governed by general environmental factors--assumes that crest cells are pluripotent when migration begins. They would then respond to environmental factors that are relatively nonspecific, in the sense that all migrating cells, and perhaps other cells as well, would be affected. If the migrating crest cells were pluripotent, the idea of specific control of migration would be untenable. Such cells would be unable to respond differentially to directional cues in the environment. Moreover, if cells were pluripotent at the onset of migration, the specializations that they ultimately manifest would necessarily be in response to developmental cues that they receive from the local environment through which they migrate, or in which they reside when migration ceases.

To discriminate between these alternatives, we have attempted to define the mechanisms by which the patterns of crest cell migration are regulated. We have tried to consider exactly how much (or how little) specificity would be sufficient to account for the pattern of crest cell distribution that we observe. In addition, we have considered the question of whether the environment can affect phenotypic expression of crest cells, and if so what role it plays in regulating cellular specialization.

2. CONTROL OF CREST CELL MIGRATION

By suitable cell marking procedures, (cf. Weston, 1967), the pattern of neural crest cell migration can be followed during normal development and under a variety of experimental conditions. Such studies have revealed that embryonic chick trunk crest cells migrate away from their source in two well defined streams--one dorso-lateral, which associates closely with the superficial ectoderm, and one ventral, which enters the somitic mesenchyme (Weston, 1963; 1970).

The cells in the ventral stream enter the somites and migrate only within and not between them. As a consequence, these crest cells ultimately acquire a segmental distribution. The distribution of the cells in the dorso-lateral stream, however, is more or less uniform and unsegmented. Thus, migration seems to be facilitated in an organized cellular environment such as the ectoderm or somitic mesenchyme. The fact that migration is favored only within existing tissue environments suggests that the discrete pathways and segmental distribution of crest cells might simply be the consequence of discontinuities in the environments which these cells enter. This idea is supported by experiments where labeled crest cells are transplanted heterotopically into regions of unsegmented mesoderm. Under these conditions crest cells migrate away from their source, but disperse uniformly (Weston, 1963).

The structure of the environment, therefore, could impose a pattern on the distribution of migrating crest cells nonspecifically (cf. Weston, 1971). This factor alone seems sufficient to account for the precisely delimited streams of cells and their initial non-random distribution. No inherent specificity within the migrating cell population need be postulated.

The localization of crest cells--especially those that form sensory and autonomic ganglia--appears to be quite precise. The extent to which general environmental factors are sufficient to cause this localization is important in defining how much intrinsic specificity exists in crest cells during their migration, and how much specificity is necessary for ganglion formation. Therefore, we performed a variety of experiments to test the proposition that the crest was a heterogeneous mixture of covertly specialized cells that underwent "directed migrations" to specific sites in the developing embryo. In these experiments, labeled neural crest in different stages of migration was grafted into unlabeled host embryos of a given age, or newly-formed crest was grafted into progressively older host embryos. The results indicated that cells remaining in "old" crest grafts migrated normally in young hosts, while migration of "young" crest cells was restricted in older host embryos. This is consistent with the idea that in older embryos the environment through which crest cells migrate probably changes its ability to support cell motility (Weston and Butler, 1966).

It seems clear, however, that simply limiting motility, and hence stopping continued dispersion of crest cells, cannot account for the precise regional accumulation of crest cells into compact ganglia. In addition to limiting motility, some developmental change in the somitic environment should also elicit a change in the mutual affinity of crest cells. Such a change, coupled with cell proliferation would probably be sufficient to account for the appearance of the cohesive cluster of cells that forms a ganglion.

One possible environmental factor that could account for these changes is the presence *in situ* of Nerve Growth Factor (NGF; cf. Levi-Montalcini, 1965; Levi-Montalcini and Angeletti, 1968). NGF is known to promote mitosis in responsive cells, and in its presence, aggregation of dissociated ganglion cells is enhanced (Levi-Montalcini and Angeletti, 1963; Weston, 1971). Moreover, NGF has been observed to prevent cell dispersion in explanted ganglia (Weston, 1971; Weston and Kaplan, 1972). Finally, NGF activity has been detected in the axial mesoderm of chick embryos just at the time in development when ganglia are forming as coherent compact structures (Bueker, et al., 1960; Winick and Greenberg, 1965).

Before we can attempt to predict what role NGF plays as a possible morphogenetic agent, however, it is necessary to consider the kinds of changes in cell adhesive interactions that might be expected to influence cell motile and morphogenetic behavior (i.e. whether cells will move at all, whether movement will be random or directional, or whether cells will clump). To this end, we have proposed a model causally relating cell social and motile behavior *in vitro* to the relative adhesive interactions of cells to each other and to their substrate for locomotion (Weston and Roth, 1969). In this model, the competition between these sorts of adhesions is operationally characterized as a ratio (cell-cell/cell-substrate), and the absolute value of the adhesive ratio is postulated to affect cell motile and social behavior. In the early embryo, the functional substrate for cell movement is likely to be other cells. For purposes of extrapolating the *in vitro* model of cell behavior to conditions which prevail *in vivo*, therefore, we assume that *isotypic* adhesive interactions (i.e. adhesions between like cells) are special cases of the cell-cell interaction in the formulation of the adhesive ratio. Likewise, we assume that *heterotypic* adhesions are special cases of cell-substrate interactions.

In the specific case of neural crest cell morphogenetic behavior, we can identify four general morphogenetic phases: 1) Initially, crest cells exist as a coherent epithelium on top of the neural tube. 2) When migrations begins, these cells relinquish contact with each other in favor of interactions with other cells in the embryonic environment. Crest cells then presumably use these cells in the environment as substrates for migration. 3) When, in older embryos, the environment apparently changes to limit cell motility, (cf. Weston and Butler, 1966), there may be a concomitant change in the relative adhesion of crest cells to somite cells, and of crest cells to each other, that may be causally related to the change in motility (cf. Weston, 1970). 4) Finally, when ganglia begin to form in the somite, crest cells may tend to relinquish adhesive contact with somite cells in favor of mutual contact interactions with each other. Specifically, we would predict that the heterotypic adhesive interaction (between crest and somite cells) should be higher than either isotypic adhesion while migration is underway. When crest cell motility is restricted the model predicts that heterotypic adhesions would be reduced. Moreover, when ganglia form, the effective "sorting out" of crest cells from somite could result when isotypic adhesions become greater than the heterotypic adhesion. Finally, if NGF plays a causal role in the observed changes in morphogenetic behavior of crest cells, exogenous NGF should effect these changes in isotypic and heterotypic adhesions *in vitro*.

Comparisons of adhesive stabilities can be made using suitable techniques (cf. Roth and Weston, 1967; Roth, et al., 1971). Preliminary results of such comparisons between somite cells and cells of nascent sensory ganglia (Weston and Schwarz, in preparation) show that in the absence of NGF, heterotypic cell adhesions are usually more stable than the isotypic adhesions. In the presence of NGF, these relationships are reversed. That is, NGF appears to reduce heterotypic adhesions and enhance isotypic adhesions. If substantiated, these results will be consistent with the suggestion that NGF is an environmental agent that could mediate the observed changes in crest cell morphogenetic behavior.

3. CONTROL OF CREST CELL SPECIALIZATION

The foregoing discussion has indicated that it is possible to account for the initial distribution in the somite, the cessation of migration (localization) and perhaps also the aggregation of cells to form sensory ganglia without postulating any specific phenotypic differences among the migrating crest cells. On the contrary, it seems reasonable to propose that when the initial migration has been completed, the crest cells still remain pluripotent and hence capable of responding to various different local environmental cues by expressing different phenotypes.

As mentioned previously, there seems to be a correlation between cellular dispersion and melanogenesis, or cellular aggregation and neurogenesis *in vivo* (Cowell and Weston, 1970; Weston, 1971). When nascent (3½-4 day) and to some extent, older (5-6 day) chick embryo sensory ganglia are explanted *in vitro*, moreover, the cells in the explants soon disperse on the substrate. Pigment cells ultimately appear in the periphery of these cultures (Peterson and Murray, 1955; Cowell and Weston, 1970; Weston and Kaplan, 1972), and under some conditions in culture, sympathetic neurons also appear (Weston, unpublished observations). In general, neurogenesis in these cultures is confined to the central area of the explant where the cells remain compact, while in all but the youngest ganglia (or under experimental conditions where the explanted ganglion disperses completely) pigment cells appear only in the periphery where the cells are dispersed. Cowell and Weston (1970) have demonstrated, moreover, that the ultimate appearance of melanocytes in these ganglion cultures requires that some developmentally significant events, involving DNA synthesis and perhaps some DNA-dependent RNA synthesis, occur during the first few days in culture. Whatever these events are, they are correlated with cell dispersion, and cannot be explained simply in terms of the dilution of some intracellular inhibitor of melanogenesis (cf. Peck, 1964; Weston and Kaplan, 1972). These results are consistent with the idea that at least some cells in nascent sensory ganglia remain pluripotent, like crest cells, and still retain the ability to differentiate in response to specific cues present in the particular culture environment.

Conversely, in our cultures, ganglion cells will undergo neurogenesis in response to exogenous NGF only when cell contact can be established and maintained (Weston and Kaplan, 1972). Thus, when Stg. 26-28 (ca. 5-day) chick embryo sensory ganglia were dissociated with 0.1% Trypsin, and the cell suspensions plated at various cell densities, unequivocal differences in differentiative behavior were observed at high and low plating densities (cf. Table 1). Neurogenesis was never observed in low density cultures whether NGF was present initially or not. At higher cell densities in the initial absence of NGF, cells tended to remain dispersed, and neurogenesis was modest. In the initial presence of NGF, however, cells quickly reassociated and underwent neurogenesis, so that ultimately much of the culture surface was covered with a network of nerve fibers, cell bodies, and glial cells. This network varied from a condition of cellular confluency to a condition of dense nerve cell clusters connected by large nerve fiber trunks. Melanogenesis occurred at virtually all cell densities both in the presence and absence of NGF. At high cell densities in the initial presence of NGF, however, melanogenesis was poorer than in controls without NGF. These results are consistent with the observations that neurogenesis occurs only in coherent clusters of crest cells *in vivo*, and in the compact central region of explanted ganglia *in vitro*.

TABLE I

The Effect of Cell Plating Density on Melanogenesis and Neurogenesis by Dissociated 5-day Sensory Ganglion Cells (From Weston and Kaplan, 1972)

Approximate Cell Density (cells/mm^2)	NM → PM		NM + NGF → PM[1]	
	Melanogenesis	Neurogenesis	Melanogenesis	Neurogenesis
1000	+[2]	±	+	++
500	+	+	+	++
250	++	-	++	±
150-200	++	-	++	-
20-130			+	-

[1] Cells were initially cultured 3-4 days in minimal medium (NM), with or without NGF. These cultures were subsequently fed with and maintained in a medium known to permit both neurogenesis and melanogenesis (PM; Cowell and Weston, 1970). Thus, the only major variable in the culture procedure was the presence or absence of NGF during the first few days of culture.

[2] Code reflects relative numbers of pigment or neuronal cells in the culture.

NGF may promote or maintain close cell interaction, but cellular interaction alone is apparently not sufficient for neurogenesis to occur. Thus, when 4-5 day sensory ganglia are cultured on agar so that cellular dispersion is prevented and the coherent ganglion structure is preserved, no nerve fiber outgrowth is observed unless NGF is present. This suggests that NGF may have two separable functions--one to affect morphogenetic cell behavior, and the other to promote neurogenesis in responsive cells. It also suggests that the environmental cues that impinge on pluripotent crest cells ultimately to direct their differentiation may involve complex temporal and spatial sequences of interaction.

4. GENETIC STUDIES OF NEURAL CREST CELL DEVELOPMENT

There are, in the mouse, a number of potentially useful genes that affect neural crest development. These genes may be identified by their characteristic pattern of pleiotropic effects. Thus, because of the known developmental fates of neural crest cells (cf. Weston, 1970), the following patterns or combinations of anomalies offer presumptive evidence that the development of neural crest cells is affected: (1) an altered pattern of pigmentation coupled with lethality or neurological or behavioral disorders; (2) inherited neurological anomalies that occur in conjunction with specific malformations of head cartilage, the face, the inner ear, or teeth; or (3) similar genetically caused morphological anomalies that are associated with alterations in pigment patterns. Using these criteria a number of genes have been identified which are putative neural crest mutants in the mouse, and many of these alleles affecting pigment, neurological, and skeletal phenotypes have been mapped with respect to linkage groups (cf. Sidman et al., 1965; Green, 1966; Weston, 1970). Examples of possible "neural crest genes" in the mouse include alleles such as Splotch (Sp, XIII), Patch (Ph, XVII), Piebald lethal (sl, III) lethal spotting (ls, V) and a number of others (cf. Weston, 1970).

Mutant-bearing organisms obviously have some gene controlled processes altered or inactivated. Suitable comparisons of developing systems in normal and mutant-bearing embryos could allow us to infer the nature of the developmental process under control of specific genetic loci. With very few exceptions, however, only the gross morphological or physiological consequences of pleiotropic mutants affecting neural crest development have been described. The time in development when the neural crest genes begin to function, the earliest detectable anomaly, the primary genetic lesions and the developmental process leading to the expression of the gross condition are largely unknown. The chances of detecting the site or time of gene action might be enhanced, however, if the presence of a given genotype can somehow be predicted in the cells prior to its gross manifestation. In order to recognize such genotypes in very early embryos or in cultured cells or tissues, some independently recognizable marker is necessary. One potentially useful marker to permit the genetic analysis of neural crest development, for example, seems to be the locus for the polymorphic enzyme isocitrate dehydrogenase (Id-1) (Henderson, 1965). This locus serves to mark the adjacent neural crest gene, Splotch, on linkage group XIII (Hutton and Roderick, 1970). Splotch (Sp), which is located approximately 10 cM from Id-1, affects coat pigmentation when in the heterozygous state and is lethal at 13 embryonic days when homozygous. Developmental lesions of the nervous system which precede embryonic death are first detectable in the homozygote at about 9-1/2 to 10 days (Auerbach, 1954). The combination of pigmentary and neurological defects suggest that early neural crest development is anomalous (Auerbach, 1954; Weston, 1970). Since IDH activity can be detected and the isozyme pattern characterized at least by nine days of development (Epstein et al., 1971), it seems likely that this enzyme can be used to predict the Splotch genotype in early embryos prior to developmental lesions characteristic of Splotch homozygotes. This can be done by means of the following mating system:

The strain carrying Sp heterozygotes is homozygous for the $Id\text{-}1^a$ allele (Weston, unpublished). Such mice must be bred to a stock carrying $Id\text{-}1^b$ and the heterozygotes from this mating ($Id\text{-}1^a$ Sp/$Id\text{-}1^b$ +) subsequently intercrossed. With this mating system, embryos showing the A (slow) form of isocitrate dehydrogenase would very likely also be homozygous for Splotch ($Id\text{-}1^a$ Sp/$Id\text{-}1^a$ Sp) and potentially moribund, while embryos with hybrid or B (fast) enzyme variants would be heterozygous carriers ($Id\text{-}1^a$ Sp/$Id\text{-}1^b$ +) or homozygous normal ($Id\text{-}1^b$ +/$Id\text{-}1^b$ +), respectively.

It seems likely that the Splotch gene affects the morphogenetic behavior of crest cells. Since we can identify the genotype of young embryos or their cells, and we can compare adhesive properties of these cells under a variety of conditions, it is now possible to test and perhaps verify this idea, and to establish the causes and consequences of this genetic lesion in some detail.

In general, the neural crest provides a developing system with a number of operationally useful phenotypic markers to characterize cell differentiation. In addition, the role of the cellular environment in promoting the expression of these various phenoytpes is gradually being elucidated. Because these cellular phenotypic markers are available, and we are increasingly able to regulate their expression *in vitro*, it seems feasible also to begin to examine the action of other such genetic lesions known to affect neural crest development. This should provide the opportunity to analyse the time of gene action, its regulation, and its immediate and ultimate consequences in a developing system normally involving morphogenetic behavior, interaction and phenotypic specialization of an embryonic cell population.

REFERENCES

Auerbach, R. (1954), Analysis of the developmental effects of a lethal mutation in the house mouse. J. Exptl. Zool. 127, 305-329.

Bueker, E.D., I. Schenkein and J.L. Bane (1960), The problem of distribution of a Nerve Growth Factor specific for spinal and sympathetic ganglia. Can. Res. 20, 1220-1228.

Cowell, L., and J. Weston (1970), Analyses of melanogenesis in cultured chick embryo spinal ganglia. Develop. Biol. 22, 670-697.

Epstein, C.J., J.A. Weston, W.K. Whitten and E.S. Russell (1972), The expression of the isocitrate dehydrogenase locus (Id-1) during mouse embryogenesis. Devel. Biol. 27 (in press).

Green, M.C. (1966), "Mutant Genes and Linkages." In: Biology of the Laboratory Mouse, 2nd Ed. (ed. E.L. Green). McGraw Hill, New York.

Henderson, N.S. (1965), Isozymes of isocitrate dehydrogenase: subunit structure and intracellular location. J. Exp. Zool. 158, 263-273.

Hörstadius, S. (1950), The Neural Crest; Its Properties and Derivatives in the Light of Experimental Research. Oxford Univ. Press, London.

Hutton, J.J., and T.H. Roderick (1970), Linkage analysis using biochemical variants in mice. III. Linkage relationships of eleven biochemical markers. Biochem. Genet. 4, 339-350.

Levi-Montalcini, R. (1965), The nerve growth factor: its mode of action on sensory and sympathetic nerve cells. Harvey Lect. 60, 217-259.

Levi-Montalcini, R. and P.U. Angeletti (1963), Essential role of the nerve growth factor in the survival and maintenance of dissociated sensory and sympathetic embryonic nerve cells in vitro. Develop. Biol. 7, 653-659.

Levi-Montalcini, R. and P.U. Angeletti (1968), Nerve Growth Factor. Physiol. Reviews, 48, 534-569.

Peck, D. (1964), The role of tissue organization in the differentiation of embryonic chick neural retina. J. Embryol. Exp. Morph. 12, 381-390.

Peterson, E.R. and M.R. Murray (1955), Myelin sheath formation in cultures of avian spinal ganglia. Amer. Jour. Anat. 96, 319-355.

Roth, S.A. and J.A. Weston (1967), The measurement of intercellular adhesion. Proc. Nat. Acad. Sci. U.S. 58, 974-980.

Roth, S., E.J. McGuire, and S. Roseman (1971), An assay for intercellular adhesive specificity. J. Cell Biol. 51, 525-535.

Sidman, R.L., M.C. Green and S.H. Appel (1965), Catalog of the Neurological Mutants of the Mouse. Harvard Univ. Press, Cambridge.

Weston, J.A. (1963), A radioautographic analysis of the migration and localization of trunk neural crest cells in the chick. Develop. Biol. 6, 279-310.

Weston, J.A. (1967), Cell marking. In: Methods in Developmental Biology (F.H. Wilt and N.K. Wessells, eds.). Crowell, New York, 723-736.

Weston, J.A. (1970), The migration and differentiation of neural crest cells. Advances Morph. 8, 41-114.

Weston, J.A. (1971), Neural crest cell migration and differentiation. In: "Cellular Aspects of Growth and Differentiation in Nervous Tissue," (ed. D. Pease) U.C.L.A. Forum in Medical Sciences 14.

Weston, J.A. and S.L. Butler (1966), Temporal factors affecting localization of neural crest cells in the chicken embryo. Develop. Biol. 14, 246-266.

Weston, J.A. and R. Kaplan (1972), The regulation of melanogenesis and neurogenesis in cultured chick embryo sensory ganglia. Devel. Biol. (submitted for publication).

Weston, J.A. and S.A. Roth (1969), Contact inhibition: behavioral manifestations of cellular adhesive properties in vitro. In: Cellular Recognition (R.T. Smith and R.A. Good, eds.). Appleton-Century-Crofts, New York, 29-37.

Winick, M., and R.E. Greenberg (1965), Appearance and localization of a nerve growth-promoting protein during development. Pediatrics 35, 221-228.

Manuscript received after December 1, 1971

SINGLE GENE TRANSLATIONAL CONTROL OF TESTOSTERONE "REGULON"

SUSUMU OHNO
Department of Biology
City of Hope Medical Center
Duarte, California USA

As each gene locus sustains a deleterious mutation at the rate of 10^{-5} per generation, an organism having too many functionally significant gene loci would exterminate itself from unbearable mutation loads. Although the mammalian genome has room for 3 X 10^6 gene loci, the actual number of loci may only be 1/50th of that.

Once it is realized that the difference between mammals and prokaryotes with regard to the number of gene loci is more likely to be 10-fold rather than 1000-fold, it becomes clear that each regulatory system of mammals should contain fewer components than that of prokaryotes. During evolution from unicellular prokaryotes to multicellular eukaryotes of increasing complexity, the necessary increase in number of regulatory systems would have imposed unbearable mutation loads unless the increase was compensated by simplifying each system. The often expressed view that each mammalian regulatory system must be enormously complicated is unsound.

On the basis of our previous findings (Ohno and Lyon, 1970; Bullock *et al.*, 1971; Dofuku *et al.*, 1971; Gehring *et al.*, 1971; Ohno *et al.*, 1971) on the X-linked testicular feminization (*Tfm*) mutation of the mouse (Lyon and Hawkes, 1970), I have reasoned (Ohno, 1971) that the response of target cells to testosterone is solely mediated by this regulatory locus. When it is not bound to testosterone, a protein specified by the wild-type allele of the *Tfm* locus stays in the cytoplasma and serves as a translational repressor of certain enzymes. Binding with testosterone or its metabolites changes its allosteric configuration so that the bound form detaches from these *m*RNAs, releasing a translational block (specific enzyme induction), and moves into the nucleus where it activates RNA polymerase I inside the nucleolar region. This results in increased production of ribosomes; hence, hypertrophy. The *Tfm* mutation represents a loss by this protein of the binding affinity to testosterone and its metabolites; 5α-dihydrotestosterone (DHT) and 5α-androstan-3α-17β-diol. Thus, testosterone elicits no response from target cells of affected *Tfm*/Y.

Two- to three-fold increase in RNA polymerase I activity has been observed in target cells of not only testosterone but also of estrogen, hydrocortisone, etc. How does a steroid hormone activate this class of RNA polymerase? Since mammalian RNA polymerases have been shown to contain a small subunit having the molecular weight of 38,000 (Blatti *et al.*, 1970), the possibility arose that the so-called 3S nuclear receptor protein (Bruchovsky and Wilson, 1968; Fang *et al.*, 1969) when complexed with DHT might activate RNA polymerase I by becoming its sigma-factor-like subunit. This in turn raised the possibility that the 3S nuclear DHT-receptor protein might, in fact, be a protein specified by the wild-type allele of the *Tfm* locus. Indeed, when not bound to DHT, this protein appears to exist in the cytoplasma (Bruchovsky and Wilson, 1968; Fang *et al.*, 1969).

Preliminary results so far obtained on kidney proximal tubule cells of +/Y and *Tfm*/Y mice give an affirmative answer to the first and a negative answer to the second possibility. After the *in vitro* incubation of +/Y kidney homogenate with 3.1 X 10^{-10} M of ^{3}H-DHT for

60 minutes at 30°C, each cell nucleus incorporated about 2000 molecules of DHT. When the solubilized RNA polymerase fraction was extracted from these isolated nuclei according to the method of Roeder and Rutter (1969), about 80% of the ^{3}H-DHT counts moved into the solubilized fraction. After prolonged (24 hrs) dialysis, up to 70% of the DHT count was lost from the fraction. The lost count represented unbound as well as loosely protein-bound DHT. The dialyzed fraction was then loaded on a Sephadex G-200 column equilibrated with 0.05 M tris-HCl buffer, pH 7.9, containing glycerol, $MgCl_2$, EDTA and dithiothreitol to protect polymerases (Roeder and Rutter, 1969). All the protein-bound ^{3}H-DHT eluted as a single sharp peak which exactly coincided with an RNA polymerase I activity peak. No ^{3}H-DHT peak was seen in the 3S region. Thus, it appears that the complex between DHT and 3S nuclear receptor protein becomes an independent entity only in high salt concentration such as 0.6 M NaCl or 0.4 M KCl (Bruchovsky and Wilson, 1968; Fang *et al.*, 1969). In the native state, it exists as a subunit of RNA polymerase I.

When the same experiment was done on *Tfm*/Y kidney cell nuclei, however, no difference was found between +/Y and *Tfm*/Y with regard to the nuclear incorporation of DHT, the amount of protein-bound DHT or the activation of RNA polymerase I by DHT as shown in Fig. 1.

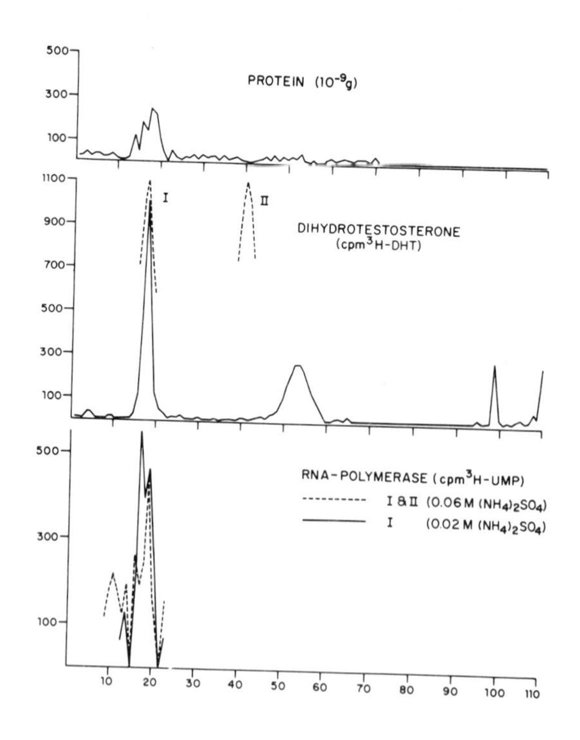

From the above, we tentatively concluded that the 3S nuclear DHT-receptor protein is not specified by the *Tfm* locus and that *Tfm*-protein is not involved in activation of RNA polymerase I.

In the case of *Tfm*/Y kidney, however, hypertrophy does not occur despite three-fold increase in nuclear RNA polymerase I activity. It may be that a protein specified by the *Tfm* locus functions only as a translational repressor, therefore, its influence is confined to the cytoplasma. If some of the ribosomal proteins in target cells are under translational control by the *Tfm* locus, increased transcription of ribosomal RNA genes in *Tfm*/Y cells would not result in increased production of ribosomes.

REFERENCES

Blatti, S.P., C.J. Ingles, T.J. Lindell, P.W. Morris, R.F. Weaver, F. Weinberg, and W.J. Rutter, (1970) Structure and regulatory properties of eucaryotic RNA polymerase. Cold Spring Harbor Symp. Quant. Biol. 35, 649.

Bruchovsky, N., and J.D. Wilson, (1968) The intranuclear binding of testosterone and 5α-androstan-17β-ol-3-one by rat prostate. J. Biol. Chem. 243, 5953.

Bullock, L.P., C.W. Bardin, and S. Ohno, (1971) The androgen insensitive mouse: Absence of intranuclear androgen retention in the kidney. Biochem. Biophys. Res. Comm. 44, 1537.

Dofuku, R., U. Tettenborn, and S. Ohno, (1971) Testosterone-regulon in the mouse kidney. Nature New Biology 232, 5.

Fang, S., K.M. Anderson, and S. Liao, (1969) Receptor proteins for androgens. On the role of specific proteins in selective retention of 17β-hydroxy-5α-androstan-3-one by rat ventral prostate *in vivo* and *in vitro*. J. Biol. Chem. 244, 6584.

Gehring, U., G.M. Tomkins, and S. Ohno, (1971) Effect of androgen-insensitivity mutation on a cytoplasmic receptor for dihydrotestosterone. Nature New Biology 232, 106.

Lyon, M.F., and S.G. Hawkes, (1970) An X-linked gene for testicular feminization in the mouse. Nature 227, 1217.

Ohno, S., and M.F. Lyon, (1970) X-linked testicular feminization in the mouse as a *noninducible* regulatory mutation of the Jacob-Monod type. Clin. Genet. 1, 121.

Ohno, S., R. Dofuku, and U. Tettenborn, (1971) More about X-linked testicular feminization of the mouse as a noninducible (i^s) mutation of a regulatory locus: 5α-androstan-3α-17β-diol as the true inducer of kidney alcohol dehydrogenase and β-glucuronidase. Clin. Genet. 2, 1.

Ohno, S., (1971) An argument for the simplicity of mammalian regulatory systems: Single gene determination of male and female phenotypes. Nature in press.

Roeder, R.G., and W.J. Rutter, (1969) Multiple forms of DNA-dependent RNA polymerase in eukaryotic organisms. Nature 224, 234.

CELL-TO-CELL CONNECTIONS

WERNER R. LOEWENSTEIN
Department of Physiology and Biophysics
University of Miami School of Medicine
Miami, Florida USA

I shall deal here briefly, with a basis for cellular communication that is present in most organized tissues (Loewenstein, 1966). It consists of a system of passageways connecting adjoining cells. Each passageway unit consists of a matched pair of membrane regions of high permeability (junctional membranes), one region on either side of the cell junction, and a seal (junctional seal) insulating the interior of the connected cell system from the exterior (Loewenstein, 1966). Little is as yet known about the structure of the passageways or even their dimensions, but their reality is well assured by electrical measurements with intracellular probes and by studies with tracer substances injected into cells. In at least one cell junction, molecules up to the order of 10^4 weight pass through (Kanno & Loewenstein, 1966); probably many parallel passageway units make up a communicating junction.

Communicating junctions form rapidly where the cell membranes come into contact. A large part, perhaps all, of the cell surface membrane is capable of making communicative junctions: when a cell junction is broken, the membranes seal; a new communicative junction is readily formed by putting the cells into contact at other spots (Loewenstein, 1967; Ito & Loewenstein, 1969). The calcium ion concentration on both sides of the membrane plays an important role in the permeability transformations associated with the making and breaking of the junction (Loewenstein, 1968b).

The functional role of junctional communication is not known. This communication, which is widespread among embryonic tissues (see Furshpan & Potter, 1968), is, for several reasons, a good candidate for the close-range cell interactions associated with normal growth and differentiation: (a) the interior of the connected cell system is continuous for a wide range of cytoplasmic molecules; (b) the system has a sharp diffusion boundary (the nonjunctional membranes plus the junctional seals) and a finite volume, and, hence, it has the potential of obtaining information on the number and position of its cell members on the basis of simple clues of chemical concentration (Loewenstein, 1968b).

In the absence of knowledge of the relevant communication signals, we have explored the possibility of an involvement of the junction in growth and differentiation by searching for defects in junctions in cells with uncontrolled (cancerous) growth. The approach was guided by the idea that if the junction is indeed conveying growth-controlling signals, interruption of the signal flow by genetic junctional defect (noncoupling) should lead to cancerous growth (Loewenstein, 1968a). It is, of course, a priori, extremely unlikely that the many forms of cancer should all have the same cause (in fact, several cancerous tissues present junctional communication, Potter _et al._, 1966; Borek _et al._, 1969; Sheridan, 1970). But since the junction is a vulnerable bottleneck (Loewenstein _et al._, 1967), we hoped that noncoupling might be a cause frequently enough to give a reasonable chance of finding some noncoupling kinds of cancer cells.

The first indications for noncoupling were found in certain malignant epithelial tumors (Loewenstein & Kanno, 1967; Jamakosmanovic & Loewenstein, 1968; Kanno & Matsui, 1968) and in certain cancerous epithelial cells in culture (Borek et al., 1969). Conclusive evidence of noncoupling has just been obtained in a cancerous strain of cells in culture, derived from a liver tumor with a method that avoids probing into the cancerous cells (Azarnia & Loewenstein, 1971). These cancerous cells fail to make communicative junctions with several types of normal cells (which make such junctions readily even with cells of different type and genus (Michalke & Loewenstein, 1971) (Fig. 1). They also appear to be incapable of junctional communication among themselves.

These new results are encouraging. It is now feasible to inquire into the correlation between the junctional and the growth defects with simple genetic techniques, and this Azarnia and I are now in the process of doing.

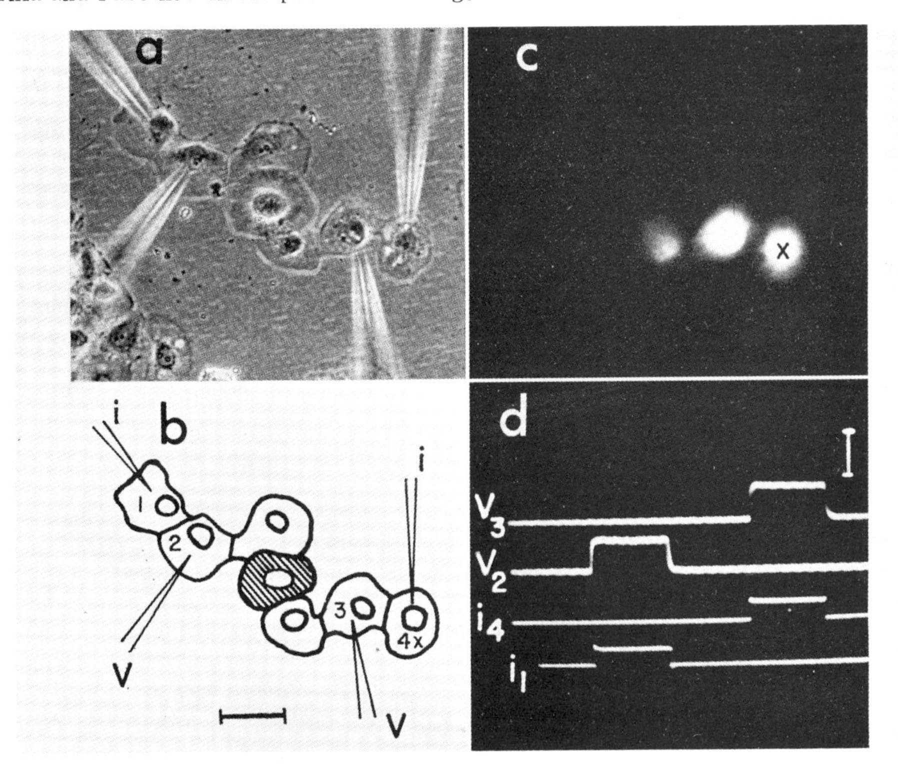

Fig. 1. A cancerous cell that fails to make communicative junction. a, phase contrast photomicrograph of two chains of normal liver cells in culture bridged by a cancerous cell (hatched in the tracing in b). Electric current pulses (i, 3×10^{-8} amp, 100 msec duration) are injected into cells 1 and 2 (100 msec delay) with microelectrodes and the resulting voltages (V) are recorded inside cells 2 and 3 (circuits are closed via the grounded exterior). In d are the corresponding oscilloscope records (voltage calibration, 500 mV). The records show that the normal cells are coupled on either side of the cancer bridge, but not across the bridge. Simultaneous with the electric measurement, fluorescein (330 molecular weight) was injected into cell 4. The darkfield micrograph of the fluorescence in c shows that the molecule passes from normal to normal, but not from normal to cancerous cell. (From Azarnia & Loewenstein, 1971.)

REFERENCES

Azarnia, R. and Loewenstein, W.R., (1971), Intercellular communication and tissue growth. V. A cancer cell strain that fails to make permeable membrane junctions, J. Membrane Biol. 6

Borek, C., Higashino, S. and Loewenstein, W.R., (1969), Intercellular communication and tissue growth. IV. Conductance of membrane junctions of normal and cancerous cells in culture, J. Membrane Biol. 1, 274.

Furshpan, E.J. and Potter, D.D., (1968), Low resistance junctions between cells in embryos and tissue culture, Curr. Topics Devel. Biol. 3, 95.

Ito, S. and Loewenstein, W.R., (1969), Ionic communication between early embryonic cells, Devel. Biol. 19, 228.

Jamakosmanovic, A. and Loewenstein, (1968), Cellular uncoupling in cancerous thyroid epithelium, Nature 218, 775.

Kanno, Y. and Loewenstein, W.R., (1966), Cell-to-cell passage of large molecules, Nature 212, 629.

Kanno, Y. and Matsui, Y., (1968), Cellular uncoupling in cancerous stomach epithelium, Nature 218, 775.

Loewenstein, W.R., (1966), Permeability of membrane junctions, Conf. on Biol. Membranes: Recent Progress, Ann. N.Y. Acad. Sci., 137, 441.

Loewenstein, W.R., (1967), On the genesis of cellular communication, Devel. Biol. 15, 503.

Loewenstein, W.R., Nakas, M. and Socolar, S.J., (1967), Junctional membrane uncoupling. Permeability transformations at a cell membrane junction, J. Gen. Physiol. 50, 1865.

Loewenstein, W.R., (1967), Cell surface membranes in close contact. Role of calcium and magnesium ions, J. Colloid Interface Sci. 25, 34.

Loewenstein, W.R., (1968a), Some reflections on growth and differentiation, Perspectives in Biol. and Med. 11, 260.

Loewenstein, W.R., (1968b), Communication through cell junctions. Implications in growth and differentiation, Devel. Biol. 19, (Sup.) 151.

Loewenstein, W.R. and Kanno, Y., (1967), Intercellular communication and tissue growth. I. Cancerous growth, J. Cell Biol. 33, 225.

Michalke, W. and Loewenstein, W.R., (1971), Communication between cells of different type, Nature, 232, 121.

Oliveira-Castro, G.M. and Loewenstein, W.R., (1971), Junctional membrane permeability: effects of divalent cations, J. Membrane Biol. 5, 51.

Potter, D.D., Furshpan, E.T., and Lennox, E.J., (1966), Connections between cells of the developing squid as revealed by electrophysiological methods, Proc. Nat. Acad. Sci. 55, 328.

Sheridan, J.D., (1970), Low resistance junctions between cancer cells in various solid tumors, J. Cell Biol. 45, 91.

Manuscript received after December 1, 1971

CELLULAR INTERACTIONS IN SLIME-MOULD AGGREGATION*

A. ROBERTSON, MORREL H. COHEN, D.J. DRAGE, A.J. DURSTON, J. RUBIN AND D. WONIO
Department of Theoretical Biology
University of Chicago
939 E. 57th St.
Chicago, Ill. 60637

1. INTRODUCTION

Cellular slime mould[1] amoebae live in damp situations feeding on bacteria.[2] When the food supply is exhausted the amoebae aggregate by chemotaxis towards centres which secrete an attractant.[2] In some species (e.g. Dictyostelium minutum) a fruiting body consisting of a mass of spores supported by a multicellular stalk is made on the spot, directly from the aggregate. In others there is an intervening period of migration, either by a freely moving multicellular organism, a slug (e.g. D. discoideum), or by a slug which produces a continuous stalk (e.g. Polysphondylium violaceum).[1]

The life cycle therefore contains stages during which the amoebae exist as free individuals and stages during which they are joined to form a multicellular organism. The life cycle includes all of the processes[4] found at the cellular level in the development of Metazoa: growth, division, contact formation and breaking, movement, secretion and differentiation. In the Metazoan the complexity of the adult is produced by many repetitions of these basic processes. Even so, development is highly reliable, which suggests that there are systems, operating at the multicellular level, to control the occurrences and courses of the basic cellular processes.[4,5,6,7] Therefore, it is important to understand the nature and mechanisms of such developmental control systems before we can begin to understand the mechanism of morphogenesis. The system controlling aggregation in the cellular slime moulds is a prototypic developmental control system.

In this paper we shall confine our discussion to aggregation, and the period of interphase[8] between feeding and aggregation, as they occur in D. discoideum. In particular we want to find out how the aggregative signal is produced and propagated, and how it controls aggregation. We therefore describe our understanding of the signal and the experimental techniques we are using for its analysis.

The attractant used by D. discoideum amoebae is almost certainly cyclic AMP.[9] At the end of interphase some amoebae begin, autonomously, to produce approximately periodic pulses of cyclic AMP. The period is about 5 min. Each pulse contains about 3 x 10^9 molecules and is short, lasting for less than 2 sec.[10] The cyclic AMP diffuses away from a signalling amoeba. Its amplitude is reduced both by diffusion and by the action of an extracellular phosphodiesterase[11] secreted by the amoebae. If a sensitive amoeba detects a supra-threshold concentration of cyclic AMP at a portion of its anterior membrane surface, it responds by releasing its own pulse of cyclic AMP, after a 15 sec. delay[10], probably from its posterior end, initiating from its anterior end a step of movement lasting for about 100 sec. towards the signal source.[12] Once signalled it remains refractory[12,13,14] to further stimulation for about 2 minutes. These features of the signalling and response mechanisms lead to the outward propagation of a wave of inward movement with respect to the original signal source, provided the field of sensitive amoeba is above a critical density[10] below which the signal cannot propagate. This critical density corresponds to an inter-amoeba spacing of about 60μ, which in turn corresponds to a diffusion time, for cyclic AMP, of less than 2 sec. Therefore, the rate-limiting step in signal propagation is the 15 sec. intracellular delay.[10]

Because each amoeba acts as a local repeater of the signal, individuals tend to move first towards their nearest neighbours rather than directly towards the original signal source.[15] The amoebae therefore form streams in which polar contacts are made.[16] The front, sensitive and motile, end of an amoeba makes contact with the rear, signalling, end of its anterior neighbour in the stream. Stream formation will occur whenever each amoeba repeats the signal, whether the signal is a pulse or is produced steadily leading to a gradient[17] of an attractant. In those species such as D. minutum which produce aggregates with circular peripheries, there is clearly no signal propagation by the free amoebae.

We have made a detailed theoretical analysis of wave propagation during early aggregation and have used three experimental techniques to elucidate the signal and response mechanisms. These are: grafting the tip,[18] which acts as an organiser of development after the early stages of aggregation, into fields of amoebae during interphase (section II); introducing an artificial signal source into the field (section III)[19]; and examining mutants selected for failure to aggregate (section IV)[13,20,21]. Combining the results of these with deeper mathematical analysis has enabled us to relate the morphologies of aggregation to the general features of the control system (section V).

Through the entire development cellular interactions play a central role, in communicating by secretion of, e.g.,cAMP and in specific and non-specific contact formation.

II. TIP TRANSPLANTS

Spemann and Mangold[22] showed that the dorsal lip of the newt blastopore is an organiser: on being grafted into a second gastrula, the dorsal lip can induce a secondary embryonic axis. Raper[18] showed that the tip of the D. discoideum slug has an analogous function, and we have now found that tips from all developmental stages are capable of acting as organisers.[23] This is important in two ways. It improves the analogy between the slime-mould developmental control system and the control systems of more complex embryos, and it gives a technique for examining signal propagation from a natural source which is easy to recognize and to manipulate.

Tips were removed from late aggregates, coni, slugs, and fruiting bodies of D. discoideum and transplanted to fields of interphase and post interphase amoebae, and to aggregates, slugs, and fruiting bodies both with and without tips.[24] The tips used in the transplants were removed by means of a capillary tube which had been drawn into a sharp point. The point was placed at the "junction" of the tip with the rest of the structure and was gently manipulated until the tip separated and could be removed. In a developmental stage in which the junction between the tip and "body" of the organism is clearly defined as in the fruiting body or late slug, the decision as to where to remove the tip was easily made. However in the case of the earlier slug or the conus, where the distinction was not as clear, the separation was made as close to the end of the specimen as possible.

Tips transplanted into fields of amoebae never began to attract them until approximately three to four hours from the beginning of interphase. The amoebae in the field became able to relay the signal from the tip at about five to six hours and to form streams towards the tip at about eight hours from the beginning of interphase. Tips from all stages showed the same ability to produce this sequence of responses. In those cases where periodicity is normally easy to observe, aggregation and fruiting, it was observed in the transplant experiments. We therefore think that the tips continue to secrete an attractant, most probably cAMP, periodically throughout development. This is good circumstantial evidence for the persistence of a single developmental control system.

As shown in the table (Table 1) tips from all stages can organize development of all stages. The graft, as shown by Raper[18] in the slug, must preserve polarity: apparently the tip can only signal backwards along the anterior-posterior axis of the organism. Further, when the graft is made into a late stage of development, the base of the tip must be pushed through the slime sheath which surrounds the slug or the early fruiting body. If the slime sheath is not disrupted, the tip cannot signal through it and develops autonomously, regulating to produce a small fruiting body. If the tip of the host is left in place, two developmental axes may emerge, or the grafted tip may lose its organization and become assimilated by the

Table 1

Results of Tip Transplant Experiments*

Field \ Tip	Aggregation Center	Conus	Slug	Young fruiting body
Vegetative				
Interphase				
Age at time of Transplant: 0-1 hours	+		+	+
2-3 hours		+		
4-5 hours				+
6-7 hours			+	+
8-9 hours			+	
Post Interphase	+		+	+
Aggregation with streams	+	+	+	+
Conus		+	+	+
Slug	+	+	+	+
Fruiting body		+	+	+

*A "+" entry indicates that a tip transplanted from the specified stage of development organized the development of at least a portion of the field into which it was transplanted. No entry indicates that the transplant has not yet been performed.

host. When a second axis was organized, the time course of organization was standard. The transplanted tip required approximately 30-40 minutes to set up its own field of influence and another 30 minutes were required for two new autonomous structures to separate. When the host tip has been removed, the grafted tip organizes the host, and normal development ensues with the grafted tip in control. In particular, the clear periodic movements characteristic of fruiting-body erection are always present.

The function of the tip as an organizer throughout development is therefore established. We are continuing this series of experiments in order to determine the quantitative details of tip function and to compare them with those of the pulser described in section III.[19] We are using tips stained red by growing D. discoideum amoebae on bacterial cultures containing Serratia mariescens. The detailed results will be published elsewhere.

III. THE PULSER

As it is very likely that cyclic AMP is the aggregative attractant for D. discoideum,[9] we decided to use a microelectrode to release pulses of cyclic AMP electrophoretically. In this way we hoped to mimic the natural signal with a system giving us precise control over signal amplitude and time course and allowing signal release from an aperture smaller in diameter than a single cell. The microelectrode technique should also make it easy to screen other possible attractants. It is particularly useful because it gives a much better defined signal geometry than the tests of Shaffer[25], Konijn[26] or Bonner.[27]

In our experiments a glass micropipette with an internal tip diameter of 2-5μ is filled with a buffered salt solution containing millimolar cyclic AMP and fluorescein. The fluorescein is used as a marker in setting up the system and in particular to allow monitoring of the bias potential used to hold negative ions within the electrode. The microelectrode is held so that its tip is within the aqueous film above the agar substrate on which amoebae are plated out. The bias is reversed periodically, the electrode tip being made negative with respect to the agar, which is held at ground potential. Therefore, negatively charged ions including cyclic AMP are driven out of the electrode. It is easy to vary the period, amplitude and duration of the cyclic AMP pulse. In our first experiments we used a 265 sec period and a 2 sec. pulse with an amplitude chosen to be approximately 100 times larger than the pulse from a single amoeba.

When the electrode was put into position in a field of amoebae that had been centrifuged 1 1/2 hours earlier there were no responses. At 4 hours after centrifugation some cells were attracted to the electrode tip and formed a loose circular cluster around it. The cluster had a diameter of about 200μ and was reminiscent of the "clouds" seen during normal interphase.[15] These probably form when one or a few cells start to secrete cyclic AMP continuously, and their neighbours have differentiated far enough to respond to, but not to relay, the signal.[17] The fixed size of the cloud shows that there is a threshold for the movement response to cyclic AMP.[17] At about 6 hours after centrifugation some signals were relayed, and by 8 hours proper aggregation was under way with streams leading to the aggregation centre controlled by the microelectrode. These results show that the competences required for aggregation develop in a specific order. They are: ability to detect cyclic AMP and to make a chemotactic response to it; ability to relay the cyclic AMP signal; ability to form intercellular contacts as in streams; and ability to signal autonomously. We are not concerned with subsequent development here, but clearly the list of competences could be extended throughout the life cycle using similar techniques.

At present we are investigating the normal frequency and amplitude range of the signal by filming responses to the pulser. The results so far indicate that the refractory period shortens during interphase from more than six minutes to about two and that the theoretically estimated pulse size for a single amoeba is correct within an order of magnitude. The range of entrainment therefore extends up to a frequency of one pulse every two minutes, while the autonomous frequency of the amoebae has a mode of one pulse every five minutes, setting a lower limit.

IV. MUTANTS

Several mutant strains which fail to aggregate or have abnormal aggregation are now known.[13,20,21] These have varying morphologies and also vary in their ability to attract and to be attracted by wild type centres.[13,20]

However, it is not sufficient merely to classify a mutant as aggregateless or aggregation-abnormal. The behaviour of each must be recorded by time-lapse cinemicrography, and the pulser can be used to determine whether or not the mutant strain can detect and relay a cyclic AMP signal. Filming has already shown that there are many types of chemotactic mutants. The pulser and tip transplants should resolve these even more precisely.

We have therefore collected and manufactured aggregateless and other mutants in order to film their undisturbed behaviour and to record their responses to the pulser and transplanted tips. In this way, we hope to establish the specific functional block in each mutant strain, to establish details of the signalling and response systems and possibly to begin a real genetic analysis of the developmental control system. The blocks so far discovered fall into four general classes: those leading to failure to make the normal movement response; failure to relay signals; failure to make autonomous signals; and failure to form normal intercellular contacts. Table 2 shows the results of our preliminary analysis of those mutants we have filmed.

Table 2

Some Aggregation Defective Mutants

Mutant	Category	Defect	Morphology
aggr 50	agg -	no movement response to or relaying of signals	single cells or clumps
73	agg -		
1	agg -	no autonomous signals	loose mounds of cells
10	aggregation abnormal	non polar movement response	normal
66	aggregation abnormal	has dispersive and aggregative phases during each signal period	

Mutant 10, for example, is not strongly attracted to wild type centres. Time-lapse films show that signal production and relaying are at least superficially normal in this mutant, but that the movement response is not. During early development, each amoeba shows pseudopodal activity around the entire periphery of its membrane when signalled. The effect is random jiggling of the cell with no net displacement each time it is signalled. The field of amoebae propagates waves of quivering. Later, cells develop a polar, but still abnormal movement response. It therefore seems that 10 has defective polarity, as regards its movement response. Normally, in aggregating wild type cells, only one end of the cell can produce pseudopods, probably because the rest of the membrane has become structurally modified during interphase. We do not know whether this lack of polarity is also reflected in the signal propagation mechanism, nor do we know whether the amoebae can make polar intercellular contacts.

V. INTERPRETATION OF MORPHOLOGIES

It is obvious that the morphologies produced during interphase and aggregation reflect properties of the signalling and response systems. We have discussed interphase and aggregation in detail elsewhere, so we shall only mention two examples here.

There are[17] two fundamentally different kinds of aggregate - those with basically circular, smooth peripheries and those with streams.[1] Streaming is a consequence of the instability of the radial inward flow of aggregating amoebae to azimuthal perturbations. It can only occur when each amoeba acts as a signal relaying source, and is thus an attractive instability. When only the centre, initially a single founder cell, secretes, the radial inward flow is stable to azimuthal perturbations and the aggregate remains round. Stream formation is diagnostic of signal relaying, be the signal continuous or periodic.[17]

Interphase morphology can undergo quite complex variations. In D. discoideum "clouds" form[15] typically during the fourth and fifth hours from centrifugation. We have interpreted these as the result of the activity of some amoebae which have developed the competence to produce cyclic AMP, but cannot yet secrete it, or cut off its production once sufficient has been made for a single pulse. Thus the competence to signal has several components which also develop in a sequence.[17]

The observed normal morphologies of D. discoideum are well reproduced in the pulser experiments, which allows us to understand them quantitatively.[19]

The various slime mould species show a range of aggregation morphologies.[1] These are discussed in our earlier paper,[17] but the important point is that one can tell by inspection whether signal relaying occurs. It must do in D. discoideum and D. mucoroides, but not in D. minutum, and it must also occur in Polysphondylium violaceum and P. pallidum, as well as in D. lacteum, D. rosarium, and Acytostelium leptosomum, as all these species have well-defined streams. D. lacteum is an interesting intermediate as it relays, but its signal is continuous. This is presumably also the case in Gerisch's streaming mutant of D. minutum.[13] In wild-type D. minutum cells are stimulated to secrete within about an hour of making contact with a centre.[13] If this time were reduced significantly, streams would form, as indeed happens in very dense cultures of wild-type cells.

Morphologies of aggregateless mutants of D. discoideum are also extremely interesting and can be related to their signalling and response capacities. The morphologies are listed in Table 2.

Finally, more subtle morphological differences among species which show streaming can be understood by combining observations on the sequence of development of competences during interphase and a theoretical analysis of the dynamics of signal propagation with the above instability arguments.[17] The density of amoebae in an interphase population which are competent to relay increases monotonically with time. When it reaches a critical value, the percolation density, long range signal propagation over the interphase field becomes possible. If the first autonomous signalling centers emerge well after this critical density is reached, the field propagates the waves densely and quasicontinuously. Many streams then form around each center as in D. discoideum. If the autonomous signalling centers emerge before the critical density is reached, signal propagation starts when it is reached and occurs along the percolation channel containing the center. Amoebae which are chemotactically sensitive but cannot yet relay move into the channel, reducing the density outside the channel, and confining signal propagation to the channel itself. This leads to the formation of one or two giant streams as in P. violaceum.

VI. DISCUSSION AND CONCLUSIONS

The observations and experiments that we have described all show that the cellular slime moulds develop using processes akin to those in the Metazoa. In particular the tip developed in the late aggregation of D. discoideum behaves like a classical organiser. Understanding the D. discoideum developmental control system is therefore important in a wider context than just that of the development of cellular slime moulds. As far as we know the slime-mould morphogenetic control system is the one best understood in developmental biology, but there is much evidence that similar systems exist in Metazoa, particularly for the control of morphogenetic movement.[4,6,13] There is also indirect evidence that neural transmitters are involved in the control of morphogenetic movement - for example in sea-urchin gastrulation.[28]

We therefore feel that the pulser provides a technique that may be quite generally useful in investigating developmental control systems, both by mimicking known organisers and by controlling the local movements of isolated cells, e.g., the migration of the chick precardiac mesoderm.

Two examples of what may be called "developmental control systems" were considered by other speakers at this symposium. These are the circadian clock of Drosophila (Benzer)[29], and the system involving the T-locus of the mouse which specifies some components of the cell surface (Bennett, Boyse, and Old).[30] Both these systems affect development in a comprehensive way and both can be analysed genetically, as can the slime mould system. It should therefore be possible to dissect all these systems into their components and to understand their mechanisms - at least in principle. In understanding how a control system works during development it is important to remember that the mechanism has evolved by natural selection acting on its products. It may be difficult to argue on solely logical grounds: there are many ways in which a particular result might be brought about during development, but only one of those can in fact exist.

The connection between morphologies of aggregation and the nature of the developmental control system is also of wider significance. We have pointed out that local signal production gives rise to streaming morphologies whereas production of the signal by the centre gives rise to smooth morphologies. Now cancer may be regarded as a disease of developmental control. The basic cellular processes involved are differentiation, growth, and division. We can conclude from the general relationships between morphological instabilities and structure of the control system that underlie the results for the slime moulds, that if cooperative effects occur within a tumor it will have an irregular periphery whereas the periphery would be smooth in their absence. By cooperative, we mean differentiation, growth, or division of a cancerous cell enhancing that of its neighbours.

Manuscript received after December 1, 1971

REFERENCES

1. J. T. Bonner. (1967) The Cellular Slime Molds. Princeton: the University Press.
2. G. Potts. (1902) Flora 91, 281-347.
3. J. T. Bonner. (1947) J. Expt. Zool. 106, 1-26.
4. M. H. Cohen and A. Robertson. (1971) Proc. Sixth IUPAP Conf. Stat. Mech., eds. S. A. Rice, K. Freed, and J. Light, Chicago; University of Chicago Press.
5. L. Wolpert. (1969) J. Theoret. Biol. 25, 1-47.
6. A. Robertson and M. H. Cohen. (1972) Ann. Rev. Biophys. Bioengineering, 1, (in press).
7. B. C. Goodwin and M. H. Cohen. (1969) J. Theoret. Biol. 25, 49-107.
8. J. T. Bonner. (1963) Symp. Soc. Exptl. Biol. 17, 341-358.
9. T. M. Konijn, D. S. Barkley, Y. Y. Chang, and J. T. Bonner. (1968) Am. Naturalist 102, 225-233.
10. M. H. Cohen and A. Robertson. (1971) J. Theoret. Biol. 31, 101-118.
11. Y. Y. Chang. (1968) Science 160, 57-59.
12. M. H. Cohen and A. Robertson. (1971) J. Theoret. Biol. 31, 119-130.
13. G. Gerisch. (1968) Current Topics in Dev. Biol. 3, 157-197.
14. B. M. Shaffer. (1957) Am. Naturalist 91, 19-35.
15. B. M. Shaffer. (1962) Advan. Morphogenesis 2, 109-182.
16. B. M. Shaffer. (1964) in Primitive Motile Systems in Cell Biology. pp.387-405, eds. R. D. Allen and N. Kamiya, New York: Academic Press.
17. M. H. Cohen and A. Robertson. (1972) Proc. 1st Intl. Conf. on Cell Differentiation, eds. R. Harris and D. V:za Copenhagen: Munksgaard (in press).
18. K. B. Raper. (1940) J. Elisha Mitchell Scient. Soc. 56, 241-282.
19. A. Robertson, D. J. Drage and M. H. Cohen. (1971) Science (in press).
20. A. J. Durston, unpublished observations.
21. M. Sussman. (1955). J. Gen. Microbiol. 13, 295-309.
22. H. Spemann and H. Mangold. (1924) Arch. Mikrosk. Anat. Entwmech. 100, 599-638.
23. A. Robertson. (1972) in Lectures on Mathematics in the Life Sciences, 4, ed. J. D. Cowan. Providence: Amer. Math. Soc. (in press).
24. J. Rubin and D. Wonio, unpublished observations.
25. B. M. Shaffer. (1956) J. Exptl. Biol. 33, 645-657.
26. T. M. Konijn. (1968) Biol. Bull. 134, 298-304.
27. J. T. Bonner, A. P. Kelso, and R. G. Gillmor. (1966) Biol. Bull. 130, 28-42.
28. T. Gustafson and M. Toneby. (1970) Exptl. Cell Res. 62, 102-117.
29. R. J. Konopka and S. Benzer. (1971) Proc. Nat. Acad. Sci. U. S. A. 68, 2112-2116.
30. D. Bennett. (1964) Science 144, 263-267.

LIST OF INVITED ATTENDANTS

A. C. ALLISON, Clinical Research Centre, Harrow, Middlesex, England.

B. ANDERSSON, Institute for Tumor Biology, Karolinska Institutet, S. 104 01 Stockholm 60, Sweden.

G. ATTARDI, California Institute of Technology, Division of Biology, Pasadena, California 91109, U.S.A.

D. BENNETT, Dept. of Anatomy, Cornell University Medical College, New York, N.Y. 10021, U.S.A.

S. BENZER, California Institute of Technology, Division of Biology, Pasadena, California 91109, U.S.A.

H. BLOMGREN, Karolinska Institutet, Institute for Tumorbiology, 104 01 Stockholm 60, Sweden.

E. A. BOYSE, Sloan Kettering Institute for Cancer Research, New York, N.Y. 10021, U.S.A.

S. BRENNER, Laboratory of Molecular Biology, Medical Research Council, Cambridge, England.

M. BURGER, Dept. of Biology, Princeton University, Princeton, New Yersey 08540, U.S.A.

G. BURNSTOCK, Dept. of Zoology, University of Melbourne, Parkville, Victoria 3052, Australia.

A. BUZZATI-TRAVERSO, Unesco, Paris, France.

M. CALVIN, University of California, Laboratory of Chemical Biodynamics, Berkeley, California 94720, U.S.A.

H. CANTOR, National Institute for Medical Research, London N.W. 7, England.

H. CLAMAN, University of Colorado Medical School, Denver, Colorado 80220, U.S.A.

J. J. COHEN, National Institute for Medical Research, London N.W.7, England.

M. H. COHEN, James Franck Institute, The University of Chicago, Chicago, Illinois 60637, U.S.A.

M. COHN, The Salk Institute for Biological Studies, San Diego, California 92112, U.S.A.

F. CRICK, Medical Research Council, Laboratory of Molecular Biology, Cambridge, England.

M. CRIPPA, C.N.R. Laboratory of Molecular Embriology, 80072 Arco Felice (Napoli), Italy.

B. CUNNINGHAM, The Rockefeller University, New York, N.Y. 10021, U.S.A.

G. DORIA, Gruppo CNEN-EURATOM di Immunogenetica, Centro Studi Nucleari della Casaccia, Roma, Italy.

J. E. DOWLING, Biological Laboratories, Harvard University, Cambridge, Mass. 02138, U.S.A.

R. W. DUTTON, Dept. of Biology, University of California, San Diego, La Jolla, California 92037, U.S.A.

G. M. EDELMAN, The Rockefeller University, New York, N.Y. 10021, U.S.A.

A. GAREN, Dept. of Molecular Biophysics and Biochemistry, New Haven, Conn. 06520, U.S.A.

R. M. GAZE‘ National Institute for Medical Research, London, England.

E. GIACOBINI, Draco Research Laboratory AB DRACO, Fack S–221 01 Lund, Sweden.

D. GINGELL, Dept. of Biology as Applied to Medicine, The Middlesex Hospital Medical School, London W 1 P 6DB, England.

G. GIUDICE, Universitá di Palermo, Istituto di Anatomia Comparata "Andrea Giardina", 90100 Palermo, Italy.

C. F. GRAHAM, Zoology Dept., Oxford OX 1 3 RE, England.

F. GRAZIOSI, LIGB, 80125 Napoli, Italy.

M. F. GREAVES, National Institute for Medical Research, London N.W.7, England.

G. A. HORRIDGE, Dept. of Neurobiology, Canberra ACT 2601, Australia.

D. H. HUBEL, Harvard Medical School, Laboratory of Central Nervous Physiology, Dept. of Neurobiology, Boston, Mass. 02115, U.S.A.

J. H. HUMPHREY, National Institute for Medical Research, London N.W.7, England.

T. HUMPHREYS, Dept. of Biology, University of California, San Diego, La Jolla, Calif. 92037, U.S.A.

M. IVERSON, Division of Experimental Biology, National Institute for Medical Research, London N.W.7, England.

F. JACOB, Institut Pasteur, Paris XVme, France.

M. JACOBSON, Jenkins Dept. of Biophysics, The Johns Hopkins University, Baltimore, Md. 21218, U.S.A.

N. K. JERNE, Basel Institute for Immunology, CH 4058 Basel, Switzerland.

E. R. KANDEL, The Public Health Research Institute of the City of New York Inc., New York, N.Y. 10016, U.S.A.

B. KATZ, Dept. of Biophysics, University College London, London WCIE 6 BT, England.

E. KLEIN, Dept. of Tumor Biology, Karolinska Institutet, S 104 041 Stockholm 60 Sweden.

S. W. KUFFLER, Dept. of Neurobiology, Harvard Medical School, Boston, Mass. 02115, U.S.A.

P. LAWRENCE, Medical Research Council, Laboratory of Molecular Biology, University Postgraduate Medical School, Cambridge CB2, England.

C. LEVINTHAL, Columbia University, Dept. of Biological Sciences, New York, N.Y. 10027, U.S.A.

W. R. LOEWENSTEIN, Dept. of Physiology and Biophysics, University of Miami Medical School, Miami, Florida 33152, U.S.A.

B. LEWIN, Editor of Nature New Biology, MacMillan Journals Ltd. London WC2 3LF, England.
T. E. MANDEL, Basel Institute for Immunology, CH 4058 Basel, Switzerland.
U. J. MCMAHAN, Dept. of Neurobiology, Harvard Medical School, Boston, Mass. 02115, U.S.A.
H. S. MICKLEM, Immunobiology Unit, Dept. of Zoology, Edinburgh 9, Scotland.
N. A. MITCHISON, Dept. of Zoology, University College London, London WCIE 6BT, England.
E. MÖLLER, Division of Immunobiology, Karolinska Institutet, 104 05 Stockholm 50, Sweden.
G. MÖLLER, Division of Immunobiology, Karolinska Institutet, 104 05 Stockholm 50 Sweden.
A. MONROY, Laboratorio di Embriologia Molecolare del CNR, 80072 Arco Felice (Napoli), Italy.
A. A. MOSCONA, The University of Chicago, Dept. of Biology, Chicago, Illinois 60637, U.S.A.
J. G. NICHOLLS, Neurobiology B, Harvard Medical School, Boston, Mass. 02115, U.S.A.
G. J. V. NOSSAL, The Walter and Eliza Hall Institute of Medical Research, Victoria 3050, Australia.
S. OHNO, Dept. of Biology, City of Hope Medical Center, Duarte, California 91010, U.S.A.
I. PASTAN, Dept. of Health Education and Welfare, Public Health Service, National Institutes of Health, Bethesda, Maryland 20014, U.S.A.
B. PERNIS, Laboratory of Immunology, Clinica de Lavoro "Luigi Devoto" Universitá di Milano, Milano, Italy.
J. D. PITTS, Dept. of Biochemistry, University of Glasgow, Glasgow W.2, Scotland.
G. PONTECORVO, Imperial Cancer Research Fund, London WC2, England.
M. C. RAFF, National Institute for Medical Research, London N.W.7, England.
K. RAJEWSKI, Institut für Genetik der Universität of Köln, D-5 Köln 41, W. Germany.
S. RIVA, Dept. of Microbiology, Gruppo Lepetit S.p.A., 20158 Milano, Italy.
A. ROBERTSON, The University of Chicago, Dept. of Theoretical Biology, Chicago, Illinois 60637, U.S.A.
W. J. RUTTER, University of California, Dept. of Biochem. Biophysics, San Francisco Medical Center, San Francisco, California 94122, U.S.A.
S. SCHLOSSMAN, Harvard Medical School, Beth Israel Hospital, Boston, Mass. 02215, U.S.A.
P. SENSI, Director of Research, Gruppo Lepetit S.p.A., 20158 Milano, Italy.

R. L. SIDMAN, Harvard Medical School, Dept. of Neuropathology, Boston, Mass. 02115, U.S.A.

L. G. SILVESTRI, Director of the Dept. of Microbiology, Gruppo Lepetit S.p.A., 20158 Milano, Italy.

M. G. P. STOKER, Imperial Cancer Research Fund, London W.C.2, England.

M. TERZI, Imperial Cancer Research Fund, London W.C.2, England.

A. TISSIERES, Dépt. de Biologie Moléculaire, 1211 Geneve 4, Switzerland.

G. TOCCHINI-VALENTINI, Centro Acidi Nucleici CNR, Gruppo Regolazione, Roma, Italy.

J. TOOZE, Imperial Cancer Research Fund, London W.C.2, England.

E. VERWEY, North-Holland Publishing Company, Amsterdam-C, Netherlands.

J. D. WATSON, Cold Spring Harbor Laboratory, Cold Spring Harbor, L.I., N.Y. 11724, U.S.A.

N. WEINER, Dept. of Pharmacology, University of Colorado School of Medicine, Denver, Colorado 80220, U.S.A.

J. A. WESTON, Dept. of Biology, University of Oregon, Eugene, Oregon 97403, U.S.A.

L. WOLPERT, Dept. of Biology as Applied to Medicine, Middlesex Hospital Medical School, London W.1 P 6DB, England.

D. YAFFE, c/o Dr. Paul Gross, Dept. of Biology MIT, Cambridge, Mass. 02139, U.S.A.

SUBJECT INDEX